The Face of the Earth

The Face of the Earth

Natural Landscapes, Science,
and Culture

SueEllen Campbell

with Alex Hunt, Richard Kerridge,
Tom Lynch, and Ellen Wohl

UNIVERSITY OF CALIFORNIA PRESS
Berkeley · Los Angeles · London

University of California Press, one of the most distinguished university presses in the United States, enriches
lives around the world by advancing scholarship in the
humanities, social sciences, and natural sciences. Its
activities are supported by the UC Press Foundation and
by philanthropic contributions from individuals and insti-
tutions. For more information, visit www.ucpress.edu.

University of California Press
Berkeley and Los Angeles, California

University of California Press, Ltd.
London, England

Library of Congress Cataloging-in-Publication Data

Campbell, SueEllen.
    The face of the Earth : natural landscapes, science,
and culture / Sue Ellen Campbell ; with Alex Hunt . . .
[et al.].
        p.   cm.
    Includes bibliographical references and index.
    ISBN 978-0-520-26926-2 (cloth : alk. paper)
    ISBN 978-0-520-26927-9 (pbk. : alk. paper)
    1. Nature and civilization.   2. Natural history.
I. Title.
    CB460.C36   2011
    508—dc22   2010052469

Manufactured in the United States of America

20 19 18 17 16 15 14 13 12 11
10 9 8 7 6 5 4 3 2 1

In keeping with a commitment to support environmen-
tally responsible and sustainable printing practices,
UC Press has printed this book on Rolland Enviro 100,
a 100% post-consumer fiber paper that is FSC certified,
deinked, processed chlorine-free, and manufactured
with renewable biogas energy. It is acid-free and
EcoLogo certified.

Contents

Introduction

This book is about natural landscapes and some of the great agents that have shaped them. It is equally about how we humans have worked to understand the world around us, both through the sciences and through the stories we produce as cultural beings. We will look at how human lives and cultures are (and always have been) intertwined with such things as tectonic forces, changing climates, the presence and absence of water, adaptation, and complexity.

Curiously, English does not have a good name for this subject or this attitude of inquiry, one that neither foregrounds nor ignores our own involvements, one that draws equally on knowledge from the sciences, the humanities, and the arts. *Natural* often means "not counting human"—but we too are natural, depending on earth, water, and sunlight to keep our animal bodies alive, subject to hunger and plenty, illness and health, heat and cold; and our emotional, spiritual, and cultural lives are inseparable from the world around us. *Landscape* may bring to mind something a gardener designs and maintains, or maybe the tradition of painting from which the word actually derives—a single view of the outdoors, cut out from the whole to be framed and represented by an artist's hand. But in this book we use these terms simply to say that we are focusing on parts of the earth that are more given to us than shaped by us—"natural land"—and that we are the ones doing the seeing and understanding—"scapes." We like the notion of "the face of the earth" because it also suggests this combination of the human and the other-than-human in a poetic, visual image—though we borrowed the phrase from the title of a geology text, *Das Antlitz der Erde,* by geologist Eduard Suess, who named the ancient supercontinent Gondwanaland and proposed the concept of the biosphere.

A human face is something people read, trying to detect signs of processes—the currents of feeling, intention, and thought, and sometimes physical illness or well-being. We think of those currents as interior, as things happening "behind" the face, but they also compose the face, producing its movement, its expressions, what it reveals about personality, experience, mood, attitude. Here we are exploring ways of reading the shapes, textures, and colors of the land, its stone, waters, winds, and the other living creatures that share it with us, both to see something of the ceaseless material processes that produce the land and to understand some of the ways we have lived on it, perceived it, and conceived of it through our human cultures, how we have found in it and assigned to it meaning, value, and emotional power. To talk of the face of the earth is to draw a deep analogy between reading a human face and reading the earth's face, and to express, through that analogy, something of the full complexity of our relationship with our natural physical environment. In a sense, we're trying to stand face-to-face with the land so that we might read and understand its *character.*

Charles Darwin used this metaphor at the beginning of the third chapter of *The Origin of Species:* "We behold the face of nature bright with gladness, we often see superabundance of food; we do not see, or we forget that the birds which are idly singing round us mostly live on insects or seeds, and are thus constantly destroying life; or we forget how largely these songsters, or their eggs, or their nestlings, are destroyed by birds and beasts of prey; we do not always bear in mind, that though food may be now superabundant, it is not so at all seasons of each recurring year." Whose gladness does Darwin mean? That of the wild creatures momentarily satisfied with food, or that of the human onlooker cheered by their health and abundance, or that of "nature" as a personified idea? All these meanings are present in the metaphor, and their combination is similar to the combination we want to capture in this book. We are concerned, that is, with how the earth works, how people feel about the earth, and—to use another metaphor—the health of the earth. And Darwin's sense of what the "face" conceals from us is also important to our book.

This subject is multiple, complex, and changeable, so we have not tried to be encyclopedic or exhaustive. We've chosen instead to look at a few aspects of a few kinds of landscapes; to concentrate primarily on the United States, Great Britain, and Australia; and to focus mostly on major traditions in those cultures. Made mobile by cheap fossil fuels and fortunate economic, cultural, and professional circumstances, we've traveled and talked with friends in other parts of the world; counting everyone who contributed, one or more of us has spent considerable time on every continent. We've also read or consulted hundreds of books, drawn from all over the library, from the sciences (especially geology, ecology, and climate science) and the arts and humanities (especially literature, history, visual art, and popular culture). Still, we've left many alluring covers unopened, many places unvisited. We've followed some of the hundreds of enticing complications and tangents we've encountered but had to leave many more unexplored until another day. In short, we have tried to offer intriguing and suggestive examples of the many ways that we and our earthly surroundings are tied to each other. We hope to offer you some new ideas and leave you with new questions.

In what follows, you will find five major chapters, each looking at one or more kinds of landscape through a different selection of scientific and cultural lenses. (To make reading easier, each of these is split into shorter pieces with their own titles.) Chapter 1 focuses on landscapes shaped by tectonic forces, including volcanoes and hot springs but much more as well: in classical terms, fire and earth. Chapter 2 considers the earth's climate as a whole and then zeroes in on landscapes shaped by the ice ages: air and water. Chapters 3 and 4 look at places shaped by the presence and absence of water, by fluidity and adaptation, wetlands first, and then deserts: water and earth, earth and fire and air. Chapter 5 brings together some of the major themes of the whole and considers them in the context of one final overarching characteristic of our planet, complexity, with examples drawn from grasslands and forests. Finally, we have included a list of our major sources and some suggestions for further reading.

Lest we lose touch with the ground beneath us, you will also find, interspersed, twenty-six short personal essays or "on the spots"—each of which speaks of what it is like to be an informed visitor to one place at one time. Some of the writers of these short pieces are scientists, more are not; some are visiting from afar, some from homes quite nearby; all are paying close attention to the particulars of individual places. Standing in a certain location, what does one see, hear, feel, and smell? What does one think about? And what do those impressions and thoughts mean to us? The answers to these questions will be partly cultural and partly personal, and we invite you to compare these evocations of the personal meaning of particular places with your own feelings about treasured landscapes. Indeed, the relation between such feelings and scientific knowledge is one of the central subjects of this book.

Finally, a note about the authors: This book was conceived, in many conversations, e-mails, and writing experiments, by SueEllen Campbell and Richard Kerridge. SueEllen became the lead writer and organizer. Tom Lynch, Alex Hunt, and Ellen Wohl agreed to write different parts. In the end, chapters 1 and 2 were written by SueEllen; chapter 3 by Richard, SueEllen, Ellen, and Tom; chapter 4 by Tom with contributions from SueEllen; chapter 5 by SueEllen with contributions from Alex; and this introduction by SueEllen and Richard. We always wanted the voices of others in the on-the-spot essays, and so it has happened: in addition to SueEllen, Ellen, Richard, and Tom, another fifteen people have written these short pieces. Their names accompany their words, and you will find information about all of us at the back of the book.

Landscapes of Internal Fire

◇ **ON THE SPOT: OVER A RIVER OF LAVA**

SueEllen Campbell

We park the cars where lava obliterates the road, then set out across the wavy, corrugated terrain of Kilauea's lower edge. The view is sweeping, simple, and subtle: to our right the clean line of the Pacific, to our left the deceptively gentle slopes of Kilauea and Mauna Loa, the most massive mountain on earth, and underfoot a wondrous textured sheen of iridescent blacks.

Despite the warm, muggy air, everyone in the group of volcanologists who have allowed me to tag along is booted, gloved, blue-jeaned, and hard-hatted, and so am I. Given all the warnings I've read, I feel a bit uneasy, so I'm carefully following the route of my most seasoned guide and watching my step on the lava. Thank goodness we aren't crossing the jagged rubble of 'a'a, so sharp and chaotic that every step might bring an ankle-breaking, skin-splitting fall. This pāhoehoe is much more level, its ropy coils most often sprawling like pools of stirred taffy or dropped and tangled skeins of yarn, but even it can be tricky when the skeins tip sideways into sinuous ravines. Now and then someone steps on a thin-skinned bubble and produces the sound of breaking glass, a reminder that these black coils might hide lava still hot enough to burn both boot and foot. A step on an especially fragile bit of crust that just happens to cover an emptied lava tube will mean bruises, scrapes, and broken bones—or immediate death if the tube is still full of molten magma. Stray too close to the shore and we might fall with a big chunk of the rock shelf in a spectacular and likely fatal splash.

Right now none of these hazards is visible. But I was also out here last night, and that was a different story. Along with dozens of others, my husband and I cautiously picked our way by flashlight far enough out over the black to see just the outside edge of one spot where a river of orange lava hit the ocean and shattered into black sand and a gaudy burst of steam. We spent a long time watching these fireworks and studying the scattered flickerings on the slopes above, spots where fresh, bright lava was advancing toward the sea, expanding the land itself—reminding us, showing us, how all such oceanic islands have been built.

Somewhere up to our left, we knew, was Pu'u 'Ō'ō, the side vent of Kilauea that has been erupting pretty much continuously since 1983. Everywhere these lava flows have left kipukas, islands within islands of untouched vegetation, shady refuges for many of Hawaii's endemic species. Where we were, on the wet, windward side of the Big Island, lava is covered in mere decades with a tangled scrub of ferns and other hardy plants, though on the dry, leeward side, the lava flows stay bare for a very long time.

Today we are aiming for a hazard: a skylight, a hole through the top of an active lava tube.

Now the leaders of our group are slowing down and gathering in our stragglers. "We're here," one says, points to our right, and leads us in a wide circle until we are facing back toward the cars. We inch forward. The air grows warmer, then hot. Something sharp and acrid hits the back of my throat and I make my breath shallow. My eyes begin to water. Wind rushes across my ears. My heart pounds. I stop moving and look—ahead a few feet, and then down.

How can I describe this?

I see a river of fire, but fire without flames, without those pale yellows, blues, and greens of a campfire, just a stunning intensity of reddish orange that says nothing so urgently as *hot*. I feel it scorching my skin. Thicker than water or syrup, it's the color of an electric stove burner on high. It pulses as it flows, the pulsing of hot coals, of blood in arteries. It's very fast and mostly silent—a quiet hiss, an occasional soft gurgle. It's maybe ten feet across, no telling how deep. Patches of darkness appear and disappear on the surface, orange cooling to black, folding under, returning to liquid. It looks like the films I've seen of liquid steel flowing out of huge vats into the forms of I beams, the outpouring of some enormous underground foundry.

This is not the familiar stuff of Earth's surface, this beautiful, terrifying river of fire. It is the material of the mantle channeled upward through magma chambers and plumes, melting as the pressure on it lessens closer to the surface. I have learned this. But something closer to instinct tells me what it is, too, something beneath or maybe before words, some recognition hard-wired into my species by our long association with volcanic landscapes. I know that just here and just now I'm seeing the inside of the planet, the world turned inside out. ◈

PROLOGUE

All Earth's landscapes are shaped in large part by structures and forces that lie below our feet, beyond the reach of our senses, and usually far from our minds. We live on our planet's surface among our human kin, things we ourselves have built, plants and other animals; time moves for us at a rate set by our own heartbeats. We can't see the spinning of the earth's core, the drifting of continents, the rising and falling of mountain ranges, and usually it takes an educated eye even to recognize the traces of such slow and mighty happenings. Sometimes, though, the ground buckles, mountains explode, water boils into the air, and our curiosity and imagination snap awake. We glimpse the inside of our globe, maybe also a different kind of time, and we begin to wonder just what it is that we stand on.

It may be that this is one of the oldest human questions. We developed as a species at least partly in Africa's Rift Valley, where the earth's surface is tearing apart and underground forces are especially visible, and many of our earliest ancestors lived among hot springs, geysers, volcanoes, and boiling lakes of lava. We know this because aridity and volcanic ash are both excellent preservatives that have helped keep for us such evidence as the footsteps of two or perhaps three walkers captured about 3.6 million years ago between two layers of ash in Laetoli, Tanzania. These hominids were probably members of the species *Australopithecus afarensis,* as was the woman now called Lucy, who lived some four hundred thousand years later in Ethiopia's Afar Depression, an even more active tectonic region farther north along the rift. Like the walkers' footprints, Lucy's bones were first on the surface, then swallowed into the earth's near depths. Brought back up into the light, such traces of our ancestors' moving bodies raise questions about both our planet and our own nature. Their brains were much smaller than ours and their thought processes different (they did not, for instance, know how to make stone tools), but they walked the way we do, upright, all ten toes facing forward, with a spring in their step from their arched feet. It is hard to imagine that they did not feel some of the things we do about the ground we share. At what point in our evolution did we begin to experience not just fear at such sights—mountains turning into giant clouds of ash—but also awe and curiosity about our world? And when did we begin to imagine things we could neither see nor touch?

Not just coincidentally, many of our earliest civilizations appeared in tectonically active areas, where volcanoes created fertile soils and where the uplift caused by colliding plates led to varied habitats and a mixture of fresh and salt water. In Indonesia and Central America, in the Andes and around the Mediterranean, in the sorts of places that most vividly allow us to glimpse and then create stories about the hidden inner nature of Earth, landscape features like volcanoes have been key shapers of human myths, religions, natural philosophies, and sciences.

Geological violence, we have often thought, must come from the actions of powerful beings who live inside or beneath volcanoes—gods, monsters, spirits. If so, we need to propitiate them, offer them sacrifices, or call to our defense other beings who are equally powerful but more benign. There are many such figures. The fire goddess Pele lives in Hawaii's volcanoes, the monster Oni in Japan's. Papua New Guinea's are home to the Kaia spirits. The Titan Typhon is imprisoned under Etna, along with his jailkeeper, the metalsmith Hephaestus (called Vulcan by the Romans), and the one-eyed giants, the Cyclops, who help Hephaestus at his forge. (The Titans were the first Greek gods, later overthrown by the Olympians. Their mother was Gaea, Earth; their father was Ouranos, Sky. Their names show up often in the context of geology.) Those who live near Etna and Vesuvius are protected by the Christian saints Gennaro and Agatha. Satan is ruler of a volcanic hell, and the lion-headed horses described in the Bible's book of Revelation breathe fire, smoke, and brimstone, which is volcanic sulfur.

Stories about characters like these recognize the real mystery and raw power of Earth's interior. They show us the contrast between small humans and mighty natural forces. Lava streaming down a volcano's slopes does look like long strands of hair, as it becomes in images of Pele. When they erupt, volcanoes do rumble, bang, throw out rocks, and shake the land, as stories about angry Oni and Typhon suggest. Fresh lava does resemble the melted metal of forges. And the heat and sulfuric vapors of active tectonic features do torment the human senses. We watch the world, and then we think about it with the same kinds of plots, images, and characters we use to think about ourselves; we apply our understanding of human motives, emotions, and actions to the world around us; we personify. The stories that result are the ones we usually call mythological and religious.

For a long time, some of us have also studied the natural world by looking for laws, processes, essences, and principles—often retaining personification for its metaphorical and rhetorical strengths but trying to set it aside as a source of literal explanations. This is an effort with a complicated history, one that we'll glance at in this chapter as we begin to explore some of the ways we have seen landscapes shaped by forces from inside our planet. We'll focus on the European tradition that is most often called natural philosophy. This tradition, generally speaking, developed (roughly between 600 B.C.E. and 1900 C.E.) into physical sciences like geology, chemistry, physics, and astronomy, the fields that we might say are most concerned with general laws and those that have most to do with landscapes shaped by tectonic forces. (The companion effort, natural history, which typically paid more attention to specific and often anomalous natural details, developed into biology and other life sciences. Only in the past two centuries has the word *science* taken on its current professional and disciplinary meaning.) We'll also look at current scientific ideas about tectonic landscapes. Along the way, we'll consider such questions as how our understanding of nature involves cultural and subjective forces—imagination, curiosity, wonder, and storytelling—as much as it does careful observations, logical deductions, and classificatory schemes. And, as we'll do throughout this book, we'll keep an eye out for places where cultural and scientific frameworks (these labels, of course, oversimplify for convenience) come together to shape our knowledge of and responses to particular kinds of landscapes.

IMAGINING THE INTERIOR

Let's look at some of the ways this long tradition has found to envision the physical and especially the hidden nature of our planet. What did the natural philosophers observe, think, understand, imagine, propose?

We must begin with the world's basic elements. For many early European theorists, these were fire, air, water, and earth, along with their corresponding qualities of heat, cold, wetness, and dryness. For such a simplification to work, of course, each of these four elements and qualities has to be understood broadly. So fire encompasses all sources of heat and light, including the sun, lightning, stars, and fresh lava. Air includes all gases, not just those we breathe. Water includes other liquids. And earth includes solids of all kinds. Some theorists added to their list of basics a mysterious substance they called ether, something others saw as a purer form of air or of fire. Ether—a word whose root means "to kindle, burn, or shine"—fills the atmosphere beyond the reach of air, past the clouds or the moon, and it might also be a key constituent of the soul. (This has always been the most poetic element, probably because of its intangibility. Consider Alexander Pope's wonderful line, "All the unmeasured aether flames with light.") Other thinkers added a similarly intangible "quintessence" (literally, the fifth essence) to describe the substance of which heavenly bodies are made; for alchemists, the quintessence was a substance latent in all things, one that might, ideally, be abstracted. And many insisted that change is essential as well—a force that affects everything else, the only thing, paradoxically, that endures. Variations on these ingredients and their interactions dominated all kinds of theories from the Greeks and Romans through the Renaissance and even into the nineteenth century, a testament to their flexibility and comprehensiveness as explanatory and organizing categories.

What made earth rise above water into hills and mountains? Do seafloor and dry land regularly change places, or has the process of drying out been continuous since the biblical Flood? What accounts for the sequence of rock layers? What created precious metals and gems, and how are they distributed? Is Earth losing heat, and what caused its heat in the first place? How might such oddities as corals and fossils be explained? And, more specifically, the questions examined in this chapter: What is inside the earth? Why do volcanoes erupt? How does rock melt? What creates earthquakes? Why are hot springs hot, and where is their source of water?

To answer all such questions, earth, air, fire, and water have sufficed in some combination. Consider, for instance, the question of what lies far beneath our feet. The planet is full of air and windy caverns: this is what Lucretius, Aristotle, Seneca, and Pliny thought. Or it is filled with water, great lakes, watery abysses, rivers, and moist vapors, said Plato, Virgil, St. Isidore of Seville, and Paracelsus—a theory that lasted at least nineteen centuries. Descartes declared it to be full of exhalations. John Leslie, in the nineteenth century, imagined it filled with light. And many have envisioned Earth as a globe of incandescent caverns and passageways of fire.

And the very center of the earth, the core? It is fire itself, said Pythagoras (500 B.C.E.), Athanasius Kircher (1665), and Christopher Polhem (1731). Something molten, imagined Empedocles, the philosopher we credit with insisting (about 450 B.C.E.) that everything changes constantly. Perhaps it is an empty space to which the rays of stars penetrated, then began growing toward the surface as metals, proposed Johann Glauber, 1661. Benjamin Franklin thought the core was hot gas, or maybe metal-rich fluid. It must be magnetic material (William Gilbert, 1600; Thomas Cooper, 1813), or heavy metals that sank to the center of an original ball of mud (Amos Eaton, 1818), or something solid that had absorbed heat as the planet passed through a hot region in space (Simon Poisson, 1835). Or maybe it was a great empty cavern (Mary Somerville, 1840); a dense cube or spherical tetrahedron (Richard Owen, 1857); fluid (James Dana, 1879); supercritical gas (Siegmund Guenther, 1884). Many of these theories involve change, too. Cores expand, collapse, or trade places with the crust. Heavy metals sink and precious ores grow outward.

Inside the planet, interactions among the basic elements cause earthquakes and volcanoes. From the time of the early Greeks into the nineteenth century, many natural philosophers attributed these events to the motion of winds through caverns and tunnels, the heating of these winds by friction and sometimes compression, and then their release. Vapors are often involved, they thought, caused by heat (both interior and solar) acting on rain, or by the "fermentation" (that is, chemical reactions) or burning of coal, bitumen, sulfur, and other minerals; sometimes the winds are simply "exhalations." More winds mean more pressure, perhaps explosions, and stronger earthquakes; ventilation shafts might be dug to help release that pressure. Winds that catch fire melt stone into magma and then propel gases, rocks, and ash out of craters and vents. Or explosive quakes and eruptions happen when underground fires encounter underground water. For some, volcanoes are safety valves that help reduce earthquakes. Both are more common near seas and oceans, where wave and storm pressures force salt water into the earth through fissures, and where the phases of the moon and tides affect the timing of tremors. Occasionally, it was thought, some surface event set off these forces: several days of lightning, a passing comet. Or perhaps internal exhalations release electricity, which in turn creates both earth-quakes and lightning.

These are relatively sober theories. Many of them—even some of those that now sound especially unrealistic—were serious attempts to revise earlier ideas to account for new observations. Other theories have been more imaginative and sometimes much more idiosyncratic, though even the strangest typically have found more than one believer. More than a few natural philosophers have wondered about the nature of minerals and stones, whether they might be somehow organic. The thirteenth-century scholastic theologian Johannes Duns Scotus believed that stones and metals are alive. An Italian contemporary, Ristoro d'Arezzo, believed that distant stars sometimes pull earth from water to create mountains and valleys. Dante Alighieri agreed (1320), and added that the mechanism includes interior generating vapors caused by those stars. Pietro d'Anghiera wrote in 1516 that a tree of gold grows out from the earth's interior, and others of his era thought base metals had been turned into noble ones, including gold, by celestial rays. For Edward Jorden (1631), minerals grow from seed, and the heat released by their fermentation creates hot springs. For John Josselyn (1674), trees of metal grow upward in mountain hollows. For others, mountains grow like plants, heated by the earth.

That Dante Alighieri took an interest in such questions reminds us that literature, religion, and science have frequently not been clearly separate endeavors. But it is hard for us now to see what we'd call science in the model in his *Divine Comedy*. This complicated and inventive vision includes an icy center of the globe in which Satan is imprisoned, fiery layers farther from the center for those whose sins are of passion, a massive funnel-shaped hole from surface to core that was created when God hurled Lucifer out of Heaven, a Mount Purgatory on whose summit perches the Garden of Eden, and other elaborate mixtures of conventional Christian allegory, a poet's wild imagination, and a highly original mixture of physical and metaphysical speculations. Many scholarly hours have been spent trying to explain such cosmographies in terms that make sense to modern readers.

John Milton's *Paradise Lost*, written some three and a half centuries later, offers a similarly intricate mixture of classical and biblical theories, with occasional traces of the newer theories of the Renaissance. Milton imagines a "wild Abyss" of Chaos, the blinding light of Heaven above it, the created universe hanging from Heaven by a golden chain, and a volcanic Hell at the bottom, one whose vivid imagery may owe

something to his visit some decades earlier to Naples, where Vesuvius
was erupting, and the nearby Phlegraean Fields with their smaller volcanic features. Satan and his companion angels fall from Heaven into a

dismal Situation waste and wild,
A Dungeon horrible, on all sides round
As one great Furnace flam'd, yet from those flames
No light, but rather darkness visible
. . . and a fiery Deluge, fed
With ever-burning Sulphur unconsum'd.

They lie at first on a burning lake, but soon enough Satan gathers his
energies to move:

Forthwith upright he rears from off the Pool
His mighty Stature; on each hand the flames
Driv'n backward slope thir pointing spires, and roll'd
In billows, leave i' th' midst a horrid Vale.
Then with expanded wings he steers his flight
Aloft, incumbent on the dusky Air
That felt unusual weight, till on dry Land
He lights, if it were Land that ever burn'd
With solid, as the Lake with liquid fire.

When Satan flies toward our universe to exact his revenge through
us on God, it is through an anarchic, confused Chaos where "hot,
cold, moist, and dry, four Champions fierce/Strive here for Maistry,
and to Battle bring/Thir embryon Atoms"— the four classical qualities, which have not yet even crystallized into "Sea, nor Shore, nor Air,
nor Fire." Although he wrote at a time when modern science was beginning to emerge and he was familiar with recent theories (such as those of
Galileo, whom he met), Milton still drew on the mythic power of these
ancient ideas and images to create his own immensely influential drama.

Combinations of theology, developing science, and imagination also
appear in many theories based on attempts to read Genesis as a geological textbook. In these cases, the Bible provides a fixed theoretical framework but little help incorporating detailed field observations.
Although specifics vary from one version to another, the story of the
Flood has been a particularly durable source of ideas about mountains
(raised up by the water's turbulence), valleys and ocean basins (created
to contain the waters), and fossil fish and shells found inland (left by
receding waters). It has been less successful at explaining such things as
volcanoes, although Mount Ararat, where Noah's ark rested while the

waters abated, is itself a very high volcano, far higher than its surroundings. One influential book from 1684, Thomas Burnet's *Sacred Theory of the Earth*, offered a more complicated story that assumed the historical occurrence of a global flood and put it in a scientific framework. The initially fluid planet, Burnet explained, settled into a solid core wrapped in layers of water and crust. Baked by the sun, the outer crust cracked, creating mountains, valleys, and ocean basins and releasing the interior waters, which rushed out as the Flood. Finally, the waters either drained back into the interior or settled into the Pacific, Atlantic, and Mediterranean basins. By the end of the nineteenth century, at least for the respected geologist, linguist, politician, and paleontologist Eduard Suess, the Flood had become a local event with local causes: an earthquake in the Persian Gulf, possibly joined by a cyclone, led to a flood in the lower Euphrates valley. Similar theories continue to be formulated today, involving such events as global or regional climate change or tsunamis set off by earthquakes or volcanoes.

Other strikingly imaginative visions of the world appeared during the same centuries, visions whose ties to the emerging earth sciences are sometimes quite tenuous. One Benoît de Maillet argued in 1720 that our world arose entirely from a diminishing sea, with mermen and mermaids becoming men and women. A prominent Frenchman, Comte Georges Louis Leclerc de Buffon, wrote in 1778 that the Atlantic Ocean was formed when Atlantis sank—more than two thousand years after Plato described the disappearance of that island. The even more prominent Englishman Charles Darwin proposed that the Pacific formed when the moon was torn from Earth. George Catlin, who is better known for his drawings and paintings of the American West and its native inhabitants than for his work on geology, thought that two currents underlay the Americas: one runs south under the Rocky Mountains, then excavates the Gulf of Mexico; the other runs north under the Andes, where volcanoes warm the Gulf Stream.

Among the more durable of these proposals, that Earth is hollow, had an influential early appearance in 1692, in an essay by Edmond Halley (of comet fame). He was wondering what makes Earth's magnetic poles move. Parts of his answer are surprisingly close to current beliefs, while other parts led to several centuries of wild speculation. Halley proposed that the planet's wandering magnetic field is caused by the rotation at different speeds of a crust and a core separated by some fluid medium . . . or perhaps by three rotating concentric spheres, each illuminated by some unknown source and capable of supporting life. In 1721, a French engineer named Henri Gautier argued that gravity should decrease with depth and become negative, and that therefore (though the logic here is hard to recapture) the earth must contain a mirror image of its surface, an interior ocean connected with the exterior one at the poles, and a hollow center. A hundred years later, partly through the unlikely conduit of Puritan clergyman Cotton Mather, whose 1721 book *The Christian Philosopher* endorsed Halley's ideas, this theory found an American adherent in the eccentric John Cleves Symmes Jr. and his promoter, J. N. Reynolds. Symmes lobbied for funding to locate and explore those polar holes, and after his death, Reynolds worked to fund a voyage to the South Seas and on to Antarctica, where he expected to find an open sea beyond a barrier of ice. This expedition did take place, with Reynolds aboard until mutinous sailors put him off the ship in Chile; one result of his travels was a story later read by Herman Melville called "Mocha Dick, or the White Whale of the Pacific."

Edgar Allan Poe made use of Symmes's and Reynolds's ideas in a couple of short stories and in his 1838 novel, *The Narrative of Arthur Gordon Pym*, which ends with notorious abruptness as its narrator-protagonist is drawn down into a stormy white chasm near the South Pole. That Symmes, Reynolds, and their hollow-earth idea were familiar to Americans caught up in the romance and heroism of exploration is also evident in the use to which Henry David Thoreau put it in the final chapter of *Walden* (1854). "What was the meaning of that South-Sea Exploring Expedition, with all its parade and expense," he asks,

> but an indirect recognition of the fact that there are continents and seas in the moral world, to which every man is an isthmus or an inlet, yet unexplored by him, but that it is easier to sail many thousand miles . . . than it is to explore the private sea, the Atlantic and Pacific Ocean of one's being alone. . . . It is not worth the while to go round the world to count the cats in Zanzibar. Yet do this even till you can do better, and you may perhaps find some "Symmes' Hole" by which to get at the inside at last.

Probably neither Poe nor Thoreau believed this polar hole to be real; it was, after all, a possibility that diminished as polar exploration advanced during the nineteenth century. But it's typical of the two writers that Poe uses the idea as an occasion for an overwrought dramatic moment while Thoreau (who was quite interested in science) includes it as an ironic reminder that we ought first to know ourselves.

Perhaps the best-known literary exploration of the hollow-earth idea is Jules Verne's *Journey to the Center of the Earth*. As scholars of such matters point out, *hollow* can mean many things, from one enormous interior space large enough to hold a second sun to a maze of smaller caverns and passageways. Dozens of writers and dreamers in the nineteenth and early twentieth centuries (including L. Frank Baum, creator of Oz, and Edgar Rice Burroughs, creator of Tarzan) were entranced not so much by the scientific theorizing as by the opportunities such schemes offered for romance, utopian speculation, satire, and cultural critique. In Verne's case, they allowed him to tell a lively adventure story while playing with scientific ideas, or, in other words, to help create the genre of science fiction. Thanks to translations by the poet Charles Baudelaire, Verne was an admirer of Poe and even wrote a sequel to *The Narrative of Arthur Gordon Pym* (titled *The Sphinx of the Ice Fields*), which leaves Pym dead at the South Pole. In *Journey to the Center of the Earth* (1864), Verne takes his own characters—a scientist, his nephew, and a local guide—into the interior not through a polar hole (though he knew about Symmes's theory) but through volcanoes. They enter via Iceland's Snæfells and exit via Italy's Stromboli. As his characters travel, debating such questions as how much hotter it will be as they descend, they find emptied lava tunnels, fossil-laden sedimentary strata, a mighty cavern with walls of coal, colorful threads of precious metals, a subterranean river, a vast central sea, light like that of the aurora borealis, a forest of giant mushrooms, a battle between an ichthyosaur and a plesiosaur, a geyser, a field of bones from all the animals that have ever lived, and more—in an amusing and wildly unlikely mishmash of modern and obsolete science and pure fantasy.

The story's playful spirit—and the shift of the hollow-earth idea away from science and into popular culture and fiction—is even more evident in the two film versions of this novel. One, made in 1959 and starring James Mason, Pat Boone, and Arlene Dahl, added not just a female character and mild romantic interest but also a few musical numbers, the ruins of Atlantis, and a memorable scene in which the explorers accidentally release a torrent of water when one of them chips a large crystal off a wall. The second, made in 2008 (with Brendan Fraser in the lead), adds not just the female character—now a feminist adventure guide—but also such elements as a passage across stones kept floating in the air by magnetism and some very large diamonds the explorers carry back to the surface. In this version, they read and carry with them Verne's novel. Neither film, fittingly, shows any interest in scientific believability.

Another recent movie, *The Core* (2003), retains some older hollow-earth elements inside a framework more closely related to modern geological theories and technology. Encased in a ship made of the wondrously hard element "unobtainium," the heroic crew descends into Earth's core to try to restore its proper rotation inside the mantle and thus fix the planet's malfunctioning magnetic field and shield. Along the way, they pass through the inside of an enormous geode; at one point, the navigator calls out, "Guys, we're dodging diamonds the size of Cape Cod." (There are such things as underground crystal caves, but they are in the crust, not farther down as these films suggest.) Up-to-date trappings aside, the makers of these movies—like Verne himself—know they are not scientists, or even natural philosophers. They are entertainers who want to tap into our curiosity about what lies below.

Certainly all these now-abandoned theories about the structure of the earth are fun to read. But reading them with any seriousness can also be very disconcerting. What begins by seeming like a very odd idea indeed may gradually come to seem quite reasonable, and as a result, what we think we know to be true may sound as fantastical as the most startling ideas ever proposed. It is more, not less, unsettling to read a bare summary of them in chronological order, such as the one offered by Susan J. Thompson's *A Chronology of Geological Thinking from Antiquity to 1899*, a fascinating catalog of ideas that defy easy organizing or simple story lines. What emerges most strongly in such a summary is the inventiveness with which we have tried to make sense of the physical world by combining observation, deduction, and imagination—and how myth, religion, and other kinds of assumptions and beliefs help shape what we see, how we think, and what we imagine.

Yet of course we want to see some kind of reasonably linear plot in this history. We hope for some gradual replacement of mistakes with truths, or even some clear sequence of major theories, perhaps with faster progress during the Renaissance or the Enlightenment, when the scientific method should come clearly into play. Alas, no such storylines are obvious. And so we need to do some heavy editing to detect the narrative lines buried among miscellaneous theories. We can see how some topics emerge as shared concerns, then fade away. Some clusters of similar beliefs last for centuries, consolidate into fairly clear alternatives, and finally settle into broad agreement. With a strong enough desire to read the past in terms of the present—as we usually do—we can even see our current understandings slowly emerging.

For instance, although new ideas based on new evidence rarely take hold as quickly as one might think, we can find some advances that are clearly driven by new discoveries and improved instruments. The eighteenth-century discovery of inactive volcanic cones (the Puys) in the center of France is a relatively early example, and developments in nineteenth-century chemistry helped determine that columnar or prismatic basalt was not a precipitate from water but a form of cooled lava. Together, these two insights helped settle a lively debate between the Neptunists, who saw geological history in terms of water, and the Plutonists, who focused on volcanism. Technological advances, not surprisingly, accelerate closer to the present. Early in the twentieth century, the development of radioactive dating methods led to much stronger work on the age of the earth and its rocks. And in the 1960s, more knowledge of the planet's past magnetic reversals combined with the technology for seafloor exploration to help solidify the theory of plate tectonics.

Other stories illuminate some of the ways science and social history are tied together. For instance, we can see the extent to which geological theories have depended on where their inventors lived and traveled, how their work was influenced by local landscapes, intellectual and religious traditions, and economic and political conditions. While the Greeks and Italians saw volcanism as central, for instance, the English long regarded it as exceptional and focused instead on sedimentary layers. One of the first Englishmen to take volcanoes seriously was William Hamilton, envoy to Naples from 1764 to 1800, years in which Vesuvius was often active. And such important geologists as Charles Lyell took advantage of renewed access to the Continent following the French Revolution and the Napoleonic wars. Improved communications helped broaden scientific perspectives, but actual field experience has always been crucial. Several Scots, who gradually came to recognize the many volcanic features in their country, combined theories of sedimentation and volcanism and markedly increased the supposed age of the earth. American scientists resisted revisions to the age suggested by Genesis longer than Europeans did, and in the early nineteenth century, the relatively secular geologists from southern states accepted longer ages than did their more religious northern colleagues.

As we try to understand this history of ideas, we should also remember that how we think is always tangled up with how we classify things, how we organize and separate them into categories, how we draw lines between one kind of thing and the next—and according to what principles, either unconscious or deliberate. In a history like Thompson's *Chronology*, attempts to order the parts of the world are everywhere, as natural philosophers ask such questions as, Are there four elements, five, a hundred? Are lightning and lava like cooking fire? Are corals stone, dead creatures, or alive? What kind of thing is a fossil, and are all rocks shaped like plant or animal parts the same kind of thing? How many kinds of rock are there; how many minerals? By what criteria should we categorize rock strata? Is basalt a precipitate? If not, why does it often appear in large, flat sheets layered with sandstones and limestones?

From today's perspective, many earlier attempts to classify the world's furnishings sound very much like the inventions of Argentine short-story writer Jorge Luis Borges. In "The Analytical Language of John Wilkins," Borges describes a Chinese encyclopedia *(Celestial Emporium of Benevolent Knowledge)* that classifies animals into categories including those "that belong to the Emperor," "those that are trained," "mermaids," "those that tremble as if they were mad," "stray dogs," "those that are included in this classification," and "those that resemble flies from a distance." Borges also describes a classification of stones into groups of ordinary, intermediate, precious, transparent, and insoluble—a list not so different from some that have been suggested in all seriousness. Scientists would certainly say that their current classifications are better than discarded ones—more accurate, more useful, more complete, based on better information. But it's also hard to imagine that we'll ever be fully and permanently satisfied that we've finished this task. As Borges notes, "Obviously there is no classification of the universe that is not arbitrary and conjectural. The reason is very simple: we do not know what the universe is."

◇ ON THE SPOT: AT THE EDGE OF AN OVERTHRUST BELT

Scott Denning

Walking along the dry bottom of a narrow canyon in the Tobacco Root Mountains, I craned my neck to look beyond the rimrock above to distant pine trees along a further ridgeline. The landscape of Sacry's Ranch would be familiar to anybody who has spent time in the semiarid West: a layer cake of sedimentary rocks eroded into rugged topography, stair steps of cliffs, benches, and slopes. Harder limestone and sandstone resists the wear of water and wind, forming cliff walls and caprock, but softer shale crumbles into gentler slopes dotted with sage and cactus.

I had spent all summer at "field camp," a rite of passage required before my senior year of college as a geology student, at a geological field station in southwestern Montana. We had learned the look and feel of well over a dozen distinct rock formations, recognizable beds of sedimentary rock mostly laid down in shallow seas over hundreds of millions of years—their makeup, their colors, the fossils they entombed, the characteristic landscapes they produced, even the way their surfaces eroded to produce slabs or gravels. But most important, we had learned to recognize the order in which these formations occurred in the tilted and folded layers of the Rocky Mountain foreland. The familiar formations were our alphabet. In the first week of field camp, I had recited their names each night in bed, memorizing them before I could recognize them in the field: Flathead, Wolsey, Meagher . . .

Hiking the hills and canyons, our professors had taken us through a series of increasingly complicated structural problems. The familiar rocks had been bent, folded, twisted, broken, uplifted, eroded, and deformed in remarkably complex ways. Huge blocks of ancient "basement rock" had been lifted skyward seventy million years ago, shouldering aside the layered sediments above. Blankets of sandstone, limestone, and shale were forced to drape across the basement uplifts, folding into complex patterns of pleats. Solid rock deformed like modeling clay.

Yet by careful exploration, on foot in the dusty hills and canyons, we had learned to recognize the shapes of the larger structures in the rocks, relying on our sure knowledge of the rock alphabet. Now, at the end of the course, I could recognize most of them from a distance from the lay of the land, the way the vegetation grew on them, even from a small sample held in my hand in the lodge at night. Weeks of hiking had made my legs strong and my arms suntanned— and fed my confidence that I could see through the arid landscapes to the underlying tectonic structures.

Why, then, could I make no sense of Sacry's Ranch?

My final mapping project seemed designed by God to frustrate— and disorient—geology students. I knew how to trace my way either upward, toward younger layers, or downward, toward older layers, to work out the shapes of the folded rocks. But here at Sacry's, the familiar sequence was cut and sliced by faults like a partly shuffled deck of cards. The layers rarely stacked more than three or four deep before they were interrupted, recognizable rocks in the wrong order: *A, B, C, H, I, J, B, C.* I could never feel certain which sandstone was which because the sequence kept getting interrupted by faults. I didn't know what to expect around the next canyon bend or over the next hill.

So I was relieved to see the massive gray cliffs looming above the trail. The Mission Canyon formation is easy to recognize: smooth limestone beds of well-cemented fossils that resist erosion, it makes cliffs or caprock. I picked my way among the boulders of a steep ravine that provided the only way a trail could climb up out of the canyon. In the mess of shuffled rock layers, it was good to have my bearings. Knowing that younger rocks would lie on top of the familiar limestone, I felt I could predict what I'd find above.

As I crested the brim of the caprock, the trail opened out onto a wide bench above the canyon below. The air was hot but dry, the sweet smell of fresh sage pungent on the breeze. I swung my daypack off and stopped for a drink of water. A pleasant mosaic of grass, sage, and ponderosa stretched into the middle distance, with further benches and rock outcrops rising to the wooded ridgeline above.

But my relief didn't last: hiking up toward the top of the hill, I became baffled yet again. I had expected to continue through my alphabet, but the slopes above the canyon didn't cooperate. As I reached the trees, I recognized another unmistakable rock formation on the ridgeline ahead, where the ground was broken into familiar ledges. Amazement began to well up in me, replacing my frustration and sweeping me up in the excitement of discovery. As I climbed out of the canyon, I was crossing older and older rocks, and the top of the ridge was the oldest rock of all, full of pebbles eroded from the crystalline craton before the evolution of land plants. Nothing had prepared me for a mountain in which my whole alphabet was presented in reverse.

Sunlight shone clearly through the trees and I crossed the ridgetop, feeling the empty air beyond. I found myself atop a high promontory with the ground dropping steeply below me for hundreds of feet into a wide valley. I recognized the ranch buildings and road far below and could see the trucks parked where we'd left them at lunchtime. The valley floor, I knew, was formed from the Cretaceous volcanic ash deposited millions of years after the folding and faulting that formed the Rockies.

As I stood on the brow of the ridge, a fresh breeze blew in my face, ruffling my clothes and hair. I felt as if I were on the prow of a great ship made of rock. At last the structure was clear to me. The mountain had somersaulted! The block on which I stood, hundreds of feet thick and maybe a mile wide behind me, had folded so steeply that it had turned all the way over and slid far across the younger rocks below. The mountain I'd spent all afternoon climbing was literally upside-down, with the youngest rocks in the canyon below and the oldest rocks here at the top of the ridge.

The overturned fold-and-thrust fault of Sacry's Ranch was an outpost of a different tectonic province: not the drape-folded foreland but the overthrust belt that stretched west from here for hundreds of miles. I was standing on the leading edge of a slice of rock that had been pleated like a tablecloth and then broken free, sliding across its neighbors far beneath the surface of the earth as the Rockies were being born. The folding and faulting must have taken place over a million years in the silent darkness and crushing depths of the earth's crust. But standing on the vertiginous edge I felt the movement in my body. The wind in my hair was the wind of our passage across the younger rocks below, and my sense of time and space, movement and permanence, deep time and youth was transformed for good. ◇

MUNDUS SUBTERRANEUS

One pleasure in reading earlier theories about the earth is trying to distinguish—in our own current terms—fact from fiction. Which ideas do we still think are reasonable? Which have we simply revised? What phenomena have we reclassified? Which have we completely discarded, and why? Sometimes, though, it is more satisfying simply to enjoy the ways conventional thinking, sober observation, and wild speculation may together create visions of remarkable creative and imaginative force. We can see, too, some surprising continuities in our metaphors and images—some of our best, though not always conscious, tools for making sense, for finding or creating order, for expressing our imaginative responses to our world.

We might linger with an especially appealing example, the work of Athanasius Kircher, a seventeenth-century Jesuit priest and scholar who came originally from Germany but lived mostly in Italy. Kircher was the sort of man for whom the phrase "Renaissance man" was invented. His work was widely disseminated and very influential (Halley knew his book on magnetism, for instance); and his many interests included volcanoes and the structure of the planet's interior. Published in 1665, his massive *Mundus subterraneus* was read right away in its original Latin by scientists and other well-educated thinkers, and in 1669, parts of it (mixed with commentary) were anonymously rendered into English as *The Vulcano's: or, Burning and Fire-vomiting Mountains, Famous in the World: With their Remarkables.* This edition followed the Great Fire of London in 1660 and a large eruption on Mount Etna in 1669, two conflagrations that appear on its title page. Mount Etna's "late Wonderful and Prodigious Eruptions" are identified there as the occasion for the edition, and a short poem suggests that London's fires were symbols of a "last, and Universal Fire"—an extension of the analogy that appears several times in the text between volcanic activities and the "very final Destruction and Consumation of the whole Universe." Perhaps, too, English readers might have seen a subtle reference to recent political turbulence in the description of "the Spirit of Sedition and tumult" that agitates the waters of Scylla and Charybdis and the fires of Mount Etna: nearly twenty years of civil war had just been settled in 1660 with the restoration of the monarchy. (*Paradise Lost,* that other literary response to revolution, restoration, and volcanic landscapes, was published in 1667.)

Kircher's geology mixes old-fashioned with surprisingly modern elements in ways that are sometimes hard to separate. His long list of the world's volcanoes and hot springs draws on two centuries of European exploration, and he knows more will be found, especially in North America. Yet the list includes several that sound like underground coal fires in Germany, in Scotland, and near Newcastle—a possible addition by the translator. And Kircher draws from the fictional account (published in 1558) of Nicolò Zeno's purported visit to a monastery in Greenland built of pumice and heated by volcanoes. A huge volcano on that island and others around the North Pole, Kircher tells us, help keep some northern waters ice-free and some northern pastures green, but they also prevent access to the Pole.

Like scientists and scholars today, he cites his sources. He has more respect than we might, though, for the theories of his much-earlier predecessors: men like Plato, Aristotle, Pliny, Vitruvius, Cicero, Seneca, Virgil, and Ovid. He quotes poetry and fables such as the stories of Vulcan's forge and the Cyclops. He identifies the volcanic spots most often associated with hell and purgatory, especially Iceland's Hekla, and, adopting what may be the most durable volcano metaphor, he describes his own explorations on Vesuvius and Etna as visions of hell, lacking only the "horrid fantasms and apparitions of Devils." Yet he is careful not to speak of such religious geology as literally true: these volcanic landscapes suggest or resemble or remind us of—but are not identical to—the fires of hell or the final conflagration. And he also explored the volcanic regions of Italy in person, which is to say he did geological field-work long before that was a common activity, examining landscapes by foot rather than only in books.

For Kircher, the interior is an organized space of water and fire acting together, both elements stored and transported in lakes and caverns linked by rivers, streams, channels, and fissures. His is an energy-conserving model of perpetual circulation, interaction, and transformation. Volcanoes and earthquakes eliminate the "super-fluous clogg of fumes and dregs of dross" and allow the earth to be "cherished and refreshed with the all-reviving Air; so serving as breath pipes both for expiration and inspiration to the whole body of Nature, or the Universe." He refers to the globe's heart, hidden and intimate bowels, entrails, nutriment, vomit, and exterior members, and he describes fire as the "life of the *Macrocosm,* as spiritous blood is of the *Microcosm.*" (Both Kircher and his translator likely knew of seventeenth-century debates about the circulation of the blood, involving the English physician William Harvey's challenge to Galen's classical theory. Galen posited two kinds of blood, one made by the liver to feed the organs, and one made by the heart to carry the vital spirits. Harvey argued for just one type of blood and a self-contained system driven by the heart.)

Kircher also writes of the "internal Oeconomy" of the earth, its "organiz'd parts." He uses many household metaphors, referring to fire-houses, ducts, conservatories, store-houses, treasuries, nests of heat, privy-chambers, and retiring places. More elaborately, in the captions to its illustrations, the book explains that while we are not to take the cross-sections of the planet too literally—"For whoever has seen them?"—the point of the pictures "was to signifie, according to the best imagination of the Author [that is, Kircher], that they are after some well-ordered and artificial, or organiz'd way or other, contriv'd by Nature; and that the Under-ground World is a well fram'd House, with distinct Rooms, Cellars, and Store-houses, by great Art and Wisdom fitted together; and not, as many think, a confused and jumbled heap or Chaos of things, as it were, of Stones, Bricks, Wood, and other Materials, as the rubbish of a decayed House, or an House not yet made." Later, combining his two major metaphors, houses and bod-ies, Kircher writes, "For the Fire and Water sweetly conspire together in mutual service, with an inviolable friendship and wedlock, for the good of the whole in their several and distinct private-lodgings, as we may so say, and hidden receptacles."

Although much of its language and many of its specifics are certainly out of date ("The Sea replenished with fatness and unctuosity . . . "), some elements of Kircher's vision are surprisingly modern. If we substitute words like aquifer, underground river, magma chamber, and magma plume for his subterranean lakes, fire-houses, and conduits, we're not so far from today's geology. We are likely to find familiar his sense of a "conservation of Nature in its perpetual and constant course" and a "constant and inviolable Law of Nature." His body metaphors suggest the notion of Gaia, the idea that the planet is a self-sustaining or living system: both an ancient and a modern notion, encompassing fertility goddesses and the idea of a Mother Nature or Mother Earth, the classical Greek Earth goddess named Gaea and the theory proposed by scientist James Lovelock in the 1960s—though Lovelock's hypothesis differs from Kircher's in its focus on the biosphere. And Kircher's household metaphors and sense of "oeconomy" are much like our own sense of ecology as a set of relationships and interactions among parts of natural systems. The Greek root *oikos* meant "house," and *ecology* is a word that has come into wide use only recently. For Kircher's English translator (according to the *Oxford English Dictionary*), *economy* would have been a word whose usage had expanded quite recently; many of the earliest quotations in the *OED* for this word are from about 1660. At the smallest scale, *economy* meant household management, including finances. At a larger scale, it was the management of the concerns and resources of any community. Most expansively, it was the method of the divine government of the world. So this word could mean the structure or organization of a domestic or social system, or of an individual body, or of the material creation. Some of these meanings are still alive. So is Kircher's understanding of a planet for which this word is appropriate.

One more thing about *Mundus subterraneus* is worth noting, and that is the book's striking engravings. These pictures helped pioneer the use of illustrations in science books and are often reproduced today in books about volcanoes and the history of geological thinking. One shows water rising from the sea into mountaintop springs and then flowing back down. Cut-out views of Vesuvius and Etna show giant fires inside those mountains. A world map indicates for the first time where volcanoes exist, an early recognition that they usually appear in long lines and curves, a feature later explained by plate tectonics. Most striking of all are two cross-sections of the earth, one illustrating the patterns of underground water and central fire, the other its internal stores of fire. (These last two were the only engravings included in *The Vulcano's*.) In both of these "ideal systems"—accompanied by text reminding us not to take them too literally—a dark round earth is framed by a paler layer of air and an outer mass of gray clouds. In each corner, a round-cheeked head blows out cones of wind. Set off against the white of the air are large mountains and stretches of ocean. In the first, water stretches inward from the surface in giant rootlike systems, with some branches ending in large bulbous reservoirs about halfway to the center; that center is a massive ball of fire, whose flames shoot out to encounter the lowest waters. In the second, eleven large volcanoes spread around the surface, each half in shadow and half in sun, each producing its own rounded clouds of flame and smoke. Here, two of the oceans support tall sailing ships. Inside, again set off by their contrast with the dark background, are twenty-four balls of flame. The largest is in the center, surrounded by others of varying sizes, with sinuous rivers of flame connecting them to one another and to the surface volcanoes.

With its repeated triangles (two in each volcano, one for each wind), its many echoing circles (the clouds, the globe itself, the balls of smoke and flame), and even the alternating and interlocking spaces of dark and light, the design of this image matches Kircher's vision of a dynamic, living planet akin to and in harmony with the larger cosmos of planets and stars. It is a planet that provokes in us the strongest sensations of admiration, praise, and wonder. "In a word," he says, "whosoever desires to behold the power of the only Great and Good God, let him betake himself to these kind of Mountains; and he will be so astonish'd and stupefied with the ineffable effects of the Miracles of Nature, that he will be constrained ever and anon to pronounce, from the most intimate and inmost affection of his heart; *O the depth of the Riches and Wisdom of God! How incomprehensible are thy Judgments, and how unsearchable thy ways, by which thou hast constituted the World!*" The English translator began his introductory "Epistle to the Reader" with something similar, offering "the most wonderful, most prodigious, and even miraculous Operations of Nature, in the Geocosm, or Terrestrial World."

Indeed, the word *wonder* appears often in *The Vulcano's*, along with *marvelous, admiration, astonishment, incredible, prodigious, miraculous, remarkable,* and so on, and, for that matter, italics and the exclamation point. *Wonder* is a word as old as the English language itself. The *Oxford English Dictionary* gives it and its variations (wonderly, wonderment) more than thirteen columns, citing examples from *Beowulf* (840 C.E.), thirteenth-century bestiaries, and many lines from Shakespeare, most famously from *The Tempest*. Shipwrecked onto an unfamiliar land, a new world, the heroine of that play, Miranda (the Latin *mirari* means "to wonder at"), exclaims, "O wonder! How many goodly creatures are there here?" For Kircher's translator and readers, this would have been both a familiar and a strong word, one referring to a particular set of physical and mental reactions that had for some centuries interested philosophers and artists alike and had frequently accompanied European experiences in the New World. Something wondrous was momentarily heart-stopping, nearly impossible to believe and nevertheless true, radically strange and real. While Kircher does sometimes speak of fear—of horror, terror, madness—when he describes his close encounters with volcanoes, his vocabulary of wonder marks a quite different response, one that captures our fascination with the sheer power of these landscapes.

Like all human emotions and ideas, wonder (and its frequent companion, curiosity) has a cultural history. What counts as a wonder has changed again and again (though volcanoes and natural springs have usually been on the list), and so have the emotional implications of the response of wonder, what it is thought to accomplish. At Kircher's moment, according to science historians Lorraine Daston and Katharine Park, most natural philosophers believed that wonder sparked curiosity, which then led to serious inquiry. Some of the most alluring wonders were things that were obscure, secret, hidden: what could be more worth wondering at, and about, than the secrets beneath our feet?

◇ **ON THE SPOT: AMONG THE AEOLIAN ISLANDS**

John Calderazzo

It's sunrise, and I'm standing alone on the wet top deck of an overnight ferry from Naples, watching Stromboli boil up from the sea. To my left, the sun is a pink egg just lifting off the water, while directly ahead, the mint-green triangle of the volcano grows out of morning haze. I feel lucky to have jumped out of bed early enough to see this.

Stromboli is growing because the ferry is approaching quickly, a matter of my changing perspective. But this illusion also replicates the past two hundred thousand years of geologic history, compressed. Here, off the southern Italian mainland, the African tectonic plate grinds under the Eurasian plate. The result has been fiery mountains such as Sicily's Mount Etna; Vesuvius, which looms over Naples; and Stromboli's island neighbor, Vulcano, which has given us the word *volcano*.

One eruption and lava flow at a time, Stromboli really *has* boiled up from the sea bottom to its present height of 3,034 feet. Over the long arc of time, rock moves constantly, everywhere, in big ways and small, in continental drift and in slow-motion increments of mountain building.

And in and around volcanoes, rock flies. Volcanoes heave themselves up into high, holy summits, like Mount Fuji and Mexico's Popocatépetl. Then they burst or crumble apart and erode back down. Their lavas flow in glowing rivers, or leap and turn with the grace of sandhill cranes in their mating dance. They breathe and roar.

Stromboli probably roars more often than any of the others. It puts out a small and contained eruption roughly every twenty minutes, and it has rumbled for most of the past two thousand years, without any world-shaking or island-wrecking explosions. This is why it has been called the Lighthouse of the Mediterranean. It is also why, as I stand on the vibrating ferry deck, the engines thrumming far below remind me of the gargantuan thermal motors churning inside our planet. As if to underscore the point, Stromboli now emits a cartoon puff of gray smoke.

Soon we're anchored offshore, and the staggered-out smoke puffs have turned yellow. Standing at the railing, I look up at huge green walls scored by blackened lava tracks and think, *God, it's nothing but volcano!* Shaded by palm and olive trees, grapevines, and sprays of red bougainvillea, whitewashed houses made of lava rock, pumice, and tuff cling to a thin rim of shore at the base of the steep volcanic walls. The island is only a couple of miles wide and holds fewer than a thousand permanent residents. It's easy to see why.

Later, I walk a beach shining with sand and cobbles as black as sealskin. Ten in the morning and the sun is already pounding. Here and there the surf breaks down pieces of very rough lava—the sea eroding the island even as the newest earth on Earth builds up from the top down. When I hear a huff, like a dragon politely exhaling, I stare almost straight up and see more smoke. It's the first eruption I've ever actually heard. Nearby, fishermen are winching a boat over the black sand. Not one pauses or glances skyward.

At dusk, as the heat of the day lifts off of the water, I sit aboard a small tourist boat anchored offshore. Suddenly, on Stromboli's summit, a red fountain of lava shoots up as though trying to join the stars. The fireworks poof apart and spatter back down into one of three cups on the summit—perhaps Sciarra del Fuoco, the Pit of Fire. Everybody claps and whoops.

During the next eruption, I notice that low clouds directly over the cauldron glow red before and after the blast. This tells me there is exposed volcanic soup churning and boiling up there.

Down here, though, the water is as glassy as can be. There's just enough light left for me to see something momentarily break the surface not far from the boat. It's a dark, smooth, rolling thing. A dolphin? A large eel of the kind my Calabrian grandmother used to fish for in nearby waters?

Or maybe it's a god or a sea monster. Charismatic landscapes tend to spark musings on the mystical, the spiritual, or the just plain scary. North of here lie Scylla, the hulking coastal rock that Homer said sprouted squidlike tentacles, and the whirlpool Charybdis, ancient eater of ships and men.

But erupting volcanoes, above all landscapes, can set the imagination on fire. Hawaii's Kilauea gives birth to chunks of rock that float and burn at the same time. When that happens, the four classic elements forming the world—earth, air, fire, water—no longer act the way you think they should. So why not give in to your fears, your dreams, your most extravagant fancies? Native Hawaiians say that Madame Pele, goddess of fire, can unleash a killer eruption with a mere kick of her heels.

Inspired by Stromboli's lava show, I'll soon take off on a personal volcano odyssey that will last years. I'll climb Vesuvius and trudge alone (and stupidly) up smoking, rumbling Mount Etna. I'll learn that the Inca sacrificed and buried teenagers on the icy summits of Andean volcanoes to appease high mountain spirits who provide crop-nurturing rain. On the Caribbean island of Montserrat, many people will tell me that their suddenly erupting Soufrière Hills volcano is cosmic payback for the sins of anthropologists who took ancient human remains from the island without asking anyone's permission.

I've always believed in the laws of science, in the assumption of a measurable reality. But here at Stromboli, as the sky darkens now over the Tyrrhenian Sea, I find my thoughts turning to deep-down things, physical and psychological. Maybe that ripple in the water was a giant squid, risen from the abyss to test surface currents and pull me down for whatever cosmic sins I have committed. Maybe the fire that burns inside this volcano, or some other, will someday light the way for me to understand my life. ◇

THE GLOBE, TECTONIC PLATES, AND MOUNTAIN BUILDING

Our understanding of what lies inside the planet has changed a lot since Athanasius Kircher's time, thanks to accumulated observations, powerful new technologies, advances in chemistry and physics, and more focused scientific inquiry. Still, many of our own ideas can—and did, when they were first suggested—seem as fantastical as any others, maybe more.

Scientists no longer say our world is made of fire, air, water, and earth. Much more wonderfully, they talk about electrons, protons, and neutrons; protons and neutrons made in turn of quarks and antiquarks held together by gluons; quarks that come in varieties called top and bottom (or truth and beauty), up and down, strange and charm, quarks that are both waves and particles, matter waves. A chemist's periodic table of elements is now a catalog of 118 elements, some not even named, and some so unstable they've barely been detected, if at all. A physicist's list of the basic forces now consists of gravity, electromagnetism (which works by exchanging photons), and the weak and strong nuclear forces (which exchange bosons and mesons). What holds us to the earth, they say, though nobody has ever seen one, are gravitons.

These marvelous—and poetic—ideas do not concern only the very small. The universe, scientists now say, began with the Big Bang and includes black holes; the earth is incredibly ancient; and we are all made from the material of exploded stars. Now as much as ever, we believe that change is the only constant, but we now believe as well that asteroids, fissure eruptions, and human activity can create mass extinctions. Mile-thick ice sheets sometimes cover large parts of the earth; climates change radically and sometimes very quickly. Earth's magnetic fields have reversed themselves over and over. Continents drift, collide, and split into different configurations.

For these ideas, imagination and poetry are as important as ever, and it is to them that we turn now. We'll look first at the current scientific understanding of the structure of the globe, from the core out to the crust, keeping in mind that every inch of our surface owes something to processes that begin deep within. Then we'll consider some places where what is inside becomes most clearly visible on the outside: mostly at volcanoes, but also at hot springs and geyser basins.

When the planet coalesced a little over 4.5 billion years ago, its densest materials settled at its center, still extremely hot and under tremendous gravitational pressure. This core, geologists now believe, is a very dense mass of nickel and iron in two layers, the inner solid, the outer molten, together comprising roughly half the distance from Earth's center to its surface. The outer core is stirred by convection currents—that is, the hotter, more buoyant matter rises to the surface boundary of the core while cooler, heavier matter sinks inward, where it gradually "freezes" onto the inner core. Amazingly, the core (or perhaps the inner core) also seems to spin somewhat more quickly than does the surface of the globe. (Edmond Halley may have been right about this.) This spin and convection combine to create electrical currents that in turn create Earth's magnetic field. With its North and South poles, that magnetic field makes compasses work and so has been crucial to human travel. It also shields us from damaging solar and cosmic rays by deflecting them, the fact that motivated the plot of the movie *The Core*.

Around the core is the mantle, a thick wrap of rock that also divides into two major layers and several transition zones. The inner mantle is partly molten, while the outer is solid. Here, too, convection currents move heat and more buoyant rock outward while colder and denser rock sinks inward. Geophysicists know these things about the core and mantle mostly because seismic waves travel differently through liquid and solid matter: when an earthquake occurs (or an underground nuclear explosion), the patterns made by the shock traveling through the globe reveal its inner structure, as an X-ray, an MRI, or a CT scan can reveal the hidden structures of the human body.

Above the mantle is the thin layer of crust, the cold, brittle rock that makes the surface of our planet. Oceanic crust is made of basalt and is typically about four miles thick; lighter and bulkier continental crusts, on average, are just over twenty miles thick. Under young mountain ranges like the Himalaya, the crust may be twice as thick, but even here it fills only half of one percent of the distance from surface to center. If we were to lay out a cross section of the globe on a hundred-yard-long American football field, with the surface at one goal line and the center of the core at the other, the thickest slab of crust would take up only about a foot and a half, an average bit of continental crust about nine inches, and a sample of oceanic crust less than two inches.

Along with the top layer of the mantle, the crust is broken into tectonic plates. (The word *tectonic,* which came into geological use at the end of the nineteenth century, derives from the Greek word for builder or carpenter. *Architect* means, roughly, "master builder.") These slabs of rock support both oceans and continents, and they move slowly but constantly in what is called continental drift, some at less than half an inch a year, others more than ten times faster. Much of this motion is driven by the heat in the mantle, which is in large part a result of radioactive decay. North America moves west-southwest at just over an inch a year, about the speed at which a human fingernail grows, while Australia moves briskly north-northeast at a little more than three inches a year. These motions are usually jerky: pressure builds up and then releases suddenly in earthquakes.

The configuration of these plates—and thus the placement, number, and shape of continents and oceans—has altered radically over the life of the planet. Here the story really makes demands on the imagination: we have to stop thinking about time and change as things that happen on a human scale and begin to think of them as very long and very slow. Over and over again, oceans open up and disappear: Iapetus, Panthalassa, Tethys. (Iapetus and Tethys are names of Titans; Panthalassa is Greek for "all seas.") Continents collide and tear apart: Pangaea, Gondwana, Laurasia. (The name Gondwana comes from a part of India; Pangaea means "all earth," from the Titan Gaea.) New land is formed; old land erodes. Smaller chunks of crust called terranes—sometimes split-off bits of continents, sometimes arcs of islands—float around on their own, then pile up together and adhere to new continents—hence the title of John McPhee's book *Assembling California.*

Individual bits of land move closer to the poles, or to the equator, from beneath sea level to the tops of mountain ranges. The Burgess Shale, home of some of the world's weirdest fossils, formed south of the equator and under the sea during the Cambrian evolutionary explosion about half a billion years ago; now it is on a very high, steep mountain slope far inside North America in the Canadian Rockies, at an elevation of 7,400 feet. (These fossils are the subject of Stephen Jay Gould's book *Wonderful Life.*) Sometimes, too, large continental masses like the interior American West and the Tibetan Plateau are uplifted to higher altitudes. As these geological patterns change, so do the circulation paths of the oceans and the atmosphere. The global climate changes in turn, and with it, the forms of life everywhere. The high Himalaya cut off the flow of moisture to central Asia, drying out the Gobi and Taklamakan deserts and creating the monsoons in South and Southeast Asia. When South America pulled away from Antarctica, the currents in the surrounding waters grew strong enough to sever the southern continent from the warmth carried by the world's oceans, and so the Antarctic ice cap formed. Much later, when North and South America came together at Panama, the Atlantic and Pacific oceans could no longer circulate at that point. One result, a difference in salinity between the two bodies, helps drive the global circulation of heat and moisture that characterizes our current climate—a topic to which we'll return in chapter 2.

Movements and changes like these are immensely complicated, and they are a challenge to understand, given the rates at which they operate and the many ways past landforms are altered by what follows. Yet in the past several decades, geologists have done much to decipher these processes. Among the basic elements of this emerging story are the three ways—or three kinds of places—in which tectonic plates intersect.

First, in places called divergent boundaries (also constructive margins, extensional boundaries, and spreading zones), two plates separate from each other and magma from the mantle wells up between them to create more crust. These boundaries are most often beneath oceans, and the very long chains of undersea volcanoes that mark them are conspicuous on ocean-floor maps. Just how this spreading happens is still being debated: geologists have thought that the upwelling magma pushed the plates apart (ridge-push), but now they think the much greater impetus comes from slab-pull, the sinking of the oldest and faraway edges of the plates, which in cooling have become denser. When the newly exposed magma is liquid, the magnetic minerals it contains line up in response to the magnetic field at the time. (Similarly, the bits of iron in a spoonful of sand will line up when they're spread on a piece of paper with a magnet beneath. Right under the magnet, they will stand on end, and when the magnet is moved, they'll move with it.) As the magma cools and solidifies, these miniature compasses become fixed in place. Because the planet's magnetic poles wander constantly, and because every few hundred thousand years, in an unpredictable pattern (and for unknown reasons), the magnetic field reverses its orientation, these natural compasses record the rate and timing of the spreading plates. This process is understood now in part because of technology developed during World War II to detect submarines. Later, scientists used that technology to map the ocean floor, noticed patterns of matching stripes, and began to think more seriously about the theory of continental drift.

Small rifts also occur within continents, sometimes left by earlier events, sometimes beginning new ones. These include the Siberian Rift that holds Lake Baikal (the deepest lake on Earth), the Rio Grande Rift in Colorado and New Mexico, California's Death Valley and Salton Sea, and the basins in the Basin and Range areas of the western United States. A much larger rift, the Mid-Atlantic Ridge, where the North and South American plates separate from the Eurasian and African plates, comes to the surface quite visibly in Iceland (which is also probably atop a hot spot, a feature we'll get to shortly). Thingvellir, in Iceland, is the valley where the world's first parliamentary meeting took place, and where, in 1000 C.E., the question of a national religion was settled by a debate about whether volcanic eruptions were reactions to sacrilegious human activities (the answer was no, and the religion that won was Christianity). Entering the valley, you walk down a sort of open-roofed tunnel along a path perhaps ten feet wide between two walls of volcanic rock—the rift itself. Similarly, to the southwest of Reykjavík, you can stand on a bridge over another such feature, with one foot on Europe's continental plate and the other on North America's.

Undoubtedly the divergent boundary that has figured most strongly in human cultural history is the one that divides the African from the Arabian and Somali plates. This complex of rifts includes the Jordan River, the Sea of Galilee, and the Dead Sea, where Moses is said to have parted the waters (perhaps a record of a time when the center of the Dead Sea—always a tenuous body of water, given its desert location—was dry). It continues through the Gulf of Aqaba and the Red Sea; comes to the surface in the hottest place on Earth, Ethiopia's Danakil Depression; and then stretches southward along the East African Rift Valley, through the homelands of our ancient ancestors. It creates the depressions for Lakes Turkana, Naivasha, Nakuru, Tanganyika (the world's second deepest lake), Malawi (the fourth deepest), and others, some of which are home to geysers and thousands of pink flamingoes. Other landscapes along this rift system are strikingly raw, stark, unwelcoming: the film of John Le Carré's novel *The Constant Gardener* ends in what is supposed to be Lake Turkana (although it was filmed at Lake Magadi), in a landscape whose toxic mineral colors suggest the story's concern with the evils of economic colonialism.

The second kind of place where tectonic plates intersect is a transform boundary or strike-slip zone, where two plates move alongside each other at different rates or in different directions, often producing earthquakes. California's San Andreas Fault is the most famous of these, the northern and landward manifestation of the split between Baja California and the rest of Mexico and the intersection between the North American and Pacific plates. As Los Angeles moves northwest relative to San Francisco at an average rate of two inches a year, pressures build up and then release in sudden bursts, some large enough to topple buildings and start citywide firestorms. This is the scientific seed of the widespread myth that one day California will fall into the ocean, and indeed the land west of the fault line may, far in the future, split off into a separate terrane. This fault serves both literal and metaphorical purposes in James Houston's 1977 novel *Continental Drift,* which links tectonic features with the story of a California family affected by the Vietnam War, the source of a significant and enduring fault line in American culture. The New Zealand Alps also lie on a complicated transform fault that's part of the longer boundary between the Pacific and the Indian-Australian plates.

Third, at convergent boundaries, two plates collide. Where the crust on both sides of the boundary is oceanic, island arcs result, as in the case of the Aleutians, part of New Zealand, and Japan. If oceanic crust intervenes between colliding pieces of continental crust, the former is first subducted (that is, it slides beneath an adjoining plate), and then the latter begin to pile up, creating mountain ranges like the Urals, the Alps, and—the newest instance—the Himalaya. This process involves faulting, folding, erosion, and metamorphism, and the result is often considerable deformation of the rock. Sometimes the bottom of the continental crust melts under all the pressure and forms pools, which then rise, cooling very slowly below the surface to make granite and other intrusive rocks such as gabbro. Because granite is relatively light, enough so that it "floats" upward, these plutons—if they're large enough, they're called batholiths—help elevate the surface even more. As the tops of such mountains erode, their granite basements continue to rise; where those basements reach the surface, they create many distinctive landscape features.

Dartmoor, in southwestern England, with its typically gray granite tors, is a pluton, part of a batholith that includes the Isles of Scilly. Much of Canada's coastal range is a batholith, and so is California's Sierra Nevada, where the shining of the pale granite led John Muir to speak of the Range of Light. Colorado's Pikes Peak, another pluton, extends just a little farther east into the Great Plains than the rest of the mountains in the Front Range of the Rockies and thus offers especially expansive views. It was that view that in 1893 led Katharine Lee Bates to contemplate her nation's expanse "From sea to shining sea" and write "America the Beautiful":

> O beautiful for spacious skies,
> For amber waves of grain,
> For purple mountain majesties
> Above the fruited plain!

Maine's Mount Katahdin is a similar feature, and so when Thoreau climbed it in 1846 and found himself stunned by the power of raw nature in its bare rocks, he was responding to the remains of an ancient tectonic convergence. In his story of this climb (told in *The Maine Woods*), Thoreau refers to the Titans, Vulcan, and the Cyclops. He quotes Milton's Satan and compares his own climb to Satan's arduous passage through Chaos. And he fears that such rocky, exposed mountaintops are too sacred, too primitive, too raw to accommodate human visitors.

One especially interesting and much older example of a continental convergent boundary is the Caledonian range. These mountains formed between 500 and 400 million years ago, when the ancient landmass Avalonia collided first with Baltica and then with Laurentia, closing the intervening Iapetus Ocean along what is called the Iapetus Suture. Avalonia (which had broken off with other terranes from the southern supercontinent Gondwana) included England, Wales, the south of Ireland, parts of the eastern United States and maritime Canada (including Prince Edward Island and Nova Scotia), Belgium, and the Netherlands; Baltica included Scandinavia, Russia, and other parts of northern Europe; Laurentia included much of North America, the north of Ireland, and Scotland. When this collision was fresh, the results would have been as obvious as the Himalaya are today. But because the earth's plates continued to drift and recombine, the traces are now hard to distinguish and the remnants are scattered far and wide. Looking carefully at such clues as radioactive dating of rocks and the distribution of fossils, geologists now find ancient Caledonian rocks in eastern Greenland, western Norway, and Svalbard; in parts of the Appalachians and the Canadian coast; and across the middle of Ireland from the Shannon Estuary to Slieve Gullion. The Isle of Man is Caledonian, as are much of Wales and the English Lake District, including what are still some of the highest peaks in these places. The Iapetus Suture line runs beneath the Cheviot Hills, where this ancient geological boundary coincides with the border between Scotland and England.

The Caledonian orogeny (a word that means "mountain building," from the Greek *oros,* mountain) has been especially well studied in Scotland, where its effects are widespread and still significant. These effects include the northeast-southwest orientation of much of the country's geology, which resulted from a piling up of terranes north of the suture and subsequent faults that developed mostly parallel to but sometimes perpendicular to the dominant directions. These structures have affected much of Scotland's settlement history, especially by helping make the Highlands difficult to traverse; they have furnished the major travel corridors; and they have controlled the shape of much of the coastline. The Midland Valley contains a volcanic arc that was part of this event (although it seems odd to call something that lasted a hundred million years an event, geologists do so) and includes today's Ochil, Sidlaw, Braid, and Pentland hills. The geology of Glencoe is Caledonian, and Ben Nevis, Britain's highest peak, is Caledonian granite topped by Caledonian lava. As these features suggest, plate collisions can have very long-lasting results, and they often contribute to later geological changes by their distribution of such things as faults and resistant rock types.

Some plate collisions are between oceanic and continental crusts. In these cases, the thinner, heavier oceanic crust slides beneath the continental crust. The oceanic basalt and the bottom of the lighter continental crusts melt, and the resulting magma plumes rise to produce mountain ranges that include both large bodies of granite and many volcanoes, such as those that make up the Ring of Fire encircling the Pacific Ocean. More than three-quarters of all earthquakes occur in these subduction zones, as do most large, well-known volcanic mountains. The list is long and full of familiar names: Etna and Vesuvius in the Mediterranean; Mont Pelée and Soufrière Hills volcano in the Caribbean; Krakatau, Tambora, and Agung in Indonesia; Fuji and Unzen in Japan; Shasta, Saint Helens, and Rainier in the United States; Popocatépetl (Nahuatl for "smoking mountain") and Pico de Orizaba (or Citlaltépetl, "star mountain") in Mexico; Cotopaxi (Quechua for "shining peak") and El Misti in the Andes; Pinatubo in the Philippines; Tarawera (Maori for "burned peak") and Ruapehu in New Zealand; and many others.

One final important tectonic phenomenon is the hot spot, where isolated plumes of magma rise from the mantle to the surface. They can be very long-lived, and in these cases, when the plates have moved over them, lines of volcanoes clearly show their progression in time. They may lie either at the edges of or well inside plates. The Azores and Iceland are instances of the former, Yellowstone National Park of the latter. The magma that heats Old Faithful Geyser is part of a line of older eruption sites stretching southwest along the Snake River plain to the Oregon-Nevada boundary, sites buried by old basalt flows and the volcanic soils that nurture Idaho's famous potatoes. (The fertility of volcanic soils underlies the classic joke postcard picturing a very long truck whose flatbed is taken up by a single potato.) The line of volcanoes that runs down, and is partly responsible for, Australia's Great Divide— along with two matching lines of seamounts off the east coast—traces another hot spot. The northernmost of these volcanoes are the oldest, just over thirty million years old, while the southernmost are only about six million years old. (Lava flows deposited sixty million to twenty million years ago also stretch from northeastern Queensland through most of Tasmania, and Melbourne is surrounded by old volcanic features.)

Two other well-known hot spots are the Galápagos and Hawaiian islands. Indeed, one way or another, whether it is through a hot spot or along a plate boundary, rising magma creates all oceanic islands, that is, any island that isn't part of a continental mass, separated only by shallow waters. (England and Ireland are the latter kind; in the last ice age, when the sea level was much lower, they were not islands.) Even coral atolls mark eroded volcanic cones, a fact first recognized by Charles Darwin during his years on the *Beagle*. So it is to plate tectonics that we owe the isolation of some of these islands, their consequent biodiversity, and thus Darwin's thinking about evolution. The hot spot beneath the Galápagos is responsible not only for current eruptions but also for a line of older volcanoes stretching toward the South American coast, some above the water, others not.

The oldest, hidden end of the Hawaiian chain, the Emperor Seamounts, begins to the north, near the intersection of the Aleutian Islands and Kamchatka's volcanic edge. Farther south, but at the northwestern end of the visible chain, lie Kure Atoll and Midway Island. Kure is so eroded that all that remains above water is a coral atoll; all that remains of Midway is two very small, flat islands surrounded by an atoll. (Because of its isolation, Midway has served as a midocean airfield, especially important during World War II, and it is now a wildlife refuge of particular value to sea birds.) Some twelve hundred miles southwest lie the younger, spectacularly eroded cliffs of Kauai, where *Jurassic Park* was filmed on the steep Na Pali Coast (*pali* is Hawaiian for "cliff"). At the southeastern end of the island chain is the island of Hawaii, the Big Island, a cluster of five volcanoes including both the active Kilauea and Mauna Loa and the inactive Mauna Kea. And beyond the Big Island, scientists are watching while another bit of Hawaii, Loihi, builds its way up from the ocean floor—as the Pacific plate continues to move over the stationary magma plume at a rate of about four inches a year.

ON THE SPOT: ALONG THE DISTURBANCE GRADIENT
Charles Goodrich

The deep forests of the Cascades seem so steady, so imperturbable. The straight boles of the towering conifers are like the pillars of a great hall, the very sign of stability and hope. Hard to remember that almost anywhere we go in the Cascades, we're walking on a volcano, a rollicking landscape that is always moving, always faulting, uplifting, eroding, and sometimes, along the sutures where tectonic plates overlap, violently exploding.

I'm traveling with a group of ecologists, philosophers, and writers to various places around Mount Saint Helens, to see how the 1980 eruption affected the plant and animal communities, how it altered the very topography of the land. We're visiting stations along what ecologists have dubbed the disturbance gradient to witness these changes and to think about what it means to live in a volcanic landscape. How do the Cascade volcanoes shape our weather, our waters, our soils? How do they enter our consciousness, influence our thoughts and dreams?

We start in deep forest, amid hundred-year-old Douglas fir and western hemlock. Here, out of the direct path of the volcano's blast, the impacts from the eruption were limited to ash fall. Ash fell heavily enough to bury some of the shrubs and forbs of the understory, but after twenty-five years you wouldn't notice. A foot of ash can smother smaller plants, but it also acts as a moisture-retaining mulch and adds nutrients to the soil. The huckleberry, sword fern, and salal have all recovered and prospered, so the overwhelming impression here is green: green leaves, green needles, and lush green moss everywhere. In modest doses volcanic ash is a fertilizer and soil-builder. The rich agricultural soils of the valleys all over the Pacific Northwest are largely volcanic in origin.

Next stop on the disturbance gradient: Meta Lake. When the mountain blew up, the forest here was leveled. Meta Lake was filled with downed trees. But now, young conifers are growing back amid a vibrant understory of willow, alder, huckleberry, and hardhack.

Amphibians are returning in numbers; the western toad, for instance, which is declining over most of its range, appears to be thriving here. The birdlife is far more varied now than it was thirty years ago, when this was deep forest. White-crowned sparrow, American robin, hairy woodpecker, and red-breasted nuthatch prosper in the openings and edges of the altered landscape, where they find more insects and seeds. Even meadowlarks, unheard of in deep forests at this elevation, are fairly common.

Finally we head for a view of ground zero, the crater itself. The trail up from Donnybrook winds through young alder and grand fir. Then the trail emerges onto a hot, crumbly, windswept ridge where, trudging around an outcrop, we come into sight of the volcano's crater and rim. Echoing with rockfall and intermittently belching smoke, the crater looks like a great rotten tooth, a broken-off molar with a steaming mound of decay in the center. This jagged-edged emptiness is hard to reconcile with my memory of pre-eruption Mount Saint Helens, the svelte, symmetrical mountain we all called The Lady.

Below the crater stretches a vast, sand-colored barrens, sparsely covered with lupine and thin stands of scrub alder and willow. This devastated landscape, the pumice plain, was buried by the collapsing mountainside and then blanketed with pumice and ash. It looks as if ten thousand cement mixers had dumped concrete crudely over the whole plain. This obliterated terrain is what ecologists were most eager to study after the eruption. With essentially all life erased, how would plants, insects, and other animals reclaim the sterilized landscape? The scientists had imagined that repopulation of plant and animal life in these devastated areas would occur (a) slowly and (b) from the less-affected edges inward. To their astonishment, the renewal of the landscape began to occur almost immediately, and it happened from almost everywhere. Survivors persisted in myriad refugia—in underground burrows, beneath lingering snow fields, behind ridges, under logs. Even in the most heavily affected areas, biological legacies persisted: logs were rafted on debris flows, seeds were buried where they could germinate, plants that had been buried were subsequently released by erosion. And in-migration began immediately. Spiders and insects parachuted in. Voles and pocket gophers dug out their burrows and uncovered old caches of seeds, some of which ended up sprouting and growing. Alpine lupines began recolonizing the pumice plain, each plant creating a little shade, holding a little moisture, and slowly building up a little soil from leaf mulch and detritus, so that each lupine itself became a refuge enabling other plants to gain a toehold on the harsh plain.

On the hike back down, I let my friends walk ahead while I loiter in a spring-watered cleft beside the trail, a lush little grotto full of wildflowers. Coming upon this wild garden, I am taken with a powerful thirst—a thirst for bright color, for intricately textured vegetation, for the cool tang of photosynthesis on the air. This grotto, just a few dozen yards from the edge of the blast zone, must have been protected by the flank of the mountain. It was probably buried in ash, but not so deeply that the shrubs and forbs couldn't come back. Its persistence in such close proximity to the blasted crater stretches my sense of the earth's vitality and creative unpredictability.

It's a temporary refuge from the dust and devastation we've been walking through for the past several days. I sit quietly for a while, thinking about the refugia in my life—my home, my garden, the parks and open spaces of our town, the library, bookstores, coffee shops, friends. The idea of refugia takes root in me. Maybe that's one fruit of this expedition: for those of us who live in volcanic landscapes—in any landscape, really—it's essential that we take care of the vital places and the nurturing relationships from which recovery and renewal are always ready to spring. ◇

VOLCANOES AND THEIR ERUPTIONS

Of course, the most obvious and dramatic effects of plate tectonics, especially for eyes that see in human time scales, are volcanoes. Volcanologists like to say that one in ten of us lives in the shadow of an active one. But if we stretch our imaginations further into the past, the percentage grows; often, we're among the remnants of ancient volcanoes even in places we don't think of in those terms. Like the motions of tectonic plates, the effects of volcanoes on our landscapes may be subtle but are surprisingly widespread.

Many factors influence the nature of volcanic eruptions and the landforms they create, including the ingredients and thickness of the magma, what it passes through on its way to the surface, whether and where it encounters water, the shape of the vent, and how much pressure builds up before an eruption. Most volcanoes are built by more than one kind of event at more than one time, too. Nevertheless, we may divide them into several general categories.

Magma that reaches the surface in the form of extremely hot molten rock (usually basalt) is called lava. When it is not concentrated into focused vents but instead emerges from extensive fissures, it can spread out in sheets to form vast low-relief plateaus, as it has done in the Columbia River region of Washington and Oregon. Both Glasgow and Edinburgh are on land that was deeply buried by lava from scattered fissures and from Edinburgh's remnant volcano, Arthur's Seat; the curve of hills around Glasgow is another remnant. Later, when Scotland was separating from the North American plate, lava sheets covered its western edge, including the islands of Mull and Skye. Some much larger flows have covered the Paraná Basin in Brazil, the Karoo in southern Africa, a large part of Siberia, and India's Deccan Plains, where lava buried an area of more than 250,000 square miles to depths up to seven thousand feet.

Large lava flows are sometimes called traps—the Deccan Traps, the Siberian Traps. This word, which comes from the Swedish (and ultimately Old Norse) for stair steps, is also used for the layers made by the tops of columnar, or hexagonal, basalt, a pattern that sometimes occurs in lava plateaus. Scotland's tiny island Staffa and Devil's Postpile National Monument in California offer good examples of these striking formations. The Giant's Causeway on Ireland's northeastern coast, another such site, was a key location for nineteenth-century geologists, who had some difficulty agreeing on the geological mechanism responsible for these shapes. This problem was linked to the developing understanding that volcanoes could be inland as well as along the edges of seas and oceans; in fact, it was when they noticed pieces of these columns in pavements and roadside posts in France that traveling geologists began to consider the possibility of a local volcanic history.

Lava can also form boiling, steaming orange lakes, as it did in Hawaii's Kilauea during significant stretches of the nineteenth century, when the islands (then called the Sandwich Islands) were beginning to attract adventuresome tourists. British traveler Isabella Bird was one of these adventurers, and her reactions in 1873 to these lakes are strikingly similar to Athanasius Kircher's in their Christian symbolism and their landscape aesthetic. Emerging in the eighteenth century and dominating the nineteenth, the aesthetic of "the sublime" (to simplify a little) regarded especially grand and imposing landscapes as signs of the insignificance of humans before the power of God or Nature. Like its precursor, the response of wonder to the marvelous (though with a stronger element of fear), the sublime had a distinctive descriptive vocabulary. Bird was fascinated by Kilauea's lava lakes, visited them numerous times, and exclaimed in turn at their beauty, at their "awful sublimity" and "mystery inscrutable," and at their resemblance to hell. These pools of fire, she said, were "the fittest emblems of those tortures of remorse and memory, which we may well believe are the quenchless flames of the region of self-chosen exile from goodness and from God." About the same time and in the same place, Mark Twain quipped, "The smell of sulphur is strong, but not unpleasant to a sinner."

When it flows, lava generally takes one of two main forms, both with Hawaiian names. One, 'a'a (pronounced "ah-ah," with even stress on each syllable; also sometimes spelled simply aa), is much thicker, contains less gas, clinks as it flows, and both moves and cools in extremely jagged chunks that will tear up boot soles and the palms of hands. The other, pāhoehoe (pronounced "pah-hoy-hoy," also evenly stressed), is much smoother, with a surface that looks like pavements of curved and snarled rope. When pāhoehoe is new, it shines like iridescent black glass, and when you walk on it, it crackles like thin glass, or thin ice. It flows in lobes: the surface of one lobe will be cool enough to contain its molten contents, at least temporarily, while an adjacent lobe will be hot enough to have a bright orange leading edge. Beneath a cooled surface, lava typically travels much more quickly in tubes, and sometimes, though it is dangerous to do so, you can look through a hole in the surface, or skylight, straight into a river of hissing liquid rock. These tubes sometimes empty out but remain open, making tunnels that you can walk through, occasionally amid ice preserved by the insulating walls. Accessible examples are on Kilauea, at Undara in Queensland, and in El Malpais National Monument in New Mexico. (*Mal pais* is Spanish for "bad land"—used here because the ground, especially the 'a'a, was so hard on horses.)

Like granite, cooled lava is quite durable and persists in many places after surrounding ground has eroded. All around the world, in fact, high points are often volcanic in origin for this reason. Many mesas (as tablelands are called in the southwestern United States) are topped with a layer of old lava, and sometimes what began as lava flowing along low riverbeds has lasted so much longer than its surroundings that it is now the tops of hills. (This is called inverted topography.) Sometimes lava lasts in the form of sills and dikes (lava intruded into horizontal and vertical cracks) and necks or plugs (lava that petrifies in a volcano's main conduit). Northern Ireland and the mainland and islands of western Scotland contain many dikes that remain from the opening of the North

Atlantic about sixty million years ago. The Breadknife in Australia's Warrumbungle National Park is a dike that formed somewhere between thirteen million and seventeen million years ago inside a large shield volcano. Notable sills include the Salisbury Crags in Edinburgh, the ground beneath Stirling Castle, and the Great Whin Sill of northern England, a jutting shelf of hard gray rock extending for eighty miles along the northern edge of Northumberland, just south of the border with Scotland. This formidable natural rampart was exploited by the Romans, who built part of Hadrian's Wall along its crest. The Palisades along the Hudson River between New York and New Jersey are also a sill.

Edinburgh's Castle Rock is a plug, one of many instances where volcanic remnants have helped protect humans against potential attackers. Devil's Tower in Wyoming is a related feature, likely not a plug but a hardened lava intrusion. This monumental landscape feature, which figures in the film *Close Encounters of the Third Kind* as a meeting place between humans and extraterrestrial visitors, has also been a key spot in some Native American stories, including the Kiowa tale of seven sisters escaping a bear: while the sisters escape into the sky to form the Big Dipper, the bear's claws create the tower's striking vertical striations, another instance of columnar basalt. New Mexico's Ship Rock is sacred to the Navajo (Dine) people, as Australia's Wollumbin (Mount Warning), the remaining center of the enormous eroded Tweed Volcano, is to the Bundjalung people.

Sometimes lava is catapulted into the sky in fiery fountains and curtains. In these cases, it cools and solidifies rapidly into a variety of solid forms. Large chunks called bombs lead volcanologists in the field to wear helmets and sometimes walk holding hands so that one can look down while the other watches the air. Much smaller pellets and long, fine strands are called Pele's tears and Pele's hair, named after the Hawaiian volcano goddess. Lightweight, bubble-filled pumice and cinders (also called scoria) result from rapidly expanding gases: cinders, which are smaller than pumice, come from relatively fluid lava and form in the air; pumice, which is light enough to float, comes from thicker lava, forms inside volcanic conduits, and then shatters into smaller parts when it hits the air. Some thick lava explodes into even smaller shards of natural glass, the main component of volcanic ash.

Volcanoes made mostly by lava flows are shield volcanoes, so called because their wide, shallow slopes resemble a warrior's shield laid down on the ground with the domed side up. Approaching Hawaii's Big Island, you see—or think you see—two broad, low hills joined by a central saddle. This vision turns out to be deceptive: they look low only because they are so very broad. For these two shield volcanoes, Mauna Loa and Mauna Kea—whose Hawaiian names mean "long mountain" and "snowy mountain"—both reach more than 13,600 feet above sea level and 30,000 feet above their base on the sea floor. Mauna Loa is Earth's most massive mountain, occupying some ten thousand cubic miles, and because of its elevation and its midocean isolation from electric lights and air pollution, the slightly higher Mauna Kea is one of the world's best locations for telescopes. The shallow slopes are built up by the flows of relatively liquid lavas, liquid enough to cover long distances, akin to the way a thin cookie batter—or a fairly thick pancake batter—spreads not into steep little mounds but instead into flattened but slightly raised cakes.

. Many other classic volcanoes are cone-shaped, with craters rather than points on top. Hill-size cinder, spatter, or scoria cones are created by small bits of basaltic lava that fall close to their source to build straight-walled triangular hills with 30 degree slopes—the angle of repose of these fragments, or the steepest slope they can naturally sustain. These often occur in clusters, and many represent one-time-only volcanoes or vents. A particularly dramatic example is Paricutín, which began erupting from a cornfield in central Mexico one day in 1943 to become an extremely rare instance of a volcano whose entire life (nine years, in this case) was watched and documented by scientists. Now it seems obvious, at least to an informed eye, that the area is a collection of cinder cones, but until that day the residents had no idea they were living among volcanoes. Pioneering modernist painter Dr. Atl bought the new volcano from the cornfield's owner and lived there, breathing its noxious gases and losing a leg as a result, while he created thousands of images emphasizing the stark, dramatic beauty of the landscape and the way that the volcano's bright lava sprays and swirling ash plumes linked the earth to the starry night sky.

Dr. Atl also created many spectacular images of Mexico's other volcanoes, especially Popocatépetl and Iztaccíhuatl, the two giants that dominate the valley where Mexico City lies. Devising his own textured, intense pigments (Atlcolors), he painted such subjects as the long triangular shadow cast by El Popo across the Valley of Mexico, its tip just touching the wide, curved horizon where the blue of distant mountains meets the bright orange and yellow of the sunrise—all seen from the rocky heights of the volcano. El Popo and Izta are examples of the mountain-size cones called composite volcanoes or stratovolcanoes—

volcanoes common in subduction zones that often stand in dramatic solitude during at least parts of their lives. They are formed by a series of eruptions (and thus layers) of lava, ash, and rubble, sometimes strengthened from within by a skeleton of sills and dikes. Not all stratovolcanoes are so cleanly conical. Iceland's Hekla, for instance, is much longer than it is thick because of its rift location. Other volcanoes lose their clarity and solitude through erosion and through subsequent eruptions that destroy large parts of the cone (as happened at Mount Saint Helens), build new cones on the shoulders of the first, or fill in the space between two cones with debris.

The iconic stratovolcano is Mount Fuji, whose imposing symmetry has inspired many Japanese landscape artists, including Hiroshige (1797–1858) and Hokusai (1760–1849). In many of their woodblock prints this sacred, snow-covered or snow-streaked volcano rises gracefully in foreground, middleground, or distance. Fuji floats over busy, colorful city scenes, beyond lakes and marshes rich in boats and birds, between foamy waves and dangling leafy branches. Its strong lines emphasize the corresponding shapes of other elements of the scene, and its height and self-containment suggest a world that is both present and also somehow beyond the ordinary. The otherworldly quality of this type of volcano has impressed other cultures as well. Mount Ararat is a stratovolcano. Bali's Mount Agung is considered to be home of the god Mahadewa and the axis of the world. Popocatépetl's eruptions just before the arrival of Cortés were seen by the Aztecs as bad omens, as they indeed turned out to be. And California's Mount Shasta, a center of New Age activity, is thought by some to contain within itself several lost kingdoms, including the legendary volcanic island of Atlantis and several others with less impressive genealogies.

Certainly stratovolcanoes are beautiful in an imposing way. But they are also objects of awe because they can be very dangerous. The layers of ash and rubble that help build them are the remnants of two violent and potentially destructive kinds of eruptions, ash columns and pyroclastic flows.

Ash columns—familiar to anyone who has seen a child's drawing of an erupting volcano—consist of small particles of varying sizes. Sorted by gravity, the heaviest ones fall close to their source, building mountains and burying the nearby countryside, sometimes deeply, as happened when Vesuvius buried Pompeii in ash and cinders in 79 C.E. This was the eruption in which the natural historian Pliny the Elder died, having rushed toward it to see better and to help; his nephew

Pliny the Younger described it; and so large ash-column eruptions are called Plinian. Ash deposits can cause delayed destruction, too. Around Pinatubo, for instance, when rain soaked the ash from the 1991 eruption, one result was a version of the deadly mud flows called by a name of Indonesian origin, lahars. (Lahars also occur when volcanoes erupt into glaciers; this is seen as one of the chief hazards posed by Mount Rainier in Washington, because of its proximity to Seattle, and by Ecuador's Cotopaxi.)

Midweight bits may travel considerable distances: ash from the 1980 Mount Saint Helens eruption landed in visible amounts on cars and windowsills nearly a thousand miles away. The lightest ash particles rise high into the atmosphere, to be distributed by the winds around the globe. These clouds pose a danger to airplanes, whose engines can be shut down by them, a scary fact discovered the hard way in 1989 when a passenger flight from the Netherlands to Tokyo via Anchorage ran into ash from Alaska's Redoubt Volcano and narrowly averted disaster—the reason Iceland's Eyjafjallojökull disrupted so much air traffic in 2010. Thick deposits of ash (and sometimes larger stones) that have solidified over time into rock are called tuff. On occasion, a cliff of tuff has served as a valuable building site, as for instance for the Early Puebloan people (Anasazi) who excavated their cliff houses in the ash of the Jemez Volcano near Santa Fe, New Mexico, some eight hundred years ago.

Ash, gas, and chemicals from really large eruptions can also cool the climate—rarely for long periods, usually just for a few years. Materials sent into the atmosphere from the massive and long-lasting eruptions that created the Siberian Traps have been implicated in an abrupt climate change that may have caused the mass extinctions at the end of the Permian period about 251 million years ago. More recently, in 1783, the fissure eruption of Iceland's Laki caused serious famine in Iceland and a variety of upsetting weather changes across Europe (some described by the pastor and pioneering natural history writer Gilbert White)—dirty sulfuric air, excessive lightning, respiratory diseases. When Benjamin Franklin deduced the cause-and-effect links, he was participating in the beginnings of modern scientific weather study and recording. A few years later, the 1815 eruption of Tambora killed at least 117,000 people in Indonesia by pyroclastic flow, ash fall, water pollution, crop destruction, famine, and disease; so many died in the kingdom of Tambora that its language disappeared.

The sulfuric acid dust veil this volcano's ash created in the atmosphere then caused Europe's and New England's "year without a summer" in

1816. Widespread crop failures and starvation followed, as did food riots in France, Britain, and Germany; crop failures in India are even thought to have been a factor in the first worldwide cholera epidemic. The crisis set off large-scale migration from Germany into Russia and America, and some historians credit poor crops in New England with speeding up the push of settlements farther into the Midwest. The many cultural traces of this event include Mary Shelley's novel *Frankenstein:* cold, gloomy weather kept her indoors and sparked the ghost-story competition that led to the novel, and the novel itself shows signs of the advance of Alpine glaciers that may have resulted from the volcano's ash. British Romantic painter J. M. W. Turner took note of the particularly brilliant red sunsets in the fall of 1815, a phenomenon also visible following the eruption of Pinatubo. And John Keats's "Ode to Autumn" recorded the relief with which good harvests were received in the first warmer year.

Pyroclastic ("broken fire") flows are even more destructive volcanic events. These are enormous, gray, expanding billows of steam, ash, and larger debris derived from a volcano's conduits and often from chunks of the mountain itself, and they flow downhill, often staying in valleys but sometimes crossing over ridges. They travel very quickly (up to sixty miles an hour) and are very hot (up to nearly 1600°F). The 79 C.E. explosion of Vesuvius involved deadly pyroclastic flows as well as an ash column, but the first such eruption to be recognized and studied by volcanologists occurred in 1902 on the Caribbean island of Martinique. Mont Pelée killed 28,000 residents of the capital, Saint Pierre, in minutes, leaving only two survivors, one protected by thick prison walls, the other by a sea-edge grotto. It was this kind of eruption that happened at Mount Saint Helens when its side blew out, looking to climbers on the summit of nearby Mount Rainier as if the entire mountain were flying toward them. Massive pyroclastic flows rushed downhill, changing the shape of the land; turning dense, damp, mossy old-growth forests into vast swaths of stripped, fallen trunks that look to this day like spilled pickup sticks; and killing nearly every person in range—though far fewer people than would have died a century earlier, before scientists knew enough to issue sufficient warnings. The rock produced by solidified pyroclastic flows is called ignimbrite, and it is often used as building material. The controversial high-level nuclear waste storage site inside Yucca Mountain in Nevada is a mixture of tuff and ignimbrite (much of it only loosely welded); one reason it is so controversial is its tectonically active location.

Sometimes large explosions empty so much magma from the chambers beneath the volcano that the ground above them collapses into that space, creating a caldera (from the Spanish for cauldron). The eruption of New Zealand's Taupo less than two thousand years ago created a caldera that now holds Lake Taupo; Kilauea's former lava lake visited by Isabella Bird and Mark Twain was in another. Scafell in England and Glen Coe in Scotland are very old calderas. The steep walls of Tanzania's Ngorongoro Crater have helped protect its extraordinary wildlife. American examples include Crater Lake in Oregon and two less visible instances near Santa Fe, New Mexico, and in Long Valley, California. Much of Yellowstone National Park is a caldera, and the lingering heat from its last superexplosion about six hundred thousand years ago is what fuels its hydrothermal features; entering Yellowstone's gates, visitors are given a map on which the caldera rim is marked. The 1883 explosion of Indonesia's Krakatau created a caldera (and a particularly deadly tsunami), as did the eruption of Santorini about 1600 B.C.E. The effects of the latter event on Mediterranean cultures may have been considerable. Geologists, archaeologists, and historians have speculated that it may have contributed to the decline of the Minoan civilization and provided the basis for numerous legends, including the story of Atlantis and possibly some of the region's many stories of catastrophic floods.

Such speculations bring us full circle, in a way, returning us with yet another twist to the age-old interplay between what we might simplistically call material and nonmaterial explanations of our world. We have seen how the earliest natural philosophers moved away from traditional methods of explaining geology with the stories provided by myth, legend, and religion and toward what would become scientific methods. Today, we may reconnect these modes in the other direction, recognizing that myth and legend may have some scientific value and using the tools of scientific geology to renew our understanding of some of our culture's most fundamental stories.

◇ ON THE SPOT: APPROACHING CHAITÉN VOLCANO

Fred Swanson

Today's the day—the start of our weeklong field expedition into an unexplored landscape, southern Chile's Chaitén volcano. Quiet for nearly ten millennia, it erupted less than a year ago, sending ash eastward across Patagonia and into the southern Atlantic, and in the months since, still restless, it has pushed up a massive dome that nearly fills its old caldera. With my ecologist buddy Charlie Crisafulli, I've come here to learn more about the ecology of volcanic explosions, and to test some of the lessons we've learned from three decades of study closer to home at Mount Saint Helens.

We've traveled hard to get here, and now, finally, poised on the bridge into the town of Chaitén, we get our first full view of the looming volcano. Just a few miles north up the Rio Blanco, it's a big, dusty mass of rubble. Puffs of steam from vents dot its sides, and a plume rises a few thousand feet straight up before bending northward with the winds. The river below us is a braid of many channels wandering across a plain of gray silt and sand—evidence of the sediment-charged floods that have moved down the riverbed, inundating the forest along its sides and then the town. For a few minutes we watch several massive bulldozers working on dikes to protect what's left of the town, but they can't really do much: it's sitting at the mouth of the gun barrel, squarely in the sights of the next flood or mudflow.

I had imagined a crucial decision at this point in our trip: crossing the bridge commits us to engaging directly with the volcano. The possibility that a new mudflow could wipe out the bridge, cutting off our supplies and our retreat, seems very real. I worry too about how close the roads—and our own sense of safety—will permit us to get to the blast zone. But of course, now that we're so close, we can't stop.

Just over the bridge we find the town of Chaitén nearly deserted. Despite the government's efforts to relocate its five thousand citizens, about seventy holdouts remain. Abandoned dogs roam the streets, and horses graze on the tall grass in the city square. Gritty gray ash lies several inches thick on every horizontal surface, except where passing cars and trucks have blown it aside. Houses close to the river are scarred by mud lines halfway up the first story. Vultures perch atop an empty A-frame, and crudely lettered slogans (in Spanish, of course)—"Madam President, you violated all the rights of a people who will not die"—protest the government's reactions.

Our destination is the northern flank of the volcano, the point of closest access. We drive through lush rain forest, then suddenly catch a glimpse ahead of completely dead forest beneath the dirty steam cloud drifting from the dome. Soon we're stopped where deep gullies cut by streams draining the northern flank slice the road. A ten-minute hike takes us through a towering forest of standing dead trees and into a blown-down forest. The world is gray except for a few sprigs of foliage sprouting on surviving plants and many hard-to-see brown, yellow, and orange mushrooms budding from rotten wood. Rocks clattering down the dome less than a mile directly upslope remind us of the instability of the scene and the potential for further explosions. The air is full of the slightly burned, organic smell of wood stewing in the fresh volcanic deposits, a scent that takes me back to the Mount Saint Helens landscape I walked the first summer after her 1980 eruption.

We get down to work. Charlie is an intensely inquisitive naturalist with a craving for critters—he knows them all. Although we're both lanky, bearded, seasoned volcano nuts, we couldn't be more different in how we set to work. Charlie focuses on collecting—plants go into the plant press, bugs in the vials, soil and new ash into the plastic sample bags. This means he hunkers down and examines a few spots intensively. Ultimately, stories will tumble out of his samples and data. By contrast, as a geologist, I like to move quickly across the land and assemble stories from observations along the way. I scribble notes and sketches, emphasizing interactions between geological processes and the biota, and jot down questions I'll think about later. So Charlie and I compromise and adopt a blend of fast reconnaissance with hourlong periods of focused site sampling.

In the blast zone, less than a year after Chaitén's big eruption, we discover many signs of life. Amid the gray, prostrate tree trunks we find bright green, knee-high clumps of bamboo and other hearty plants sprouting from rootstocks buried by less than a foot of rock debris left by the blast. Some species of toppled trees are sending out a few green sprouts, but other species don't have this capacity. The blast zone is loaded with dead wood, both from before the eruption and because of it, so I'm not surprised to see many types of mushrooms, fruiting bodies of the extensive network of fungal mycelia spread throughout the feast of rotting wood below the surface. I'm reminded of the small pebbles I saw dangling from spiderweb mycelia on the wall of a shallow pit I dug in blast deposits ten days after the climactic eruption of Mount Saint Helens. I realized then that this fungus, adapted to germinate after heating by forest fires, had penetrated a vast blast-zone area in less than two weeks. Maybe the same thing happened here at Chaitén.

Cross-country travel in this landscape is made very difficult by the endless fascinating distractions and the physical challenges of a forest laid flat. Usually my long legs are a big advantage in moving across rough ground, but in the parts of this volcanic area most littered with fallen tree limbs and stems of shrubs, we're almost swimming through the tangle. Our shovel, packs, and camera gear are repeatedly snagged, bringing us to awkward and sometimes risky halts. And in short order we're covered with ash shaken from every stem and leaf we touch; we're as gray as the landscape. The chance to catch a breath is welcome; to trade muscle work for mind work in this fascinating landscape is a treat. Why is the bright red flower here? How much clatter are we hearing from rockfall from the dome; is it giving us a warning? What is the fate of that expanse of forest on the distant ridge—is it dead or will another crop of foliage emerge and green the landscape again? We're not very disciplined; if we were, we might miss something very important. This stuff we call fieldwork taxes all our skills and energies.

In the slant of evening light, Charlie and I finally call it a day, exhausted but abuzz with thoughts of the live volcano, the changing ecosystem, and the people of Chaitén. This wild, noisy, fragrant, dangerous landscape has really focused my attention. And as I've learned from other field adventures, today's images and ideas will provide a lifetime of reflections. ◇

HOT SPRINGS AND GEYSERS

One simplified way to describe what happens in volcanoes is to say that this is how heat escapes from the inside of the planet, as part of the very long, very slow cooling process that began when the universe first formed. Heat always wants to rise, even when it encounters barriers. Kircher's illustration of the fire houses of the interior captures this fact. His diagram of the subterranean interactions of fire and water captures another related fact: as the natural philosophers thought (though not exactly in the ways they thought), internal heat does encounter underground water, and water may increase the explosiveness of volcanic eruptions. Heat also alters water, and we experience the results on the surface in the form of such features as hot springs and geysers.

Some warm and hot springs derive from geothermal gradients—that is, water that descends far enough into the earth to become heated simply by increasing pressure and temperature and then rises quickly enough to retain the extra heat. Most hydrothermal features, though, occur in places where the earth's crust is relatively thin, where plumes of magma have worked out from the mantle toward the air, where tectonic and volcanic activities have occurred, though not necessarily recently, and where the surface layers are porous enough that over many centuries rain and snowmelt can seep deep into the ground, be heated, then rise. Both types have inspired a persistent combination of attitudes. We have noted and sometimes feared their potential dangers. We have appreciated their practical benefits. We have associated them with supernatural powers. We have found them pleasurable. And we have thought of them as healing.

To the Romans, hot springs were simultaneously medicinal, divine, and luxurious. The most important Roman hot springs area lay west of Naples in the volcanic Phlegraean Fields, which contained the entrance point to the underworld for both Homer's Odysseus and Virgil's Aeneas. It also contained an extensive system of developed hot springs at Baiae. Excavations often turn up shrines and votive items there, as well as evidence that this place was used as a resort and healing center. In his *Natural History,* an early compilation of the European world's accumulated knowledge, Pliny the Elder noted that the minerals in the waters of hot springs were good for muscles, paralysis, and internal irrigation, though he also warned of the dangers of staying too long in their heat. The aesthetic and technical care with which the Romans decorated such places can also be seen in the excavations of Pompeii and Herculaneum, whose hot baths were first fueled and then buried and preserved by the same subterranean fires.

Another ancient example is the hot waters of Bath, in England, probably the world's best-known instance of water heated by geothermal gradients rather than by magma. Celtic Iron Age people worshipped their god Sul there. Later, the Romans (who developed hot springs across their empire) built elaborate baths (the Aquae Sulis) and an adjoining temple to Sulis Minerva, in a typical association of hot springs both with their own gods and goddesses and with local deities. The English word *bath* is an old one, already in use in Old English to designate hot springs in general, and this town specifically, at least by the eighth or ninth century; its earliest uses listed in the *Oxford English Dictionary* are for these meanings. Indeed, nearly all the Germanic languages have some form of this word, including *bad,* a frequent element in European place-names, such as Baden and Marienbad ("Mary's bath," another link between religion and geology). *Spa,* similarly, which had entered English by the sixteenth century, originated as the proper name for some hot springs in Belgium that were known for their healing properties.

A similar mix of attitudes is evident among the various Native American cultures whose members have visited or lived near the thermal areas of what is now Yellowstone National Park. Some Crow people saw them as sacred sources of power. The local Shoshone thought the waters helped aching bones and swollen joints, skin problems, ulcers, indigestion, and even gallstones. Sometimes they buried important people in the hot springs, where, depending on water chemistry, the bodies might be dissolved rapidly by acids or eerily encrusted in pale mineral deposits. In harsh winters, people found warmth in the vicinity of these waters. The Nez Perce reputedly cooked food in them, another practical use of natural heat—though a risky one, as the ground at the edges is likely to be a thin and breakable barrier between feet and extremely hot water.

Natural hot springs and their artificial imitators (including hot tubs, sometimes called spas) continue to be associated with healing and hedonistic pleasure. "Taking the waters" or "taking the cure" in European and American hot springs has been a popular activity in recent centuries for treating specific ailments like arthritis, for vaguer varieties of invalidism, and—at least as important—for gathering people with money to spare and time to fill. President Franklin Roosevelt spent a good deal of time (and died) at Warm Springs, Georgia, where the mineral springs were considered therapeutic for polio, and Hot Springs National Park in Arkansas preserves a nineteenth-century American health resort. Similarly, genteel eighteenth- and nineteenth-century Britons and Europeans frequented Bath and the spas of the Continent. And the thousands of hot springs in Japan, called *onsen,* are traditional spots for social gatherings and for healing baths—not just for humans, but also for wild snow monkeys (macaques), who gather, swim, and play in some mountain *onsen,* especially during the cold winters.

Among the many fictional works that feature hot springs are two of Jane Austen's novels, the early *Northanger Abbey* and the late *Persuasion.* Both are set in Bath, where some characters drink the waters and soak in the baths to treat "gouty" constitutions and rheumatic fever, while others gather for dances, concerts, tea, and gossip. At the Pump Room, where a pump-fed fountain provided mineral waters to drink, everyone "paraded up and down for an hour, looking at everybody"—the backdrop of social convention, pretension, gossip, and triviality against which the sincerity and good hearts of Austen's heroines and heroes shine. Ford Madox Ford's 1914 novel *The Good Soldier* is set largely at the German hot springs resort of Bad Nauheim, where several key characters have gone to treat "heart" ailments, which are sometimes literal but always metaphorical. This location functions largely to set up the hypocrisy of the characters—whom the narrator calls "the good people," first seriously, later ironically, and finally bitterly—and their prewar, late-empire, moneyed society.

But hot (and warm) springs and the pools they form are not the only hydrothermal features, just the most common ones. The form of these features is determined by the specifics of their plumbing: the amount of water available, the pathway it takes to the surface, and the chemistry of the rocks it passes through. Plain hot springs occur where there is plenty of water and no impediments to its rise to the surface. Fumaroles (also called steam vents and, when sulfurous, solfataras, after Italy's Solfatara volcano) occur where there is just enough water available to create steam, but not enough to create a pool. Some fumaroles hiss quietly; others are loud like steam engines. Mud pots occur where a relatively small amount of acidic water dissolves the rocks through which it travels. These surface pools of thick gray and brown bubble and splat like chocolate fudge cooking in a saucepan and occasionally splash hot mud for considerable distances. Geysers are found where the plumbing has two additional characteristics: it is lined with something that holds high pressure (almost always silica-rich geyserite, also called siliceous sinter, which derives from rhyolite, one product of volcanoes) and it is at some point significantly constricted. With these two conditions in place, very hot water collects and pressures build until a chain reaction begins. A bit of water boils, the pressure drops, a bit more boils, the pressure drops more, and suddenly a full eruption occurs, releasing, finally, enough water, heat, and pressure for the cycle of intermittent eruptions to begin again.

Any number of transitional and hybrid hydrothermal features exist as well, such as continuously boiling springs (perpetual spouters or pulsating springs) and pools that occasionally splash, bubble, or overflow (intermittent springs). And the minerals in the waters create a wide array of associated features. Sinter cones, for example, are pale, amorphous, and sometimes fantastical piles that build up around the spouts of active or former geysers. Spring-formed travertine terraces like those at Mammoth Hot Springs in Yellowstone would, if they were green, look like large-scale models of the densely terraced mountain areas you see in the Andes and Himalaya, but instead they're various tones of white, yellow, and rust. Rarely, the pressure inside a geyser will build so high that it will explode rather than erupt, creating a hydrothermal crater or blowout, sometimes several acres in extent. Often many of these features occur together: a geyser basin will also contain hot spring pools, fumaroles, and mud pots, as well as various intermediate forms. And although their underground networks are not well understood, it is clear that they are often linked with each other. Thus interfering with one hydrothermal feature (as in drilling for generating electrical power) will very likely affect others nearby. Nor are they geologically stable, as hydrothermal areas are generally subject to earthquakes, which of course change structures underground.

The temperature of these waters varies considerably from outlet to outlet. Underground, where especially in the case of geysers they are under intense pressure, they can reach temperatures over 400° Fahrenheit. Their mineral contents and acidity vary, too, since both derive from the rocks through which they pass. Since the 1960s, an astonishing variety of thermophilic (heat-loving) and thermoacidophilic (heat- and acid-loving) bacteria, algae, and viruses have been discovered and studied in Yellowstone's hot pools and geyser runoff areas. Each takes advantage of a particular microhabitat of temperature and acidity, and each contributes differently to the kaleidoscope of greens, yellows, reds, oranges, blues, and browns that characterize this landscape. One such bacterium, *Thermus aquaticus,* provided the enzyme that made DNA fingerprinting possible. It is also used in testing for HIV and in producing pharmaceuticals. Some of these microorganisms are hazardous to human health; others have been put to work as deicers and to help remove lead paint and clean up mine tailings.

Geysers are the rarest, most vulnerable, and most spectacular of hydrothermal features. They also have the most interesting name, one that comes from Iceland, home of the only geysers known to Europeans until the nineteenth century. (Homer's *Iliad* describes what may be one near Troy, but that reference is not clear). Meaning "gusher" (and related to *gos,* "eruption"), Geysir is the proper name of a single large geyser, one described in writing at least as early as 1294. (In Icelandic, the word is pronounced "gay-seer" or "gay-zer"; the English say "geezer"; Americans say "guy-zer"—all with a hard *g* and an accented first syllable.) This Great Geysir remained a tourist attraction until early in the twentieth century, when it stopped erupting (except very rarely), either because earthquakes changed its underground structure or because it became clogged by visitors who provoked it into action by throwing chunks of turf or soap into it to increase the internal pressure. Now tourists instead watch the eruptions of Geysir's neighbor Strokkur ("churn").

The nineteenth century saw a significant widening of the world of geysers known to science, most dramatically with the discoveries in New Zealand and Yellowstone; most of those on the Kamchatka Peninsula were found in the mid-twentieth century. Although exact counts are impossible, current active or dormant geysers in the world number some nine hundred to one thousand (definitions vary, even the firmest of them leave gray areas, and the earth itself changes all the time in geyser areas). By any calculation, Yellowstone's ten thousand thermal features include more than half of the world's geysers. Kamchatka is the next-richest location, with about two hundred; New Zealand and Chile have about forty each; Iceland now has about twenty-five. As in other smaller geothermal areas, many geysers in Iceland and New Zealand have been destroyed by drilling to generate electricity.

Stories about Yellowstone's hydrothermal wonders trickled back to the East Coast throughout the nineteenth century, where they were met mostly with incredulity. The earliest known reference is in a letter sent to President Thomas Jefferson in 1805 with a drawing made on a buffalo pelt by Native guides, one that pictured "among other things a little incredible, a Volcano." Other early accounts (many by trappers) described "boiling fountains," "boiling springs," "steamboat springs," "boiling lakes," "spouting springs," and "water volcanoes." The word *geyser* took hold gradually and roughly in tandem with confirmation and finally belief. By 1874, summing up the now-complete process of word adoption, the visiting Earl of Dunraven wrote, "If I am taken to task for using the term 'Geyser' as I do, strictly speaking, the proper name of a certain spring in Iceland, it has gradually come to be considered the generic term applicable to all springs of a similar character." In the rest of his account, he used the word without the capital letter, as a common noun.

In 1870 the Langford-Washburn-Doane Expedition described and named Old Faithful, and it is this expedition that has been popularly credited with the idea of making the area the world's first national park. Certainly Yellowstone's geysers were key in that decision, as were the descriptions published by Nathaniel Langford, first in *Scribner's Monthly* and later as a book. Even more influential were the published accounts of the 1871 Hayden Survey, accompanied as they were by William H. Jackson's masterly photographs and the heightened realism of Thomas Moran's oil and watercolor paintings. In another echo of Athanasius Kircher's vocabulary of wonder, Ferdinand Hayden noted in a report he submitted to Congress, "This whole region was in comparatively modern geological times the scene of the most wonderful volcanic activity of any portion of our country. The hot springs and the geysers represent the last stages—the vents or escape-pipes—of these remarkable volcanic manifestations of the internal forces. All these springs are adorned with decorations more beautiful than human art ever conceived, and which have required thousands of years for the cunning hand of nature to form." Published in 1876 in a handsome book of chromolithographs with a text by Hayden, Moran's watercolor of Castle Geyser was among the first color images of Yellowstone to be seen by a broad public. It features a sky of roiling grays, a horizon of dark ridge and forest, a lumpy cone of white, pale blue, tan, peach, and green, and a tall spout of white spray and steam. In the foreground, a round cobalt hot-spring pool is framed by the odd and distinctive browns, purples, greens, and oranges of thermophilic microorganisms living their unlikely lives in the runoff channels of geyser and pool.

Langford, Hayden, Jackson, and Moran were but the first of countless writers, photographers, artists, scientists, and ordinary tourists to find themselves fascinated with Yellowstone's geysers and geyser basins. Albert Bierstadt's painting *Geysers in Yellowstone* hung in the White House long enough to inspire President Arthur to visit the national park, and Ansel Adams made a luminous series of photos of Old Faithful in 1942. In the 1934 series of what are widely considered the most handsome U.S. postage stamps ever issued, the National Parks series, an elegant blue-and-white image of Old Faithful represents Yellowstone. A WPA poster pictures Old Faithful in bold vertical streaks and billows of white against a sky that contains its own dynamic vapors. (The WPA, the American Works Projects Administration, designed to alleviate the Great Depression, included funds for artists and writers.) Old Faithful has long been a kind of iconic image for the national park, and early visitors could take home stereoview cards of this and other hydrothermal features. Today's tourists will find a wide choice of images of this geyser—on coasters, place mats, greeting cards, key chains, refrigerator magnets, mugs, candle holders, snow domes, Christmas tree ornaments, T-shirts, needlepoint kits, and much more.

Old Faithful is also the geyser most often witnessed, by some two million people every summer—sometimes, in late July, by five thousand at a time. Park rangers tell of visitors who don't know or can't quite believe that the eruptions are natural, not manmade. One family is said to have called from some distance away to ask whether the eruption could be held for their arrival, and visitors have asked whether it had been affected by a large electrical blackout in the Northeast. Some find the landscape terrifying and expect the volcanic forces underground to burst forth into catastrophe at any moment. But most questions are some version of "When will it erupt next?"—and indeed Old Faithful's predictability is fascinating, suggesting as it does a deeply hidden but reassuring order in a place that certainly looks like earthly chaos. Although it no longer erupts roughly once an hour as it did before a large earthquake in 1959, Old Faithful is still quite regular, with intervals that depend on the duration of the preceding eruption, and so the next eruption time is always posted, with a plus or minus range of ten minutes.

At first there is nothing to see, just a thin strand of vapor rising from what looks like a small, pale volcano in the middle of a wide, pale, shallow mound. To the east the backdrop is dark—a low hill created by an old lava flow, a forest of lodgepole pines, a sharp line marking the stopping point of the great fires of 1988. Big black ravens stalk around on the ground, croaking; a bull elk with giant antlers wanders among plumes of steam. In the winter, these geyser basins fill with billows of steam, the shaggy coats of bison grow white with a thick layer of frost, and sometimes, when conditions are right, the air fills with ice crystals that glitter in the sun.

Soon small spurts of water emerge, rising a foot or two into the air, then subsiding back into the earth. Underground, pressures are changing rapidly as water on the verge of boiling pushes hard against the narrow spot in the channel—some twenty-two feet down, it's just over four inches across, as a camera probe discovered in the early 1990s. All at once, a bright white fountain of water and steam roars high into the air, occasionally expanding so quickly that sonic booms occur inside the column. The plume rises past a hundred feet, maybe nearly two hundred feet, as up to 8,400 gallons of water shoot high, drift sideways, and fall back to earth to stream quietly off the mound toward the nearby Firehole River.

There is drama in this event, as there is in smaller or less regular eruptions nearby. Yet the aesthetics of such a place are hard to accommodate, combining as they do startlingly intense colors and sun-catching fountains with acrid, sulfuric air, barren ground, and pervasive signs warning of deadly danger. Perhaps this is why images and reactions vary so widely. Hayden reacted with wonder. Moran captured some of the landscape's fundamental strangeness, its paradoxically delicate and eerie beauty. Bierstadt's image is dramatic and brooding, while Adams's photos are all bright whites against darkness. The WPA poster is simplified and cheerful. A character in Sylvia Plath's early short story "The Fifty-ninth Bear" thinks of Yellowstone's "vile exhalations" and sees its ground as a frail "shell of sanity and decorum" separating him from "the dark entrails of the earth." In sharp contrast, John Muir remarked that geysers, like earthquakes and volcanoes, "each and all tell the orderly love-beats of Nature's heart."

Like other active volcanic and tectonic sites, hydrothermal landscapes are insistent reminders of the potential instability of the ground beneath our feet, the constant shifting of what we tend to think is stable, what we wish to believe is permanent, though we know it is not. Here the insides of the earth rise to the surface among oddly unpleasant smells and colors that seem somehow unnatural but are entirely natural, offering us glimpses of what is normally hidden. And here we come face-to-face with what we have so long seen as the most fundamental elements of our world: water and fire, earth and air, and perpetual change.

2

Climate and Ice

So much depends upon the heat within our planet. How much more must it matter that we are bathed in the warmth of that much larger fire, our sun? So big it could hold a million Earths, hotter than we can really imagine (almost 10,000° Fahrenheit at its surface), our sun sends us the energy that makes life possible—all earthly life, save for the few odd creatures who find their fuel instead at ocean-bottom volcanic vents. Plants mix the sun's energy, its light and warmth, with soil, air, and water to create wood, leaves, flowers, fruits, seeds—food, in turn, for animals and for us. When we eat, whatever we eat, we are swallowing the sun's fire, mixed with earth, air, and water. That fire warms us, too, gives us vitamins that make us healthy, improves our moods, lets us see where we are and where we are going, offers us one of our best metaphors for understanding and revelation, for those moments in which we are enlightened, when we see the light.

The sun also drives both weather and climate, forces that—like the fire beneath our feet—control much about Earth and all its living creatures, including ourselves. Most often this happens without our conscious recognition: we think about the weather all the time, of course, but until recently we haven't thought much about the climate. And we're likely to assume that we can't do much about any of these great forces, other than to pray, make offerings, maybe try seeding a few clouds to direct precipitation; like insurance companies, we typically think of droughts or hurricanes as "acts of God." But we have realized quite recently that while we can't truly control climate any more than we can control earthquakes or volcanoes, we can change it—and that now we are actively, though not on purpose, changing it in ways and at speeds we are scrambling to understand. So it matters that we all understand at least some of the story of Earth's climates past, present, and future. Yet this is a difficult story. It is complex and often highly technical. It unfolds over periods of time that are much longer than our brains easily comprehend and often in ways that are hidden from our bodily senses. Not least, it is both emotionally and imaginatively demanding.

In this chapter, then, we'll begin by looking at the main ingredients and processes in the climate system and the atmosphere. Then we'll turn our attention to places of cold and ice, and we'll consider some of our stories—both scientific and cultural—about these landscapes.

◇ ON THE SPOT: UP AND DOWN THE HIMALAYA

Ellen Wohl

The Himalaya, weather makers of a hemisphere, define a landscape of immense power. They are so high and so broad that they deflect the air masses that govern climate. They block the flow of cold air moving south toward India. Tropical storms rushing inland from the Bay of Bengal hit this wall and dump their voluminous rains on the foothills and Middle Himalaya, reaching the highest peaks only as sparser snowfall. Beyond the Himalayan rain shadow, only scattered grasses grow across the Taklamakan, the Gobi, and Tibet.

Climbing into these mountains is a progression toward colder and drier. Farm fields and small villages cover the plains at the base of the mountains. Sweat comes easily in the humid air, no matter the season. As the land rises into mountains, farms cling to the steep slopes with narrow terraces that follow the undulations where creeks cut into the hillsides. Each terrace has a higher rim downslope that holds water in the rice paddies. Men guide water buffaloes dragging wooden plows before planting, and they beat the harvested stalks of grain with hand flails while their wives and children toss the grain up from broad, flat baskets to winnow chaff from seed. The air is pleasantly soft and warm during the day, when broad-leaved banana trees provide shade. With the cooling of evening, people sit in the doorways of huts thatched with banana leaves. Bees living in a section of hollow log suspended from the roof grow quiet as uncountable stars appear in a sky undimmed by the reflection of electric lights. Despite the steepness of the terrain, the landscape feels benign, noticeably tamed by the terraced fields and dense network of foot trails.

Passing from the Middle Himalaya into the higher regions, you find more challenging terrain. The long, narrow terraces give way to little patches of field squeezed in with much labor upslope of house-size boulders resting precariously on the mountainside. The vegetation shrinks back to shrubs and then to sparse grasses, and it is not apparent what the thick-shouldered, long-haired yaks find to eat as they wander in search of pasture. Waist-high stone walls mark off the small fields, brown splats of yak dung plastered against the stones. Dried dung forms the fuel for smoky fires in this landscape of little wood, as well as fertilizing the cauliflower and potatoes that grow here. People move about indoors and out wrapped in many layers of clothing, and washing is too painful to be practiced frequently in this land of limited water and continual cold. Silver jewelry set with greenish-blue turquoise and red-orange coral forms the brightest hues in the land's subdued browns and grays. Any journey, no matter how short, requires climbing up or down, usually both. Water does not come to your house in a pipe; you must seek it out via a steep climb and carry it home on your back.

These mountains are river makers, too. The Indus, Brahmaputra, Ganges, Mekong, and Chang Jiang all rise from melting glaciers that rim the margins of the Tibetan Plateau. Although they might seem inconsequential amid the peaks that rise far above, the rivers help to shape those peaks. Cutting downward as tectonic energy forces the Himalaya upward, the rivers drive the adjacent hillsides beyond the point of balance. The oversteepened slopes give way in landslides that fill up the valley bottom, creating wedges of sediment that the rivers must carry downstream before they can continue eroding the underlying bedrock. Nothing about this back-and-forth between rivers and hills is steady or gradual. The hills release millions of tons of sediment in a single landslide. Melting glaciers release water that ponds behind dams of ice or sediment until the accumulating pressure bursts the dam and the rivers roar downstream in outburst floods that gouge out the base of the hillsides. Water in its solid state tears down the Himalaya, too. Geologists describe the effect of valley glaciers plucking up bedrock and bulldozing sediment as a glacial buzz saw that cuts into the mountains.

I never forgot the presence of these forces when I worked in the Himalaya. I went there to study glacier-outburst floods, and it took no subtle detective skills to trace the paths of floods long gone. Huge white scars of erosion marked the height of the floodwaters and boulders larger than a municipal bus lay stacked atop one another where the floodwaters had momentarily slowed in a wider portion of the valley. Each day I climbed up and down, up and down, my legs heavy with fatigue and the cold air sharp in my lungs. I felt like a human ant, always having to tilt my head at a steep angle to look fully up or down the slope on which I stood. People and yaks, footpaths, little villages all disappeared into the vertical landscape at a short distance.

I spent a month in the Himalaya, feeling myself begin to grow as lean and muscular as the tough, short-statured people who spend all their years climbing these slopes. As I finished the work and hiked back to lower elevations, a great storm reached the mountains. I waited out two days of rain, huddled into my sleeping bag in a little hut, watching the nearby river change from the cloudy pale blue of glacial meltwater to a thick brown. The rain stopped, the sun returned, and I continued downward, glad to be hiking again. When I reached the grassy airstrip at the little village of Lukla, the air was noisy with helicopters. Tourist trekkers who had returned to fly back to Kathmandu from this spot filled the hotels and overflowed into temporary tent camps, everyone talking excitedly about what had happened. The helicopters carried living and dead bodies being evacuated from a series of avalanches that had killed more than sixty people at higher elevations. Smoke rose from nearby hilltops where Nepalese killed in the avalanches were being cremated. The steep topography had transformed the benign and nourishing rains of the middle elevations to deadly snowfalls on the higher peaks. The Himalaya, named from the Sanskrit for "abode of snow," had once more made the weather. ◈

HOW THE CLIMATE WORKS

Climate scientists like to say that energy in must equal energy out. That is, however much energy we receive from the sun, we must radiate the same amount back into space. If it were otherwise, we would long since have either frozen solid or boiled dry and burned to cinders. They know how much energy we receive; how much bounces right back out, reflected by clouds, ice, and other shiny surfaces; and how much stays in our climate system, absorbed by rock and soil, plants and animals, the vast oceans, and the air itself. But when they do the math, they find that Earth's average temperature ought to be 0°F—a far cry from the much balmier 60°F of our measured global average surface (and near-surface) temperature.

Why the difference? Partly it's because the atmosphere is structured vertically so that the warmest, thickest parts are closest to the surface, while much colder parts float higher above; the average of all parts is the necessary 0°F temperature. If the near-surface air warms, as is happening today, higher levels cool to keep the balance. More important, at least for surface creatures like ourselves, is the behavior of certain particles and molecules in the lowest layers of the atmosphere. Some of these interfere with the energy coming in (most of which arrives in the short waves our eyes have evolved to see)—dust and soot, ash and sulfur-laden gases from large volcanic eruptions, and lower, thicker clouds, which reflect more heat back out into space than they hold in. Other particles, in this case mostly molecules, ignore incoming solar energy but catch and absorb it on its way back out, when it is in the form of longer waves we can't see. These molecules act like miniature heat lamps, sending some of this energy back down to the earth, where it provides, remarkably, more heat than the sun itself does. High, thin clouds behave this way, including those we create through jet contrails. So do the chief greenhouse gases, water vapor (H_2O), carbon dioxide (CO_2), and methane (CH_4), which together, in what we call the greenhouse effect, warm Earth's surface to a level that can sustain life. This simple fact of physics is also why we are now poised at the edge of considerable climate disruption. Take some of these heat-trapping molecules away and we get colder; add more of them, more heat lamps in the sky, and we get hotter.

What is not so simple is the behavior of carbon, that C in CO_2 and CH_4, the star player in one of Earth's main systems. Carbon in the air doesn't just stay there; instead, this shape-shifting trickster moves around from place to place, combining and recombining with other kinds of atoms, now air, now earth, now a living creature, now a volcano's hidden fire, now dissolved into the deepest water of an ocean. Carbon atoms move into plants, for instance, where they become leaves and seeds, and then into plant eaters of all sorts, where they enter skin, lungs, hearts. Then, when death and decay arrive, they move back out of these living tissues: every year in the autumn; over decades as larger bodies disintegrate; sometimes over centuries and millennia, as when plant parts are preserved from decay by being frozen in permafrost soils in the far north, or by being drowned, as in the bottoms of bogs. Carbon atoms also move between air and rock, as some dead creatures end up petrified—swamp plants becoming coal, sea creatures with shells becoming limestone. Over a long enough time, these atoms see the air again, too, when carbon-bearing rock makes its way to the ocean floor (sometimes directly, sometimes with long detours to the tops of mountain ranges) and then is carried very, very slowly into the mantle on subducting plates, turned to magma, and released again to the surface as lava, ash, and carbon-rich gases. By mining and burning coal and other fossil fuels, we have added a much faster route to move such rockbound carbon atoms back out into the air.

One more important thing to note about the structure of the atmosphere is how extremely thin it is, especially the bottom layer or two, where practically everything of importance to us occurs. Although there is no sharp outer edge, heights of roughly 60 or 75 miles are sometimes regarded as effective boundaries to what we consider atmosphere— roughly the elevation where auroras are formed. These are distances one can easily drive in a car on a highway in an hour or less. Farther out, we are in what we consider space: the International Space Station orbits about 215 miles up, in the thermosphere, and the space shuttle is typically about 185 miles high. The meteors we see as shooting stars usually burn up between 30 and 50 miles above the ground, in the mesosphere. Below that, the stratosphere rises to about 32 and holds the protective ozone layer. And the lowest layer, the troposphere, is about 10 miles high at the equator, where the air is warmest and most expanded, about 7 miles in the midlatitudes, and about 5 miles at the poles, where the air is coldest and most dense. The troposphere's importance to the surface of the earth is enormous: tropo- is from the Greek for "turn," and this is the sphere that changes, the space where virtually all weather occurs. Together, the troposphere and stratosphere account for 99.9 percent of the mass of the atmosphere.

By the time we're talking about the part of the atmosphere we're ever likely to visit, we usually think in feet rather than miles. Commercial airplanes typically fly some 30,000 to 40,000 feet up (roughly 6 or 7 miles), where they're above most turbulence and where the cold air increases fuel efficiency, but where if the plane's cabin pressure fails, passengers need extra oxygen to remain conscious. At the top of Mount

Everest, at an elevation of 29,035 feet (about 5.5 miles), hardly anyone can manage without extra oxygen. Everyone who climbs into air that thin (about 31 percent of the air at sea level) risks death from high-altitude pulmonary or cerebral edema, two results of oxygen depriva-tion. Everything above about 25,000 feet is called the "death zone," where oxygen is so thin that human bodies can't metabolize food and fail in other ways. Someone who lives at sea level would likely die fairly quickly about 20,000 feet up; humans can permanently adapt only to about 18,000 feet. Pilots, who must stay fully alert, need pressurized cabins or supplemental oxygen above 12,000 feet, sometimes less, and travelers and hikers are likely to find themselves winded or with head-aches much lower than that. Think of it this way: if you could turn the atmosphere sideways and ride across it on a bicycle at, say, a leisurely pace of ten miles an hour, you'd be past all the air that's thick enough to breathe in perhaps twenty minutes.

It is said that when the first human in space, cosmonaut Yuri Gagarin, looked out his spacecraft window back at Earth, he felt frightened, not of falling like Icarus from the sky, but because he suddenly could see how thin our atmosphere is, what a fine skin protects our breathing planet from the unforgiving vacuum of space. If you have a clear view of a horizon, you can see this too, in a more mundane version, at dawn or dusk, when the earth casts its shadow on its own coat of air to make a darker layer—an "earth shadow"—across the bottom of the sky. Turn your head sideways to let your eyes perceive with fewer expectations, and the layer becomes even more pronounced, that disconcerting line between earthly life and the remaining universe.

The second set of complications we'll look at—though much more briefly than they deserve—results from two facts: that the sun's energy falls on Earth's surface unevenly, and that the globe spins. The sun shines more brightly, with more energy, during the day than at night, in summer than in winter, at the equator than at the poles. That energy doesn't stay where it lands but gets redistributed around the surface, still unevenly, by air and ocean currents, winds, cold and warm fronts, and storms.

Because the sun's rays fall most consistently near the equator, the air there is hot and humid. This hot air rises, releasing its moisture as it does, and then moves toward the cooler poles. Because it is deflected by the spin of the earth (this is the Coriolis effect), the air also moves sideways. Then it descends in a band centered roughly 30 degrees lati-tude north and south, creating a hot, dry zone of high pressure—the zone of many great deserts, including the Saharan, the Arabian, and the

Australian deserts. The trade winds complete these large atmospheric circuits, flowing steadily westward and toward the equator, picking up moisture where they cross the oceans. The trades come back together again in the intertropical convergence zone, the ITCZ, a space that roughly coincides with the equator, though it moves a little north and south with the seasons (and over longer periods with such things as the growth of large ice sheets), carrying heat and precipitation with it and affecting, for instance, the line between the Sahel and the Sahara. When the trades converge, there is nowhere for all that air to go but up, and so it boils upward in enormous clouds—sometimes so high that they reach into the stratosphere—and then releases enormous torrents of rain. Thus the ITCZ is also the zone of the monsoons (from the Arabic word for "season"), where the world's most impressive rainstorms and thunderstorms occur and where we find the great tropical rainforests of Amazonia, Indonesia, and Africa. These giant circuits of air, energy, and moisture are called Hadley cells.

Paradoxically, between waves of storminess, the winds of the ITCZ weaken or die, the water grows glassy, and sailors speak of this region as the doldrums, a place where they may find themselves becalmed like Samuel Taylor Coleridge's Ancient Mariner:

> Day after day, day after day,
> We stuck, nor breath nor motion;
> As idle as a painted ship
> Upon a painted ocean.
>
> Water, water, every where,
> And all the boards did shrink;
> Water, water every where,
> Nor any drop to drink.

Coleridge interprets this event with a typically Romantic mix of natural fact and ethics: the Mariner's ship falls idle in response to his thoughtless killing of an albatross, one his shipmates (and more metaphorically the poet) see as having led them out of the icy traps farther south.

Similarly, at the outer edges of the Hadley cells, where the prevailing winds change direction, there is a narrow strip where the air is frequently still. These are the infamous "horse latitudes," where ships' sails went slack and, the story goes, sailors had to throw their horses overboard, either to preserve water or because the horses died from thirst. This, too, is the site of the Bermuda Triangle, a legendary (but unreal) site of vanishings, and the Sargasso Sea between the Azores and the

West Indies, where calm waters (skirted by currents) hold thick tangles of sargassum seaweed, to which eels migrate to breed—perhaps the origin of the "water-snakes" Coleridge's mariner blesses unaware, redeeming himself. The sargassum and the becalmed water together make a powerful image of entrapment whose resonance Jean Rhys called on for the title of her 1966 novel, *Wide Sargasso Sea*. This tale of British colonial sensuality, exploitation, madness, and paralysis is a classic "prequel," telling the story of the doomed first wife of Edward Rochester, the man who would later marry Charlotte Brontë's heroine Jane Eyre.

North and south of the Hadley cells, another matched pair of very large atmospheric circulation systems dominates the midlatitudes with prevailing westerly winds. In the Southern Hemisphere, with little or no land to interfere, these westerlies become the "roaring forties" and "screaming fifties," where gales make sea travel notoriously hazardous. In the Northern Hemisphere, where their speed is moderated by the alternating patterns of water and land (including the occasional mountain range), they control the weather of Europe, northern Asia, and much of North America. Their importance in England's climate, for instance, is everywhere visible in its literature. One early instance is the surviving fifteenth-century poem "Western Wind," with its haunting combination of wind, rain, and a traveler's yearning for home:

> Westron wind, when will thou blow?
> The small rain down can rain.
> Christ, that my love were in my arms,
> and I in my bed again.

This zone of westerlies is characterized by waves in the atmosphere, sometimes small and sometimes large enough to see on television weather maps. Carrying either a fairly steady series of mild, moist breezes or an alternating series of high and low pressure zones that can bring more extreme conditions, including major storms, these waves move farther north, then farther south, changing from day to day, from season to season, and again, it seems, with long-term climate change. Their strongest thread—the jet stream—is high above the ground, where the difference is sharpest between cold polar air and warmer air coming from the equator. A plane that catches the jet stream on its way from Tokyo to Seattle will save significant fuel and time.

When European mariners figured out how to cross the Atlantic and Pacific in the fifteenth and sixteenth centuries, it was because they discovered that they could drop far enough south to sail westward on the trade winds, then slide or tack sideways into the westerlies, which moved them back eastward. (Winds are named for the direction they come from; ocean currents for the direction they are going.) The westerlies allowed ships like Columbus's to return straight home from the Americas to Europe; they also make it faster to fly from San Francisco to New York than back. The same navigation method worked in the Pacific: to Asia with the trades, back to America with the westerlies. (The word *trade* originally meant "track," and the trade winds were so named because they blow along a regular track. The word's uses for occupations and for commerce share this origin.)

The other major actors in the redistribution of solar energy about the globe are the oceans. Equatorial water is warmer than polar water, and this temperature difference helps drive a vast system of currents that run both along the ocean surfaces and deep along their floors. In the largest pattern, warm surface water moves into the North Atlantic, where it cools and sinks to the bottom of the ocean. Then it flows back south, around Africa, into the Indian Ocean and North Pacific, where it rises, now warmer, then turns again to flow back toward the Gulf of Mexico. A complete cycle on this great conveyor belt takes about a thousand years, and since ocean water is a major absorber of atmospheric carbon dioxide as well as heat, this period is an important factor in predicting the timing, effects, and duration of climate change.

Many other important patterns of air and ocean movement exist, as do various interactions of air, water, topography, and even vegetation: the global swirl of energy is complex. Some large systems that operate on irregular timetables of several years have significant effects on climates, such as ENSO, the El Niño–Southern Oscillation, an interaction of air and ocean temperatures that brings drought or flood, productive or barren fisheries, life or death, not just to its South Pacific neighbors Australia and South America, but also to much of the rest of the globe. Similarly, a shifting pattern of high and low pressure systems connected to the Arctic Oscillation can lock in benign or turbulent patterns of weather for Europe, Siberia, and North America. Another very important factor is the ability of the oceans to hold heat (and thus moderate temperature changes) and generate water vapor, so the distance from a land area to an ocean helps determine its climate. Places farther from oceans are drier (as are places downwind of mountain ranges) and more variable in temperature, as in Siberia, central Canada and Asia, the Gobi and Mojave deserts, and the semiarid steppes of the Tibetan Plateau, Mongolia, and North America. Living things contribute to the climate system as well. Thick plant cover reduces evaporation and increases transpiration, by which plants release water vapor and oxygen into the air, so some lush places create their own humidity, clouds, and rain.

This is as good a moment as any to think about the difference between weather and climate, one that's especially important to understand in light of scientists' predictions about climate change. The word *weather* appears very early in English (the *Oxford English Dictionary* cites a use from about 725 C.E.), is thought to derive from a root word meaning "to blow," and from very early on implies temporary conditions. *Climate* comes from the Greek, where it first meant "slope" (as in incline), next the slope of the earth from equator to pole, and then a latitudinal zone, which became its earliest meaning in English. Soon it also meant a region considered in terms of its weather (as did clime), and the *OED* quotes Shakespeare's *Winter's Tale* as its earliest use of the word with its current meaning: "The Clymat's delicate, the Ayre most sweet." In short, these two words have always distinguished between temporary and characteristic conditions.

Scientists today say that climate is the statistical envelope in which weather occurs. More colloquially, we might call climate the background we generally take for granted, weather what we experience day-to-day. A very hot July or deep snow in February will be big news in London but not in Chicago: in London, such weather is not typical of the climate; in Chicago, it is. One more analogy: climate is the clothes in your closet, weather is what you put on in the morning. If you move from Ottawa to Delhi, you won't know exactly what you'll want to wear next October 15, but you'll certainly have to replace much of your wardrobe. Indeed, without moving anywhere, you may be wearing your coats and sweaters less often than you used to.

Climate is substantially predictable, particularly on a global level. Next January in New York will be colder than next July; it won't rain as much this year in Alice Springs as it will in Hilo; add more carbon dioxide to the air and things will warm up. Weather, by contrast, is dominated by turbulence and chaos, highly dependent on small differences in initial conditions: that proverbial butterfly flapping its wings in the Amazon that affects the rain two weeks later half the world away. The weather in any particular place is therefore predictable only a few days out, certainly no more than a couple of weeks. Atmospheric scientist Kerry Emanuel offers the analogy of two leaves placed side by side in a flowing creek. They may travel together for a few feet, but it won't be long before their paths diverge, and they will end up tracing quite different routes downstream. Who knows why: a slightly different swirl of water, a larger or smaller rock in their path, a subtle gesture of curled leaf-flesh. But—here's the climate analogy—both leaves will, in the end, move downstream and downhill, not up.

There is a gray area between weather and climate, and it's one that contributes to the difficulty most of us are likely to have truly grasping the nature of climate change. How many weeks, or years, of drought constitute climate rather than weather? What if your average annual precipitation remains the same, but now you see rain or snow five times a year instead of twenty, and each rare storm is four times more severe: climate change or weather? At what point does a string of longer growing seasons and fewer subzero winter nights become a shift in plant hardiness zones? Scientists are careful to explain that no one storm is caused by or evidence of global warming, though intense storms may very well occur more often in the future; nor does a chilly year prove that the climate is not really warming at all.

U.S. government statistics use a thirty-year rolling average in an attempt to see past normal variations and bring the larger picture into view: to see climate rather than weather. A disadvantage of this method, though, is that it can mask changes that locally are quite abrupt and severe. This masking effect contributes to the problem ecologists call the "sliding baseline," in which changed conditions come to seem normal; only those with special local knowledge register the shift. If you've never seen a healthy salmon run or known the calls of skylarks as the background music of ordinary pastureland, you may not truly grasp what has been lost. Even for people of an age to remember, it is hard to reassemble in one's mind a world without the Internet, or e-mail, or mass air travel. Similarly, it is hard to remember the ecological or climate conditions of one's childhood, feeling sure that one is not exaggerating and mythologizing. These uncertainties and perceptual slippages are the sorts of things that happen in the translation gap between statistical windows and individual events, between what can seem to be an abstract big picture and the concrete, particular world we experience every day.

THE GHOSTS OF CLIMATES PAST

Numerous times over the past 4.5 billion years, Earth's climate has changed radically. Scientists are keenly interested in this climate history partly for its own intrinsic interest, which is considerable, linked as it is to the evolution of life and practically everything else about this globe. But they are especially interested now because developing computer models that can do a good job of predicting requires testing them against the record of past changes, and because those changes can give us some idea of what we may be seeing in the decades and centuries to come.

So what makes the climate change? Atmospheric scientists answer this question with two words: forcings and feedbacks.

Forcings are relatively easy to understand. Some are very large-scale and slow, such as the movement of tectonic plates, where changes in the configuration of land and water alter the flow of both oceanic and atmospheric currents—by opening and closing passages for water between land masses, by building mountains and uplifting large areas, by shifting land between equator and poles, into and out of the warmest sun. Sometimes, when uplift is greatest, the weathering of rock can take more carbon out of the air than is being put in, which leads to cooling; sometimes plates move apart at seafloor seams especially quickly, releasing more carbon into the ocean and then into the air than is being taken out, which leads to warming.

Somewhat more quickly—though still very slowly by human standards—the climate responds to forcings produced by our relationship with the sun. Earth's orbit changes shape. Sometimes it is more elliptical, sometimes rounder, and these changes alter our distance from the sun and the length of our seasons, on cycles of about 100,000 years. The energy that reaches specific parts of the globe also varies according to the tilt and wobble of our axis, which again slowly change the distance different places are from the sun at different seasons—on cycles of 41,000 years for tilt and around 22,000 years for wobble. Acting together, these three effects go by the name of the Milankovitch cycle (for Milutin Milankovitch, or Milanković, the mathematician who figured them out), and they control how much heat reaches which parts of the planet at what times of year. Most crucially, given our current continental positions, the Milankovitch cycle determines how much energy reaches the Northern Hemisphere during its summer. If enough summers are too cool to melt the previous winter's snow, ice sheets and glaciers build up. The energy the sun puts out also increases and decreases over eleven-year cycles that are visible to us in sunspots, which increase with a hotter sun. This is a forcing that can raise or lower Earth's average temperature by something on the order of a twentieth of a degree Fahrenheit. Under just the right circumstances, even such small changes may nudge the climate toward relatively short-term change.

Finally, there are major forcings having to do with the content of the atmosphere. Very large volcanic eruptions (or giant asteroid hits) can cool the planet for a few years by interfering with the solar energy that reaches Earth's surface. So, on a steadier basis, does the sulfuric haze that we produce when we burn coal. And, crucially, carbon dioxide and other greenhouse gases warm the atmosphere. The relationship between temperatures and carbon dioxide is complex, but they do change in tandem and in what is likely a two-way cause-effect relationship: one goes up, so does the other. When carbon dioxide levels rise, as we've seen, its molecules absorb and reradiate more heat energy, raising temperatures. When temperatures rise for other reasons, carbon dioxide increases because of things like faster vegetative decay. Melting permafrost, for instance, releases the carbon from long-buried and frozen peat.

All of these changes are complicated by feedback effects, and here's the rub: these are much more complex, less well understood, and thus currently under intense scrutiny. Indeed, along with future human decisions and actions, feedbacks are the source of much of the uncertainty in climate change forecasts. Here the vocabulary of climate scientists can be confusing to nonspecialists. Some feedbacks are "positive," which means not that they are good (as in the power of positive thinking) but that they increase or amplify primary effects, thus tending to move a system out of balance. That is to say, a positive feedback intensifies the effect that causes it and encourages current behavior. Other feedbacks are "negative," which means they lessen or counteract the effects that caused them, thus tending to return a system to its former state. For instance, warmer air holds more water, which in turn traps more heat against the earth: a positive feedback. By contrast, more carbon dioxide in the air typically makes plants grow more quickly, which in turn makes them pull more carbon dioxide out of the atmosphere: a negative feedback. Both can get complicated over longer time spans. Wet enough air releases rain and becomes somewhat drier and cooler; faster plant growth also speeds up plant death and decay.

Another especially important positive or amplifying feedback is the albedo effect (*albedo* is Latin for "whiteness"). Just as on a hot day a white shirt is cooler to wear than a black one, snow and ice reflect much more solar energy back out of the climate system than do land, plants, and open water. On the one hand, more snow and ice means more cooling, and thus still more snow and ice; on the other hand, more melting means more absorption, more warming, and thus more melting. (Adding darker dust—such as smog and fine desert sand blown onto snow surfaces—also speeds up melting.) This albedo effect is partly responsible for today's disproportionate warming in the Arctic and the Himalaya, where temperatures have in recent years risen much faster than they have elsewhere. To add another layer of complication, if enough Arctic ice melts, large amounts of cold fresh water may flow into the North Atlantic, which in turn may slow down the ocean's circulation and lead to cooling, a negative feedback following upon a positive one. This set of responses seems to have had much to do with the past fate of glaciers, ice sheets, and sea ice in the high latitudes. Such feedbacks are often not linear, which means the effects don't stay proportional to the causes, and they may involve thresholds or tipping points. Ice is solid as it warms degree by degree, and then all at once it hits its melting point and becomes liquid. A rock slab high on a canyon wall waits, waits, waits, maybe moving a hair at a time, and then falls all at once. Thus melting ice particularly concerns scientists trying to predict the pace of climate change. Nobody really knows how rapidly the melting-warming-melting feedback loops will occur, and many of the mechanisms associated with those loops are only now becoming visible.

Of course, it is not easy to figure out what Earth's climate was like in the past, especially long ago. But paleoclimate science is remarkable for its array of ingenious research tools. Some of these tools are strikingly complicated and technologically demanding, but as researchers gradually check their results against others and refine their own techniques, pictures of past climates are slowly becoming more sophisticated.

One large set of tools involves sediments, both on land and on the floors of lakes and oceans. Land sediments are the oldest, as ocean floors are always swallowed up sooner or later by tectonic movements. But cores drilled in ocean floors, where the oldest materials are between 180 million and 200 million years old, also cover very long periods. In some cases, layers of sediment alternate with layers of volcanic basalt and other igneous rocks that can be dated for periods of more than a hundred million years through rates of radioactive decay. Uranium (which decays into lead) and potassium (which decays into argon) are especially open to this dating process. The history of Earth's magnetic reversals allows scientists to date some iron-rich sediments. Other layers can be dated by the fossils they contain, both large and tiny. Palmlike and crocodile-like fossils in the Arctic, for instance, make it clear that the planet was much warmer about 60 million years ago. Pollens and plant and animal plankton species respond to climate conditions while they're alive, and they record those conditions by their presence in sediments after they're dead.

Land sediments can preserve ancient dunes and patterns of windblown dust that come mostly from periods of global cooling and drying. Seafloor sediments can retain cobble carried from inland out to sea by icebergs during times of warming. Lake bottom sediments (especially from glacial lakes) can sometimes be dated by annual layers called varves, in which grains of pollen are remarkably identifiable and can tell us what plants grew nearby: warm-weather oaks, cold-weather spruce, wet-weather ferns, dry-weather shrubs. Lake bottoms also retain the similarly informative bodies of tiny freshwater crustaceans called ostracods. Pollen and plant bits are preserved for many years in the middens of pack rats—masses of organic materials solidified into a kind of amber by pack-rat urine—and in deep peat deposits in bogs and fens. Bits of organic material preserved in such archives can often be dated by the decay of carbon 14, an unstable version of carbon created by cosmic rays hitting nitrogen in the air.

Countable, datable, and climate-sensitive layers also occur in caves, where surface water dissolves carbons in soils and redeposits them underground to create records of such things as monsoon intensity over periods of several hundred thousand years. (More exactly, these are carbonates, which differ in their molecular structure from organic carbons, though they still contain carbon atoms.) Layers occur in corals, where summer growth is faster than winter and species flourish at differing water temperatures and depths, and where growing corals absorb uranium, which decays at fixed rates into thorium. And layers occur in the trunks of trees, where more warmth and moisture at particular seasons produce faster annual growth and thicker, lighter tree rings. These records can be spliced together to make longer histories. Great Basin bristlecone pines have produced highly reliable chronologies extending roughly nine thousand years into the past.

Bodies of beetles can indicate temperatures during their lifetimes. The edges of fossilized leaves point to their climate preferences: sawtoothed for cooler climes, smooth for warmer. For periods recent enough to include the presence of humans or our ancestors, evidence comes from such things as the shapes of jawbones, the patterns of tooth growth, styles of ax and arrow point and pottery, and most recently DNA. The species and DNA of head and body lice have been used to figure out the development of clothing and the increasing size of human communities. Soil and air chemists, archaeologists, paleobotanists, several kinds of geologists, and seafloor explorers may find themselves collaborating to establish climatic conditions in particular places and moments of the past.

And of course for the most recent millennia, since humans have known how to write, historians can add to the mix of information with records of droughts in Akkadia, famines in Egypt, cherry blossoms in Japan, France's wine harvest, sea ice around Iceland and Hudson Bay, and violent storms on England's coasts. Famous diarists like Samuel Pepys, Gilbert White, and Henry David Thoreau let us know about the dry heat that came before the Great Fire of London in 1666, the noxious haze that followed the eruption of Iceland's Laki (and Japan's Asama) in 1783 and 1784, and many details about the timing of plant behavior around Concord, Massachusetts, in the middle of the nineteenth century. Although it's startling to learn that only in 1802 did anyone realize that clouds could be studied systematically—and named—as a way of better understanding and predicting weather, certainly in the past two hundred years, proliferating instrumental climate records have been indispensable for climate historians.

Perhaps the most intriguing bodies of paleoclimate data come from ice cores. These tubes of old ice are drilled with considerable difficulty and skill from mountain glaciers in places like Peru and Kenya and from ice sheets in Greenland and Antarctica; they are then carefully preserved in cold storage at sites around the world. They contain surprising amounts of information, not just about past climates but also about the composition and behavior of the ancient air itself. We can learn about periods up to ten thousand years ago from mountain glaciers, more than a hundred thousand years ago from Greenland, and eight hundred thousand years ago from Antarctica (where even earlier times are likely to be retrieved in the coming decade). Ice records seasonal climate changes with lighter and darker layers that carry information about local snowfall amounts and temperatures, as well as about more distant conditions. Local information can have global import; Greenland's climate, for instance, seems to be strongly linked to that of much of the rest of the world, at least in part because of the global ocean conveyor and the air circulation that accompanies it.

Ice cores mark such events as large volcanic explosions whose ash extends long distances and shows up very clearly as dark layers, along with deposits of the sulfuric acid volcanoes emit. Greenland's cores show very clear layers for many eruptions: Vesuvius in 79 C.E., Saint Helens in 1479, Laki in 1783–1784, and many more. The dates of these layers can often be checked against written records, of course, and for older events where the volcano source is known through such things as the chemical signature of the ash, against dates drawn from radiometric methods. This is another opportunity to cross-check accuracy. The ice cores also contain layers of dust whose original locations can sometimes be ascertained, and whose presence indicates wind patterns and aridity, two signs of global climate conditions. Saharan dust is found in Greenland, Patagonian dust in Antarctica, and Chinese dust in the Americas. Ice shows layers from forest fires and from the haze of the Great Smoky Mountains. It holds windborne pollen. It holds salt spray (another indicator of strong winds) and other tiny ocean debris, such as shells. Ice cores hold traces of hydrogen peroxide, something the sun creates in small amounts with an annual pattern, and isotopes created by cosmic rays, signals of changes in our atmosphere's chemistry, in the sun's activity, and even in the universe's production of cosmic rays.

The frozen water itself, too, holds in its molecular makeup a very valuable indirect sign of global temperatures and sea levels: its proportion of two natural oxygen isotopes, one lighter than the other, which distribute themselves according to their weight in atmospheric water vapor, the tops and bottoms of ocean waters, and snowflakes. The proportions of these isotopes in both air and water vary significantly according to temperatures. Since annual ice layers can be counted, oxygen isotope information can be matched with, or checked against, changes that scientists expect to see from the orbital cycles we touched on above. These same isotopes also become parts of corals and of several key ocean plankton species, and thus of the climate records in seafloor cores—yet another chance to compare different kinds of records.

Most amazingly, ice cores preserve tiny bits of the actual air from many thousands of years in the past. As older snow becomes covered and compressed by newer snow until it turns into ice, small pockets of air from between snowflakes stay put in the form of bubbles. With care, scientists can preserve that air and test it for its contents—how much carbon dioxide or methane it contains, for instance. There are a couple of complications. There's a lag time, since until the ice becomes totally solid, air can still mix between the surface and buried spaces; air molecules from inside these bubbles are thus younger than the water molecules that surround them. And at a certain depth and pressure, the air is actually absorbed into the crystal structure of the ice, where it is somewhat harder to retrieve. Carbon dioxide, as we have seen, tracks global air temperatures; methane tracks the abundance of wetlands, such as in the warm tropics and subtropics and cool taiga and tundra. Both are additional important evidence of past global warmings and coolings, and again, evidence that can be compared to that gleaned from other sources. All this and the intense imaginative pleasure of having literally in hand the very molecules of air that were once breathed in and out by woolly mammoths, giant kangaroos, saber-toothed tigers, the walkers who left their footprints in that volcanic ash in Tanzania, teenage Neanderthals, the earliest members of our own species, Jesus and Muhammad, Shakespeare and Jane Austen: how wonderful!

So we're ready now to look at the history of our planet's climate, the tale that scientists have been piecing together from all these sources. Because we are ourselves living in a warm moment during a long glacial epoch, we'll touch on some of the warmer periods in that history but give most of our attention to the story of ice on Earth.

We need not linger long with most of this history, since there is less and less clear evidence the further in the past we go, since the earliest climates have least to do with our current world, and since in any case ice ages have been remarkably rare. (Although scientists generally speak of "glacial epochs" that last millions of years and much shorter "ice ages" that last tens of thousands of years, here we'll call them both by the more common but less exact term "ice age.") There seems to have been at least one quite early, a little over two billion years ago. The next are intriguing but very controversial, occurred between some 850 million and 550 million years ago, and collectively go by the name of Snowball Earth (or, more mildly, Slushball Earth). Scientists think that at several times during this period most of the planet was frozen, including locations near the equator, and that we were rescued from permanent lifelessness by feedbacks involving the creation of carbon dioxide by volcanoes. Another icy period followed between about 460 million and 420 million years ago.

The next ice age (sometimes called the Karoo) stretched from about 325 million to about 240 million years ago. This time the cold may have been triggered by the evolution and proliferation of land plants during the preceding period, which would have increased oxygen and decreased carbon dioxide levels (because of photosynthesis). It also may have been a response to the burial of the ferns and other ancient plants in the so-called coal swamps or coal measures, which tied up even more carbon dioxide underground—hence the name Carboniferous applied to much of this cold period, which also saw the creation of seafloor limestone that is now exposed on land in places like Ireland's Burren. Many coal deposits have alternating layers of coal and sedimentary formations, layers that may have resulted from the rises and falls in sea level associated with the retreats and advances of ice sheets; many coal swamps were coastal wetlands that would have been flooded and exposed as ice formed and melted. At the same time, continents were colliding (to produce Pangaea and raise the Appalachians, among other mountain ranges), which would have increased the uptake of carbon dioxide by uplift weathering. As we've seen, if you take a lot of carbon dioxide out of the atmosphere, you force a cooling.

By contrast, the next really notable climate period was about 100 million years ago, when the planet was especially warm, even more so than it has been during most of its history, with very high carbon dioxide levels, sea levels at least 660 feet higher than today's, and, evidently, no permanent ice anywhere on the planet. This was roughly the second half of the Cretaceous, so called because *creta* is Latin for "chalk," and chalk is one kind of calcium-rich limestone that was laid down in abundance by those high, warm seas. England's iconic chalk downs and cliffs—including the white cliffs of Dover—began during this period as accumulations of tiny dead sea creatures (mostly foraminifers, one of the best sources for oxygen isotope dating, and algae), and it is this past climate that makes possible the giant English landscape carvings such as the Cerne Abbas giant and the Uffington white horse. During this period, warm, shallow seas covered most of the center of North America and cut Eurasia in two. Most of the Middle East and southwestern Eurasia were underwater, too, as were other large chunks of what is now exposed land around the globe. Dinosaurs and plants like breadfruit trees lived quite comfortably on the poleward sides of the Arctic and Antarctic circles, kept warm at least partly by the carbon dioxide produced by particularly intense global volcanic activity and unusually rapid tectonic seafloor spreading. (All the new lava pouring out onto seafloors helped raise sea levels, too.) This period also produced much of the world's oil, including that in the Middle East, where shallow seas were fertile incubators for the algae that later became oil.

◈ ON THE SPOT: ON THE BURREN

Gerald Delahunty

I find it hard to believe it is high summer as I leave Kilfenora to enter the Burren—the sunshine I've enjoyed all the way from the other side of Ireland has turned into a steady rain. I have decided to stop at Poulnabrone dolmen, and as I dawdle along hoping that the downpour will pass, I glimpse among the roadside grasses occasional flowering brambles, wild marjoram, and purple hints of tufted vetch flowers, frail tendrils curling from their leaves.

I pull into the Poulnabrone car park among the other soaked tourists and sit in my car looking around, still hopeful. Although I am only a few hundred feet from the dolmen, it is just a vague landwhale against a bullet-gray sky. I feel as if I'm trapped in a grainy black-and-white photo.

My first visit to Poulnabrone was more than thirty years ago, long before it became one of the most famous dolmens in the country, enticing tourist buses onto roads built for donkey carts. Although they no longer park on the side of the narrow boreen, the buses still unload hordes of visitors, who sometimes climb on the dolmen rocks to be photographed and to gaze at the surrounding Burren. This time, the rain seems to be sparing the monument that indignity.

As I sit trapped, I amuse myself making up translations for the name Poulnabrone. I play with *poll na brón.* I've seen it translated as "pool of the quern," though I don't have the local lore, the *dinn seanachas,* to make much sense of that. Certainly "pool of sadness" or, as Yeats might have said, "well of sorrows," seems just right for the dreary day and the loneliness of the surrounding karst. There's not a house to be seen, just a single tree left to its fate on the bare limestone terrace a hundred yards or so beyond the monument.

The rain isn't letting up, but I can't leave without making my way, as always, to the dolmen. The archaeologists who excavated it in 1985 tell us that the portal tomb may have been built about 3200 B.C.E. and so is one of the earliest of the hundreds of such landmarks on the Burren. Its capstone is a daring, slim, rectangular slab of gray limestone, elegantly tilting down from its pair of six-foot-tall, west-facing portal stones. They also tell us that the human and animal remains they found in and under the dolmen appear to have been gathered there from earlier resting places, so it probably wasn't originally built as a tomb. The monument may have been a place of ritual or perhaps a symbol of ownership of the surrounding land. We can only speculate. But these monuments surely testify to the thousands of years that people have lived on the Burren, in spite of its name (from the Irish word *buirinn,* "rocky place") and its initially inhospitable appearance, its wracked, bare, limestone terraces scoured by ice sheets, rain, wind, people, and their beasts. The bones and the items found with them—a stone ax head, a bone pendant, quartz crystals, a scraper, and arrowheads—are our clues to the short, hard lives its builders led.

I am thankful for the asphalt path that leads to the monument. I've hiked over similar bare limestone grikes (long gashes in the limestone eroded by thousands of years of rain) and clints (the harder limestone remaining between the grikes) on Inish Mór in the Aran Islands and know that a slip can crack an ankle. I also know it's too late in the season to see the carpets of blue spring gentians or the spring orchids, and though I had hoped to see swaths of summer flowers, harebells, roses, and poppies, I settle for the white and yellow petals of a few mountain avens still blooming in the grikes.

Like many other Burren plants, the avens are an Arctic-alpine species that thrives here, far from its normal range and climate. Mean annual temperatures are not cold (about 50°F) and there is lots of rain, some forty to sixty inches a year. But while it's hard to believe on a soggy day like this, the winds are constant and desiccating. Plants do well here if they keep close to the ground, where the rock holds warmth, and if they have waxy leaves and petals to preserve their moisture. If I checked with a thermometer, I'd find the air inside these mountain avens a few degrees warmer than the surrounding air. Each small flower is a spa for pollinating insects.

Nor is the Burren's climate Mediterranean, though plants normally at home in the Atlantic-Mediterranean region also grow here. The delicate and delicately named maidenhair fern lives deep in the grikes, which trap the moisture and heat it needs. The several I find glisten with a wet bright green sheen.

The Burren is the western end of a broad belt of limestone several thousand feet thick that stretches across the midsection of Ireland. Elsewhere the limestone is covered by later rocks and soil. But where I walk, that soil was scoured away about thirteen thousand years ago by the last ice age. About a quarter of the Burren is bare limestone, almost pure calcium carbonate. The remainder is lightly covered with soil dropped off the backs of retreating glaciers. This glacial till is basic and rich in nitrogen-fixing bacteria and mycorrhizae, perfect for orchids—and indeed the Burren is home to twenty-two of the twenty-seven species found in Ireland. In fact, it's home to more than 80 percent of the nine hundred or so plant species native to the island, surely remarkable for a swatch of only about

175 square miles, no more than half a percent of the country's land. The three-quarters of the Burren that was not scraped down to the rock has enough soil to support a rich grassland with hazel scrub and small ash woods that shelter liverworts, mosses, and ferns. And almost all the butterfly species found in Ireland can be found here, though right now they're staying dry and out of sight..

The surface grikes and clints are matched by massive underground erosion. Beneath the crotchety surface of the Burren are splendid cave systems. Aillwee Cave leads through underground caverns decorated by dazzling calcite formations, past a waterfall, and on to an underground river and lake system. Near Doolin, the Green Holes of the Hell Cave Complex extend out under the Atlantic to form the largest such undersea cave complex in temperate waters. The food supply is so rich there that dahlia anemones drape the cave walls and common prawns grow to the size of small lobsters.

Only one Burren river, the Caher, makes it all the way to the sea. All the others disappear beneath the karst. Many Burren lakes similarly disappear down central sinkholes, draining waters accumulated in winter and spring, leaving rich summer grasslands in their wake. Such lakes are called turlochs in Ireland. In winter they give refuge and food to migrating wildfowl, whose droppings provide phosphates and nitrates that fertilize the summer pasture.

I make my way, wet and a bit disappointed, back to the car park, now dreary and almost empty. I get back into the car and turn the heater up full, hoping to dry out as I head toward Galway and on to Mayo, where I plan to stay the night. The road winds down from the plateau on which Poulnabrone sits, passing rock walls that lay claim to small meadows and wild hazel groves, in a pattern that the Poulnabrone builders would recognize as their own. ◇

OUR ICE AGE

Starting about 55 million years ago, the planet began gradually to cool, as continents and oceans drifted toward their current locations. The rise of the Himalaya and the Tibetan Plateau may well have contributed to this cooling, since it would have disrupted air circulation and led to much increased weathering, and thus a decrease in atmospheric carbon dioxide. Other changes caused by plate movements have also been implicated, such as the development of the Drake Passage, which created open ocean all the way around Antarctica, making it colder, and the closing of Panama, which linked North and South America and cut off that Pacific-Atlantic ocean circulation route. By 35 million years ago, there was permanent ice in Antarctica; by somewhere between 7 million and 3 million years ago, in the Andes, Greenland, and Alaska. By roughly 2.5 million years ago, there were continentwide ice sheets and midlatitude mountain glaciers, and we were fully started on the last ice age, the one we are still in and will be as long as there are significant bodies of permanent ice on Earth. This is called the Quaternary, the Paleocene, or the Pleistocene ice age (though technically the Pleistocene begins a little later). But we might just as well call it the Human ice age, as it is the period in which our species split off from our ancient relatives and we became what we are today. And it is this ice age that has shaped the surface of much of the world we know.

During this long epoch, glacial and interglacial periods have alternated at least seventeen and maybe more than fifty times. (Because the counting process is not straightforward, scientists come up with different numbers.) Over and over, ice sheets have grown and melted, temperatures cooled and warmed, and so the ice edge moved north and south, melting and freezing, now colder, drier, and windier, now warmer and wetter, pulling along with it the climate and many of the world's ecosystems, whole landscapes of plants, animals, human families, and cultures. The major coolings have been slow, taking tens of thousands of years, while warmings have been rapid. For at least the past 800,000 years, it has taken about 90,000 years to cool, 10,000 to warm.

We must note one more key fact about these icy times. While these past changes seem gradual when we look at them from a distance, scientists discovered in the 1990s that at least the most recent ice age was actually filled with abrupt climate fluctuations. Global temperatures changed sometimes as much and as quickly as ten or fifteen degrees Fahrenheit in a decade or less. The reasons are still being unraveled, and new theories and data appear regularly; likely culprits are various feedbacks and tipping points, especially those involving ice. One of the scientists involved in this discovery, Richard Alley, tells the story in a fascinating book called *The Two-Mile Time Machine: Ice Cores, Abrupt Climate Change, and Our Future* (2000). The graph he presents of temperatures in central Greenland, for instance, shows eleven sharp spikes of 20°F between 70,000 and 20,000 years ago, seven of them between 40,000 and 20,000 years ago, along with many smaller rises and falls. Between roughly 20,000 and 18,000 years ago, temperatures stayed low. Then they rose abruptly though jaggedly some fifteen degrees Fahrenheit, up ten, down eighteen, up twenty-seven, and then, more gradually, up another seven, at which level they settled down in a dense saw-blade high on the graph, where they have stayed for the past 10,000 years. The teeth of that saw span only about seven degrees Fahrenheit, and they indicate what is by far the longest "settled" climate period for at least the 80,000 years covered by high-resolution ice core records.

Glaciers are imposing things: turbulent seas of cliffs, peaks, and chasms, white where nothing has obscured the winter's snow, dark where rocks have fallen from adjacent hills or are only now emerging from a melting surface, sharp crevasses glowing like enormous sapphires in their sun-lit depths, lumpy ridges of rock and gravel along their sides, perhaps a cold, dark lake at the foot covered with chunks of floating ice, piercingly brilliant in the sunlight or eerie in fog or low clouds. Geologists catego-rize these large, land-based bodies of ice by size and location. Valley or mountain glaciers, including small niche (or pocket) and cirque gla-ciers as well as delta-shaped piedmont glaciers, are topographically con-strained; ice sheets (the largest), ice fields, and ice caps are not. All form when annual accumulations of snow exceed what melts or evaporates. As snow persists and deepens, the buried layers turn into increasingly dense ice—ice that flows outward from the center of pressure and down-hill. According to the shifting balance of gains and losses, the resulting glacier will expand or recede.

Now, at the beginning of the twenty-first century, glaciers cover close to 10 percent of Earth's land surface, mostly in two giant ice sheets over Greenland and Antarctica. Greenland's is nearly as large as Mexico, and in places almost two miles thick; Antarctica's is about one and a half times the size of the continental United States and up to nearly two and a half miles thick. Together, these two areas of ice contain enough water to raise global sea levels something like 230 feet. Ice fields and caps remain in Iceland, Alaska, Canada, and Patagonia. Glaciers still exist in most of the world's high mountain ranges, where they provide water to millions of people: nearly 20 percent of the world's people depend on ice and snow melt for their water. Nearly all—something like 98 percent of all mountain glaciers, and more than 90 percent of all glaciers—have been melting at alarming rates.

Although few of us will ever stand on one of the great ice sheets, sometimes it's possible to look out an airplane window on a flight over the North Atlantic and see Greenland. A vast central plain of white buries whatever past and future landscapes might lie beneath. Outlet glaciers flow toward the ocean with their flow patterns as conspicuous as if they were giant petrified rivers, as in a way they are. Scattered spires of rock rise out of the ice, stark black against stark white. Steep-walled bright blue or blue-green coastal waters are full of icebergs that have broken loose from parent glaciers, their scale impossible to discern from high above. Many mountain glaciers, too, are fairly accessible to drivers, hikers, and boat passengers, as in Switzerland, New Zealand, Iceland, and many of the national parks in the Cascades, Alaska's coasts, and the Rockies, including some ice fields between Banff and Jasper in British Columbia.

To step back in time would be to encounter quite a different case. A mere 18,000 to 20,000 years ago—a sliver of geological time, long enough to erase roads and cars but not walkers or boats—the portion of Earth's land covered in ice was much higher, not a tenth but a third. At this Last Glacial Maximum, two often-connected ice sheets blanketed about half of North America: the Cordilleran to the west, the much larger Laurentide to the east. Ice covered virtually all of Canada; all of New England, New York, Michigan, and Minnesota; most of Ohio, Illinois, Indiana, Wisconsin, and North Dakota; and good proportions of Pennsylvania, Nebraska, South Dakota, Montana, Idaho, Washington, and Alaska. The Missouri and Ohio rivers offer a rough trace of the southern ice edge, not by accident but because the ice pressed them into their current courses. South of the Cordilleran ice sheet in the American West, glaciers covered most of the higher mountains.

Ice sheets covered all of Scotland, much of Ireland, and England as far south as Yorkshire, Norfolk, and the Midlands. (The maximum extent of the ice in Britain came earlier, about 450,000 years ago, when it reached Bristol, Oxford, and north London, redirecting the Thames and the Severn as it moved.) They covered Scandinavia and much of northwestern Europe, a large part of Siberia (though not as much as one might expect, since much of it does not receive enough precipitation for glacier formation, as was also true in interior Alaska), and the major mountain ranges of Europe, Asia, South America, New Zealand, and Tasmania. Smaller mountain glaciers occurred (and in a few cases still linger, though not for long) in such widespread places as Taiwan, Japan, Morocco, Ethiopia, East and Central Africa, New Guinea, Australia's Snowy Mountains, and South Africa's Drakensberg.

With so much of the planet's water locked up, sea levels at the Last

Glacial Maximum were more than three hundred feet lower. Coastlines were much farther out than they are today, and land areas now divided by water were continuous. Ireland, England, and France were connected, for instance, as were Alaska and Russia, Australia and New Guinea. The great ice sheets were thick and very heavy, nearly three miles deep above Hudson Bay, roughly a mile above Boston, two miles over Scandinavia. Their weight pressed the continental crust beneath them hundreds or even thousands of feet down into the mantle. Around the ice-free edges of these depressions, the earth's crust sometimes bulged in response, and meltwater gathered between glacier and bulge (or between moraine and receding ice) to form large lakes, including significantly larger versions of North America's Great Lakes. Of course, temperatures were lower, with specifics depending on latitude and air and water circulation. Estimates vary, but global average temperatures likely averaged about 10°F colder than recent decades, with some regions about 40°F colder. Britain may have been about 21°F colder in the winter, about 14°F colder in the summer.

Immediately next to the ice, the earth was barren, cold, and windy, without enough plant life to support animals, a sort of polar desert. Beyond this band, tundra stretched a good distance before conditions became mild enough for grasslands or woodlands to survive, grass in drier areas, trees in wetter. Farther away, most of the world was drier, since cold air holds less moisture, and deserts were larger. But altered weather patterns brought extra rain and snow to some areas and created numerous large lakes in areas that are now very dry indeed. Lake Bonneville was a much-expanded version of today's Great Salt Lake in Utah; its former shores can be seen around Salt Lake City in the form of elevated benches, on one of which sits the University of Utah. Lake Lahontan covered much of northern Nevada, including the Black Rock Desert, site of the annual Burning Man festival, a celebration of what its founder called radical self-expression. Lake Chad was as big as today's Caspian Sea, which itself was about twice its current size. Everywhere, plants and animals moved as the climate changed, retreating into smaller protected refugia, where species could try to outlast harsh periods, and then advancing back out across larger areas. In the eastern United States, the main movements were south and north; in the West, they were up and down, into and out of high-elevation sky islands. Different species moved at different rates, and so the makeup of ecological communities changed with each shift. Some scientists think that even in the faraway rain forests of Amazonia and central Africa, the ice age forced the thickest jungles into much smaller pockets. This newly fragmented habitat, they speculate, caused new species to emerge, leading to today's high diversity.

In the vast areas of tundra and steppe in the Northern Hemisphere, large mammal populations were quite different from today's, including giant short-faced and spectacled bears, saber-toothed and scimitar cats, mastodons, woolly mammoths, dire wolves, steppe bison, cave bears, and North American cheetahs, lions, horses, tapirs, and camels, all now extinct—as well as larger numbers of the surviving saiga antelope, wild reindeer and caribou, and musk oxen. Other large Pleistocene animals lived farther from the ice, including the cattle-ancestor aurochs, woolly lions, marsupial lions, a giant wombat, and other creatures that would seem purely fabulous if we didn't know they were once real. (The fossils of the La Brea Tar Pits in Los Angeles are one of the best sources of information about such ice age fauna from milder climes.) Tundra plant species grew far to the south, even on mountains now far too warm and forested.

LANDSCAPES SHAPED BY ICE

We might well think that since all this ice is so long gone, it has almost nothing to do with today's world. But this is far from the case, for glaciers and ice sheets are among the planet's greatest earth movers. Let's turn our attention now to some of the ways they have shaped our landscape—sampling, as we go, the rich vocabulary for ice-linked landscape features that is available to speakers of English.

As the planet warmed and ice melted, sea levels rose everywhere, covering old coastlines, forcing their inhabitants farther inland, and burying the evidence of what is likely to have been a significant proportion of ice-age human habitations. Along what are known as drowned coasts, where oceans advanced into glacier and river outlets, the lower reaches of freshwater rivers became tidal saltwater estuaries like Boston's Charles River, Chesapeake Bay, and London's Thames. Freed from the weight of the ice, the land rebounded. In the past eleven thousand years alone, parts of North America and Scandinavia have risen roughly a thousand feet, and in some places the land is still rising. Along Canada's northern coast it is still not hard to find fresh-looking rebound beaches or terraces, stair-stepping out of the Arctic Ocean more quickly and recently than the postglacial rise of sea levels. Juneau, Alaska, is still rising about three quarters of an inch a year; when it has finished its isostatic uplift,

Hudson Bay will no longer be a significant body of seawater. And, of course, the earth's plants and animals had to adjust to the altering climate: colonizing newly exposed ground, moving poleward to follow familiar conditions, adapting to new sets of companions, new ecological conditions, or, in some cases, moving toward extinction.

The retreating ice sheets and glaciers left their traces everywhere they'd been. Over more than half of North America, nearly all of Britain and Ireland, and much of northern Europe and Asia, as well as in most of the world's great mountain chains, ice-created features shape much of what we see today.

Over relatively flat terrain, ice scrapes off loose surface material, often down to bedrock and beyond, sometimes excavating giant holes, such as those that contain Canada's Great Slave and Great Bear Lakes. In hills and mountains, glaciers typically gouge long, very deep U-shaped valleys or troughs, often with tributary branches either at the same level or higher on their walls, creating some of our most spectacular landscapes. Many of Scotland's glens are such troughs, as are the notches of New England's White Mountains. So is Yosemite Valley, whose tall waterfalls spill out of the hanging valleys carved by smaller branch glaciers. Fjords or fiords (this word is Norwegian) mark the troughs where glaciers reached the ocean; in Scotland, these are called sea lochs. Inland, finger or ribbon lakes fill corresponding troughs, as, for instance, on New Zealand's South Island, over much of Scotland, and in the Finger Lakes area of New York. In mountainous but only partly glaciated areas like the Colorado Rockies, it is often easy to see where the glaciers stopped, as the valley floors will suddenly narrow from ice-shaped Us to water-shaped Vs.

Higher-elevation glacier-carved forms are also thoroughly named, thanks to mountaineers. Bowls scoured out by the highest glaciers are cirques (from French for "circle"; commonly used by Americans), corries (from Gaelic coire, for "kettle" or "cauldron"; typically used in Scotland, Ireland, and sometimes Wales), cwms (pronounced to rhyme with rooms; a modern Welsh word with old roots, cognate to coomb, a small valley or hollow), and ravines (used in the White Mountains, as in Tuckerman's Ravine on Mount Washington, which also has a cirque called the Great Gulf). In the Northern Hemisphere, these bowls tend to face northeast, where the snow lingers longest. Their steep sides are headwalls, and the small, clear, round lakes they often hold are tarns (once a word used in northern England, with Middle English and Old Norse roots). Sometimes a chain of cirques and tarns will cascade down a long mountain slope, forming paternoster lakes, which resemble rosary beads and can catch the sunlight like a string of mirrors.

Narrow ridges that remain between cirques or parallel ice streams are called knife-edges or arêtes. Where enough of them appear together, serrated or saw-toothed ridges result—the origin of the Spanish word *sierra*. A single ridgeline spire left after the tops of glaciers have cut away all its surroundings is called a horn. A U-shaped pass across a ridgeline is sometimes called a col, a word from the Latin-based dialects of the Alps (and before that from the Latin word *collum,* meaning "neck"). In all these cases, where the rock is sufficiently hard (as is granite), the scouring of the ice leaves glacial polish and striations of various types (until normal weathering erases them), signs that helped nineteenth-century geologists deduce the former presence of glaciers, the direction of ice flow, and the number of glacial advances.

Glaciers and ice sheets also carve an array of streamlined hill-size features. When they encounter a resistant outcrop or knob, they may round off that rock and leave in its down-ice shelter or lee a trail of debris, making a crag and tail feature like the one that holds Edinburgh Castle and the Royal Mile. Over other large rock surfaces, the ice may carve whalebacks (also known as rock drumlins or tadpole rocks), smoothing the surface into a relatively gentle near slope and a steeper far side. Whalebacks often occur in pods and may be superimposed on each other. Similarly, in another version of this stoss-and-lee form, a glacier may smooth the surface it encounters first, and then, through freeze-and-thaw action working on cracks and faults in the rock, pluck rock from the far side to create a steep, jagged lee: sheepbacks or roches moutonnées. Very large examples exist at Deeside, England, and Lembert Dome in Yosemite's Tuolumne Meadows is another.

The result of all this excavating, plucking, grinding, scouring, and polishing is a tremendous volume of loose debris, varying in size from large boulders to the tiny particles of clay, silt, and rock flour that turn glacial streams their characteristic milky texture and hue. This debris is called till or drift, and ice carries it considerable distances before depositing it in a wide variety of ways. (Geologists may use *till* to mean unsorted and unstratified debris, *drift* to include both till and other stratified varieties.) Wind carries the smallest particles far from the ice and deposits them to make loess (a German word meaning "loose"), the deep, fertile soil that covers the central plains of North America. Loess is often visible in river bluffs, as at Council Bluffs, Iowa, on the Missouri River, or at Natchez, Mississippi. (It may also derive from deserts, as it does in central China.) Closer to the ice front, wind deposits larger, sand-size grains, creating fields of dunes that may slowly become thinly covered with vegetation and persist for thousands of years, as in Nebraska's sand hills.

Just at the edge of melting ice, outwash plains and fans of drift form porous soils that dry out quickly and acquire vegetation that may be quite different from surrounding types. Some large, recently formed outwash plains, as in Iceland, Arctic Canada, and Alaska, are called sandar (the word is Icelandic; the singular is sandur). These broad swaths of cobbles, gravels, and sands are covered with active, unstable channels called braided channels caused by large but irregular discharges of water, including the giant floods called jökulhlaups that occur when ice dams burst and release the waters from large lakes beneath or at the edges of ice sheets. (This is another Icelandic word: *jökul,* pronounced, roughly, "yokel," is "glacier," and *hlaup* is "leap.")

Much larger versions of these floods have been strong shapers of the land as well, especially during times of major ice melting, such as at the ends of glaciations. Some ice-edge or proglacial lakes become extremely large, and so when their ice dams fail, the floods can be truly catastrophic. Between thirteen thousand and fifteen thousand years ago, for instance, Montana's giant Lake Missoula, which covered some three thousand square miles (including its remnant, Flathead Lake), burst its bounds at least twenty-five and perhaps some forty times, sending its waters westward toward the Pacific Ocean with nearly incredible peak flows. (Estimates are 40 to 60 cubic kilometers an hour; imagine a bathtub three and a half miles square and a mile deep, draining in just one hour.) The Channeled Scablands of eastern Washington are one result; another is the fertile agricultural soil of Oregon's Willamette Valley, where a secondary dam allowed rich sediments to collect. Even larger lakes formed elsewhere, with one of the largest being Lake Agassiz, which at one time or another (though not all at once) covered large parts of Minnesota, North Dakota, Saskatchewan, Manitoba, and Ontario and held more water than all of today's Great Lakes combined.

One of the more obvious, though smaller, effects of ice sheets and glaciers is the long hills they create, moraines. They build these hills by pushing rock in front of them (sometimes even deforming bedrock, as with the chalk rafts in the coastal cliffs at Sidestrand in Norfolk, England, and the Dirt Hills in Saskatchewan) and by dumping debris along their sides (for lateral moraines) and at their ends (for terminal moraines), where they may form long arcs of till, marking the farthest advance of ice lobes and sometimes subsequent shorter advances. The two tails of Long Island mark such advances, as do the moraines of Cape Cod, Nantucket Island, and Martha's Vineyard. It is because it cuts through a terminal moraine that the stretch of the Hudson River called the Narrows does such a good job of protecting the waters in New York Harbor; the "hole" of Wood's Hole, Massachusetts, has the same origin. At the bottoms of mountain glaciers, terminal moraines may act as dams for piedmont lakes like Jenny Lake at the base of the Grand Tetons in Wyoming. And a medial moraine, one running along the center of an ice flow, is evidence of an upstream nunatak (this word is from Greenlandic), an isolated peak or spire that reaches above the ice.

Sometimes, when glaciers retreat, pieces of stagnant ice remain, become covered with drift, melt more slowly, and create depressions in the outwash surface. If the proportion of buried ice is large, the result is kame and kettle topography, a hummocky terrain of small mounds and small dips. If fewer chunks of ice remain, isolated kettle holes form. These can be quite deep and as much as two miles across. When the bottom of the ice is below the water table, lakes (often called ponds) form, and the shallower of these lakes may evolve into bogs. The most famous kettle is surely Walden Pond, whose depth (158 feet) and striking clarity were key symbols for Thoreau's vision of a life led "between the earth and the heavens," partaking of both.

Other glacial deposits create kames, from debris that enters holes in ice-sheet surfaces to form small mounds, and kame terraces, which form along the sides of ice sheets where meltwater runs between the ice and valley walls, as along Loch Etive in Scotland and the Connecticut River Valley in New England. (The word kame, or kaim, from Scotland and northern England, is a form of comb, a long, narrow, steep-sided ridge, so named for a resemblance to a cock's comb.) Sometimes, especially in Scotland, kame is used as a synonym for esker (from Irish eiscir), the sinuous ridges of drift that collected in under-ice rivers. Eskers, which are also called whalebacks and horsebacks, may be quite long, sometimes scores of miles. (They don't quite match geologists' categories, but these and other similar animal names are common for glacier-created shapes—in what might suggest traces of a sense of landscapes and glaciers as animate entities.) In the vast tundra spaces of Canada, where they are highly visible from the air, eskers help bush pilots navigate, and Arctic wildlife find them useful routes for travel and for denning, as they are easy to dig in and well drained, in contrast to their surroundings, which are often quite soggy. For the same reasons, they have long served as human travel routes and, along with other drift deposits, as sources of gravel for road building, fill, and clean building material. Boston's Back Bay was filled with the help of esker gravel.

Ice sheets also deposit drumlins, smooth, streamlined, elliptical hills shaped much like inverted teaspoons. With their higher and somewhat steeper slopes facing toward the source of the ice and their gentler slopes on the far side, they reverse the alignment of whalebacks and other stoss-and-lee forms. Drumlins consist of drift with a high percentage of clay and occur over softer kinds of bedrock, but their exact means of formation are still debated. They usually appear in so-called swarms, with notably large swarms in west-central New York state (some ten thousand drumlins cluster between Syracuse and Niagara Falls), the Boston area, and several places in Canada. Boston's Bunker Hill, of Revolutionary War fame, is a drumlin, as are Telegraph Hill and many of the islands one sees in the harbor when approaching Logan Airport in a plane.

Ireland is especially rich in drumlins, and indeed these (and other) glacial forms have played a significant role in that island's history. The word itself comes from the Gaelic and Irish *druim* (also *drum* or *drym*) for ridge (or *rigg*); it was introduced into geological usage in the early nineteenth century by Irish scientists, who initiated the study of these features. The large drumlin belt that stretches across Ireland from Strangford Lough and Belfast to Donegal Bay, along the political border between Northern Ireland and the Republic of Ireland, has been a barrier to travel, settlement, and communication since prehistoric times, with its confusion of small hills and chaotic drainages, its dense complex of lakes and channels, its many bogs, and its frequently poor farmland—all features caused by glaciers.

Other common evidence of vanished ice takes the form of individual rocks that have been transported away from their original source, sometimes a source that can be indisputably identified. These rocks, known as erratics, may be fairly small and move short distances, or they can be quite large (the Okotoks erratic near Calgary weighs some 118,000 tons) and move hundreds of miles. New York City's Central Park has many, as does Yosemite National Park, but America's single best-known erratic is certainly Plymouth Rock, the symbolic keystone for the English settlements in New England. The study of erratics has also been important to the development of geology, especially glaciology. In 1835, American geologist Edward Hitchcock argued that they were evidence for the biblical Flood. (The term *drift* for smaller debris left by glaciers and meltwater is a remnant of that theory.) Once it was noticed that New

England's erratics were always south or southeast of their sources, that theory had to become more complicated. Perhaps, it was suggested, the Flood was caused by an upheaval in the Arctic Ocean. Then the fact that many boulder trains were fan-shaped, spreading south and southeast of their parent rocks, indicated that the direction of the moving force varied, better evidence for ice than for floodwaters. And the presence of smaller erratics on the summits of New England's tallest mountains showed that an ice sheet had once covered everything, another blow to the flood model. The name *erratic*, in this case, refers not to irregularity or unpredictability, but to the word's earlier sense of wandering, as in knights errant and erratic stars (that is, planets)—and indeed, these are stones that have wandered, though not under their own power. Ironically, some erratics were in fact moved by water, carried by icebergs, such as those around the Willamette Valley in Oregon, where they were deposited by the Lake Missoula floods.

Taken together, the importance of all these features to the lands once covered by the great ice sheets can hardly be overstated. The contrast between glaciated and unglaciated landscapes is particularly evident in the eastern United States. North of the dividing line, lakes and ponds are abundant, as are waterfalls and rapids (good sources of power for mills), marshes, bogs, fens, and swamps; soils are typically thin, rocky, and hard to farm; exposed bedrock and boulders are everywhere. All this results from glacial carving and the disruptions caused by glacial depositions. To the south, on the Mid-Atlantic piedmont (not the coastal plain, which has quite different geology), natural lakes are nearly absent, as are bogs and swamps; boulders appear only in stream bottoms, large areas of exposed rock are rare, and soils are made of slowly eroded bedrock. These striking differences also baffled geologists as long as they believed in the biblical Flood, but they made clear sense once ice sheets—and ice ages—became part of our understanding of the planet's history.

◈ ON THE SPOT: IN THE CHANNELED SCABLANDS

Mark Fiege

I'm standing on a ledge of volcanic rock overlooking an arid coulee. Beneath me, sere grasses surround a green hay field freshened by arcs of water that shoot from wheeled aluminum sprinklers. The sun is hot and bright. My pale eyes squint in the glare, and my pink, freckled skin grows red and sore by the minute. The air smells of hot, dry grass spiced with sagebrush. Now and then a slight breeze provides a moment of cool. Around me are knobs, bumps, buttes, and cliff faces of dark volcanic rock. Up close, the outcroppings appear scraped, scoured, and sculpted. From a distance, they look like choppy undulations on a stormy lake. The contrast between aridity and evidence of water is striking: mind-boggling, catastrophic quantities of water, enough to fill a couple of Great Lakes, rushing through this land in a matter of days, over and over at the end of the last ice age, in what may have been the world's largest-ever deluges, the Lake Missoula floods.

The scablands have a biological veneer that defies their uninviting name. A hawk soars overhead; its cry, high and clear as it falls off, says *cheeeer*. When the breeze subsides, I listen to the buzzing of bees and flies and the chirping of birds. The ground underfoot is crispy with bunchgrass, cheatgrass, lichens, and dry, dormant mosses. The lichens on the rock are shades of pale green, yellow, brown, and gray. The moss is brown, bulbous, pushed up, and crunchy. I picture the moss when nourished by the rains of spring or fall: soft, cushy, and green, a living sponge. Now, under my feet and even under my touch, it crumbles into dust. Amid the moss, lichens, grass, and loose rock, in pockets of soil in an otherwise stony world, grow clumps of dull sagebrush and purple vetch. Below me, wild plum, wild rose, and other shrubby plants catch water and provide cover for wildlife. I startle a deer, and it leaps from its shady spot and bounds away.

Stubborn weathered remnants of human activity are evident on the rocky landscape, too. Shotgun shells—faded red plastic and tarnished brass—lie at my feet. Ravines dropping down into the coulee are now the graveyards of rusted pieces of metal and tangles of wire. An old barbed wire fence runs along a nearby ridgeline. Unable to penetrate the stony ground, the fence's gray wooden posts require knee braces and the stabilizing weight of loose volcanic rocks. In a few places, the fence posts stand in rock-filled rusty barrels.

The loose rocks are the story writ small. Tumbled over and over in the powerful current of those ancient floods, they are rounded, with all of their sharp edges worn away. I reach down and pick up one. Warm to the touch, it is roughly the size and shape of an ostrich egg. It is dense and heavy for its size, but it hefts nicely, and it makes me think it might be good for cracking nuts or skulls. At arm's length, its surface looks like pictures of the moon's surface, sort of gray-brown, rough, and covered with innumerable holes or pockmarks. The holes are the remains of gas bubbles in the rock when it was molten, but they remind me of lunar craters. The underside of the egg, the side that nested in the thin soil, looks slightly different. Just above the soil line yet below the curved overhang—and thus sheltered from the sun—is a ring of green-yellow-brown-gray lichen growths. Each growth is about the diameter of a quarter, and together the growths make me think of flower blossoms. Where stone touched soil, the egg has a whitish mineral coating or crust, as if water, heat, and time caused calcium or lime to precipitate. In places the coating gives the underlying dark rock a bluish tint. The ovoid rock is aesthetically pleasing to me, and I want to name it. On the central Snake River in Idaho, similar stones, some of them huge, are known among geologists as *melon boulders* or *melon gravel*. If I could name the type that I hold in my hand, I would call it *egg stone*.

If the egg stone is the story writ small, landforms are the scablands biography writ large. Coulees, like the one that I'm looking at, run through the region, each a former flood channel. In some places the coulees are almost narrow enough to throw a rock across; others (like Grand Coulee, which gave the famous dam its name) are enormous and run for miles. Sometimes coulees drop at ghost falls—Dry Falls, for example—the height and shape of which suggest the sound and fury of crashing water but which now stand eerily silent. Other falls, such as Palouse Falls, carry a trickle of water and only approach their full potential at times of maximum runoff. In other places the coulees lead to enormous holes, like Hole in the Ground, the very bottom of which collects intermittent or ephemeral water in small lakes and ponds. The holes are the greenest, most verdant oases in the scablands, the habitat of lush vegetation and waterfowl and the sites of hayfields and ranches. Yet these are the exception to the vast acreage of dry rock faces, loose stones, coulees, and arid-land vegetation that most define this tough volcanic landscape.

In some spots, the scablands have an architectural quality that derives from the intersection of geological forces. Volcanic basalt that cooled quickly formed densely packed columns that the swirling water then exposed. The columns remind some observers of tree trunks or posts in the palisade of a fort; they make me think of an ancient Greek architect—or Greek god—gone mad. Perhaps the most striking manifestation of the scablands story, however, appears not in the coulees, but on the edges of the flood zone, the places where the water was most shallow and slow. At the end of the Pleistocene, wind picked up fine dust (loess, often pronounced "lerce") left behind by the retreating glaciers and then deposited it in countless humps or hills tens of feet thick. These humps formed the Palouse, an exceptionally fertile region that farmers later converted to wheat farms. The water that made the scablands wiped away a good portion of the Palouse, but on the flood edges, where the current lost volume and momentum, the destruction was only partial. Here loess humps, many of them planted in wheat, stand on a floor of mostly volcanic basalt, the flood line at the base of the humps still clearly visible. Islands of arability in a landscape of stone, these stranded piles of soft earth objectify a profound drama at once volcanic, aeolian, and hydraulic.

Finally I turn to leave the coulee. Taking my egg stone with me, I walk back to the car, where my wife and daughter wait. I'm hot and burned and we need to be in the Puget Sound country, the wet side of the Cascade Mountains, by nightfall. We follow the winding roads in and·out of the coulees, the tires rolling on a surface made of—what else?—crushed volcanic rock. We stop for a few minutes in a small town. "What do you call this place?" I ask the postmaster and then a store clerk. "Scabrock country," each answers. Driving into a sun that now hangs low in the sky, I think of children scrambling over volcanic rock, their knees and elbows and hands bearing the crusty marks of their bloody encounters with the residue of catastrophe. ◇

ICE-AGE HUMANS

Our ties to the ice age are much deeper than the shapes of the land. Indeed, it now seems clear that our evolution, dispersal, and flourishing have been tied at key moments and places to the advances and retreats of the great ice sheets.

Some time about 10 million years ago, the climate in Africa began to change, and lush forests turned into drier woodlands. Between 7 million and 6 million years ago, the first hominids appeared, capable both of climbing trees and of walking upright on the ground, perhaps in response to a fluctuating climate. Anthropologists date the emergence of the genus *Homo* and, later, the first species that clearly resembled us (*Homo ergaster*) to the period between 2.5 million and 1.7 million years ago, when the global climate was cooling rapidly and glacial cycles were increasing in frequency.

By then, in a cooler, drier, and more seasonal climate, forests and woodlands were shrinking while grasslands expanded, bringing a better supply of nutritious tubers and large herds of herbivores. With this shift, antelope species changed from smaller numbers of woodland dwellers to large numbers of grassland dwellers, and it may well be that our own predecessors responded similarly to the reduction in our own main habitat. This is speculation, and will probably remain so, given the inevitable scarcity of evidence. Still, some have proposed that our increasing brain sizes and long periods of immaturity, which would both ordinarily be evolutionarily negative (larger skulls make birth more dangerous, and delayed maturity makes easier prey), became possible only with a shift into grassland living, a shift that freed up hands from climbing, for instance, and made them available for devising better and better tools. About one million years ago, as glacial cycles became longer and more severe, the first major wave of humans moved out of Africa, likely taking advantage of the spread of familiar grassy habitats, and, in some places, travel routes made easier by shrinking oceans.

Some have also suggested that human evolution may have followed the pattern of punctuated equilibrium, in which multiple evolutionary lines appear and disappear with relative suddenness in response to some external circumstances—like population fragmentation caused by climate stress. Several distinct groups of *Homo* did emerge and vanish during this period of significant and often abrupt climate change. Depending (as always) on how one counts, these include *Homo habilis, H. ergaster, H. erectus, H. neanderthalensis,* and perhaps others, including *H. heidelbergensis,* which seems to have been closely related to the Neanderthals, and the recently discovered and controversial *H. floresiensis.* All these groups are now extinct, except our own, *Homo sapiens sapiens.* In the context of climate, the European Neanderthals are especially interesting. Appearing about 400,000 years ago and lasting until about 24,000 years ago, by which time they were sharing the landscape and some of their DNA with *Homo sapiens,* they lived through glacial advances and retreats, moving south toward the Mediterranean coast and farther north, along with the changing weather and climate-sensitive edible plants and prey animals. Physically as well as culturally, they were a quintessential ice-age people, adapted to cold with their massive trunks and short limbs.

Many anthropologists believe that between 80,000 and 50,000 years ago, well into the last glacial period, a second large-scale dispersal from Africa occurred, this time of modern humans or *Homo sapiens*, all the while with their populations growing, shrinking, and moving with shifting ice and climate. They—or, by now, "we"—seem to have first left Eurasia for New Guinea and Australia, which we seem to have reached at least by 40,000 years ago, likely earlier. We arrived in western Europe by about 40,000 years ago and in northeastern Asia (which was much colder) some 5,000 years later, finding in those arid, grassy steppes herds of large mammals that offered food for human predators as well as hides, sinew, bone, and antler, essential materials for clothing, housing, and tools, including weapons. Throughout this time, we need to remember, the climate continued to fluctuate, spiking sharply up and down.

From early on during these cold and unsettled millennia, humans were making art: telling stories, playing music, carving, and painting. Concrete evidence is strongest for visual images, a good many of which survive, with European instances the best documented and dated. Other very early sites certainly exist elsewhere, likely everywhere that *Homo sapiens* lived. The ages of such sites are hard to tell, but sometimes they can be radiocarbon-dated through such things as the charcoal in black pigments or wooden painting sticks left nearby, or when portions of the art objects have become buried in layers with other organic and thus datable materials. We made portable art, decorated items small enough to carry around, to wear, or to include in burials. And we made rock art, both on protected walls outdoors, such as surfaces beneath rock overhangs, and inside caves, occasionally far deeper inside than most people would have traveled under ordinary circumstances.

One wonderful example is the Grotte Chauvet in southern France, found only in 1994 and radiocarbon-dated to between 30,000 and 32,000 years ago, when that graph of Greenland temperatures shows the end of a very large drop and then two or three smaller peaks. This cave includes paintings in red and black (and some very small ones in yellow), engravings, and finger tracings, some just sketches, others quite finished in appearance, of some fifteen animal species (some now extinct), including an unusual number of top predators like lions and cave bears. Chauvet holds paintings of one or two human or part-human figures, stenciled handprints made by blown pigment, and some geometric designs, including constellations of dots that turned out on close examination to be palm prints. Many of the animals—or, more precisely, their heads and necks—appear in groups arranged in receding or rising series, as though they were lined up next to one another but stretching away from the viewer—a pride of lions, a small herd of horses, groups of rhinos and aurochs. Two rhinos fight each other, their horns locked, one body twisted with force (this image is rare in being a scene); an owl looks over its shoulder (the earliest known representation of this bird); the lions seem to watch alertly for prey.

The lines and shading of some of these creatures are strikingly evocative and naturalistic. Their eyes and mouths and ears seem wholly capable of seeing, tasting, and hearing. The art here shows such skill, vision, and sophistication that its discovery forced a revision of art history away from a notion of a gradual—and later—evolution of style and technique. Here, as elsewhere, the painters (whose cultures anthropologists call Aurignacian) took advantage of the natural conditions of the cave (which is in limestone laid down during the warm Cretaceous), the natural colors of the walls, the shapes of alcoves and bumps. Horses in the Grotte Chauvet, for instance, sometimes seem to be walking out of alcoves into the wider spaces. Visitors to this and later caves such as Lascaux often experience the art as wholly three-dimensional, so effectively did the painters work with the shapes of the earth. The Chauvet cave also holds numerous bones of cave bears, along with other signs of bear presence such as paw prints and scratch marks. In one spot, which some researchers call a shrine, a large bear skull was found placed on a square rock.

Interpretations of all these images vary. Many scholars (including Jean Clottes, the director of scientific study of the Grotte Chauvet and author and editor of several beautiful books of photographs of Paleolithic cave art) think at least some of them, maybe most, were related to shamanism, a spiritual tradition in which some individuals are believed to be able to travel between worlds through visionary and literal transformations from human to animal. This is one of the earliest and most widespread forms of human religious belief and practice, one especially common around the cold North, where it has survived into our own time. There's something especially gripping about the images deep inside caves, these scholars remind us, perhaps because they have stayed so long hidden from the outside world, sometimes until very recently indeed, and perhaps because walking or crawling deep into a cave feels like moving beyond the ordinary surface world into another realm where it is hard to see and sometimes even to breathe, where we might find ourselves stepping on piles of animal bones or startling a live animal awake, where we might well feel we are entering a separate but adjacent reality.

One intriguing and quite different interpretation has been offered by

Dale Guthrie, who spent many years considering these representations from the points of view of a zoologist (more precisely, a paleozoologist specializing in ice-age mammals like mammoths), an evolutionary biologist, an artist, and a hunter. In a quirky but compelling book, *The Nature of Paleolithic Art* (2005), Guthrie points out that many cave images demonstrate very sophisticated observations of animals, such as those that might be produced by people for whom hunting was central, and that many are of unglamorous but edible subjects such as rabbits, fish (often clearly salmon), birds on the backs of grazers, and even maggots (like those that infect reindeer and make good food). He notes, too, that many images are sketchy and unskilled (such as all artists-in-training produce); that there's good evidence (from handprint and footprint sizes) that many of the visitors to these caves seem to have been male children and adolescents; that many images suggest the sexual fantasies of adolescent males; and that children and adolescents, especially boys, are still likely to explore such places as caves.

Although these two interpretations are radically different (and there are others), both recognize and preserve the power of these images. How can we help but be moved by contact with a tradition this long, one lasting tens of thousands of years? Adolescent graffiti, skilled natural history illustration, hunting magic, key elements of initiation rituals or shamanic ceremonies: we don't know what these ice-age images were for their creators. But for us they are at the very least a spur to our imagination. However enigmatically, they suggest the mental, aesthetic, natural, and spiritual lives of these people who were our kin, though their world and lives were so different from ours.

Most deglaciation occurred between 15,000 and 10,000 years ago, though because the great North American ice sheets were so large and stretched so far inland, away from the warming influence of ocean waters, it wasn't until about 6,000 years ago that they finally disappeared. As the climatic conditions changed, landscapes, ecologies, and human lives changed, too.

During the coldest parts of the ice age, the Bering Land Bridge—or Beringia—linked Siberia to Alaska. The word *bridge* suggests something narrow, but this one was more than nine hundred miles wide and to creatures traveling it would not have seemed like a bridge at all. It may have been too cold for comfort (or for herbivores and their hunters), or it may have been part of a vast "mammoth steppe," something between grassland and tundra, a sort of refuge from the ice. This period between about 13,000 and 10,000 years ago is the window most anthropologists believe allowed humans to move from Asia well into the Americas, with routes continuing south either along the coast or between the Cordilleran and Laurentide ice sheets, as these southern passages became gradually more suitable for humans with a moderating climate, melting ice sheets, and returning vegetation. Others argue that earlier migrations might well have occurred along the expanded Pacific coastlines, where ice cover was not continuous, and where any evidence of human presence has since been drowned by rising sea level. (Still other theories involve earlier boat travel along the Atlantic ice edge and across the southern Pacific, but there's as yet no evidence strong enough to make these ideas dominant.) By about ten thousand years ago, the bridge was severed by rising oceans and foot travel ended. When the next immigrants arrived—the Yup'ik, Inupiat, and Inuit who stayed in the Arctic—it was much later and by boat.

About the same time that humans moved into the Western Hemisphere, more than fifty North American mammal species became extinct. Ecologists and anthropologists are still debating the relative roles in these extinctions of ecological stress caused by climate change, a theory supported by extinctions in Eurasia during the same period of change, and human hunting pressure, a theory supported by evidence of similar extinctions following human arrivals elsewhere. It may be that the two forces acted together. Climate, for instance, might have reduced animal populations to smaller numbers and squeezed them into refugia, where the human hunters they had never before encountered could more easily kill them.

Meanwhile, back in the Old World, humans were busy domesticating plants and animals, developing agriculture, and building long-lasting settlements and the first cities. These changes mark the shift from Paleolithic (which began with the first hominid tool-making) to Mesolithic and then Neolithic cultures. They were punctuated, perhaps even driven, by the cold snap of the Younger Dryas—the most severe cold snap since the Last Glacial Maximum, which may have been caused by a sudden influx of water from the giant Lake Agassiz into the North Atlantic that shut down the world's ocean circulation. Lasting about 1,300 years, from 12,800 to 11,500 years ago, this period (named for a small alpine and Arctic flower that expanded its range with the cold) was a worldwide curtain call for the great glaciers and ice sheets, involving large temperature drops, a general expansion of ice, and a return to ice-age conditions farther to the south. Winter temperatures were especially harsh: with the prevailing weather coming across a frozen North Atlantic, northern Europe's temperatures were about 40°F colder than today's, and summers a more moderate 10°F colder.

In his book *The Long Summer: How Climate Changed Civilization* (2004), anthropologist Brian Fagan lays out this story. At the end of the ice age, humans dispersed as more temperate ecosystems spread. Their diet gradually changed from one based mostly on the meat of large animals (many of which were vanishing) to one based on smaller game and wild plant foods. These groups of hunter-gatherers were still small and mobile; they could and did move easily as climates and ecosystems changed. But in the especially warm, moist period that preceded the Younger Dryas, conditions in the lands east of the Mediterranean, in the Fertile Crescent, allowed and encouraged them to increase their numbers, spread out over larger territories, and begin to depend on certain abundant and nutritious wild plant foods. Key among these were grass seeds (wild wheat and rye) and pistachios and acorns from a long belt of nut-rich groves that stretched from the Negev Desert up across Syria to the Euphrates River. Along with their bordering ecotones (places where two kinds of ecosystem meet and access to plant and animal diversity is highest), these groves offered reliable but labor-intensive sources of energy. Thus human populations could grow but became simultaneously more fixed in place.

Then the Younger Dryas brought a thousand-year drought. The nut groves retreated, the deserts expanded, and it became much harder to acquire food. The humans clustered around water sources and built more compact and denser communities. By harvesting and saving the plants whose seeds they found most convenient and filling, and by gathering and herding the more docile sheep and goats, they domesticated local grains and animals and invented agriculture. The result, Fagan says, was a new way of life forged under climate stress—more settled, more attached both physically and culturally to specific places, more complicated, less flexible. With this came a new kind of culture, with new attitudes about the land itself, about human ancestors, and about the invisible worlds of the supernatural. The new way of life spread rapidly once milder weather returned at the end of the Younger Dryas and has continued, with only local interruptions, into our own time.

The relation between the Younger Dryas and developments in human society is one of many that have intrigued not just Fagan but a growing number of writers concerned with the interactions between human and climate histories. Disavowing the ancient but thoroughly discredited tradition of making direct and simplistic links from climate to cultural (and, worse, racial) characteristics, these writers have among their sources considerably more sophisticated and nuanced information about all aspects of this question—and an acute sense of its complications. The challenges of putting the pieces together across traditionally separated realms of knowledge are considerable. The task involves moving outside one's own field of expertise into others in which one is an amateur, shifting from a familiar set of premises, key questions, research methods, vocabulary, and so on into what can seem like an entirely different language. Uncertainties and disagreements multiply. Things archaeologists disagree about or can't say for certain are added to the complications faced by paleoclimatologists, linguists, biologists, art historians, geologists, and many others. These writers come from backgrounds and university departments of anthropology, history, climatology and meteorology, biology, geography, environmental studies, and (as in this book) literature and cultural studies. Finding their books in a large library will take you on a tour of the whole building.

The basic premise of this emerging inquiry is simple enough. Climate has so much to do with everything else about the nonhuman world that it must also have something to do with the human world. And there is one other shared starting point. There is much evidence to suggest that many episodes of significant climatic change correspond (either loosely or closely) to episodes of significant change in human cultures, societies, economies, and so on. Some of these correspondences may be purely accidental; others may involve chains of reverberations, ripples, and many-threaded entanglements; some may be direct relationships of cause and effect, maybe one-way, maybe multiple and reciprocal.

For reasons no one yet understands, the climate since the warm-up after the Younger Dryas—that is to say, for the past ten thousand years—has been unusually stable. Recall that graph of temperatures in central Greenland, and how the jagged peaks and plunges level out during these most recent millennia. No one really knows why, though one intriguing (and controversial) suggestion is the Ruddiman Hypothesis: that human activity has been changing the climate since the beginning of agriculture, inadvertently raising carbon dioxide and methane levels through forest clearing and rice growing, enough to counteract what climatologist William Ruddiman says would otherwise have been the beginning of a shift toward the next glacial advance. Whatever its cause, this stability has clearly been crucial to the history of human civilization, which has come to rely on settled agriculture, domesticated animals, and cities, all of which depend on a fairly steady climate. Even small disruptions, changes on the order of a degree or two, not ten or fifteen, have created considerable trouble for human societies. So it's not just that we're lucky to live in a stable climate period. Without that bit of luck, the history of human life would have been profoundly different.

One plot that seems to be tentatively emerging (most explicitly in Fagan's *Long Summer,* but also in Jared Diamond's *Collapse*) goes like this: Salubrious climate conditions allow human populations to grow and expand to sizes and into areas that less favorable conditions would not support. Then the climate changes in ways that disrupt something basic like agriculture. Temperatures fall or rise, frosts move earlier or later, storminess or wildfires increase, more or less rain falls in larger or smaller episodes. Crops fail and famine and disease follow, people drown or flee from floods, commerce is disrupted, competition rises for resources, populations abandon some settlements and migrate into areas that may already be inhabited. All this increases the chances of conflict. Climate changes need not be—indeed, rarely are—the only ingredients in these social changes, but they're important, sometimes as catalysts or triggers, sometimes as last straws or as stress multipliers. (It's important to remember that these physical changes are not uniform around the world, even if a temperature change is global. More moisture in western Europe, for instance, is likely to coincide with more drought in the southwestern United States. When the jet stream or monsoon shifts its path, some places get wetter, some drier.) And as may have occurred in the Younger Dryas, such stresses can also be productive, compressing populations into cities and thus creating new civilizations, forcing innovations in efficiency (food storage, irrigation) and enriching cultural linkages through trade and migration. At such moments, we might say, the processes of evolution select for human adaptability and ingenuity.

Several periods since the Younger Dryas have begun to receive similarly concentrated attention for their possible links between climate and human history. One is the global cooling that occurred when the Laurentide ice sheet collapsed 8,200 years ago, an event whose reverberations included a four-century drought in the North American plains, the separation of Britain from the European continent (and flooding of what were likely many settlements in what had been a productive river valley), and the creation of the Black Sea from what had previously been an agriculturally rich lake valley—this last at a speed that makes it one candidate for the catastrophic flood that appears in so many stories from the region, including the one in Genesis.

A second is the period before 5,500 years ago when the Sahara contained enough lakes and streams to support spread-out groups of hunters, who domesticated wild aurochs into cattle they could herd. These people created many hundreds of thousands of rock paintings and engravings of crocodiles, elephants, hippos, rhinos, and ostriches, along with cattle, boats, and humans. When the monsoons shifted south, sources of water shrank and disappeared, forcing these people to move. Some went eastward into the Nile valley, where they may have helped form Egyptian civilization. During the same period of diminishing rain, the Nile valley was widening and filling with fertile silt, partly because the level of the Mediterranean was rising, partly because the timing and severity of floods was changing. Similar episodes of human expansion, centralization, and then collapse during oscillations between moisture and drought occurred elsewhere in the cultures around the Mediterranean between five thousand and three thousand years ago, including the Sumerian and Akkadian empires.

A third example, closer to our own time (enough so that we'll shift to noting dates as "C.E." rather than "years ago") is the change of climate between about 900 and 1300 C.E. that brought steady moisture and warmth to Europe. The new conditions nourished an abundance sufficient to support such expensive projects as Gothic cathedrals; allowed the Vikings to travel to and settle Iceland, Greenland, and Newfoundland; and encouraged the Thule to expand their home territory from Alaska all the way to Greenland. This is often called the Medieval Warm Period. In sharp contrast, the same centuries brought drought to much of the rest of the world, including North America. Researchers have linked shifts in population on the plains with this climate shift, including, for instance, the collapse by about 1300 C.E. of Cahokia in southern Illinois, the general decrease in settlements in the northern plains, and the increase farther south in places like the Texas Panhandle. Prolonged drought clearly played a major role in the collapse of significant settlements like those in Chaco Canyon in northwestern New Mexico (deserted between 1150 and 1200), where the line between semiarid and desert is never more than just out of sight and agriculture depends on capturing and delivering every possible drop of rain and flake of snow. It may also have contributed to the collapse of the Classic Mayan sites in Central America. Several different and interlocking climate-system processes and relationships seem to be implicated in these megadroughts, and it seems increasingly likely that similar events will occur in our not-so-distant future.

ON THE SPOT: ON THE ARCTIC TUNDRA

Ellen Wohl

It takes time to realize that I am in a desert after I fly into the western end of Alaska's Brooks Range in late July for a backpacking trip along the Wulik River. The Wulik is north of the Arctic Circle, so I am prepared for cold, wet weather and perpetual daylight. My expectations are met; the only flesh that I expose is hands and face, light rain falls for much of every day, and the sky never grows dark. And yet, average annual precipitation here is less than ten inches. Only gradually, as I start to understand the implications of individual details of the landscape, do I perceive the tundra as desert.

Not a tree is to be seen. Shrubby willows that rarely reach more than head-high outline the course of the Wulik. Even tinier birches rise just to my knees on the drier uplands. The ground-hugging vegetation creates a spacious landscape of long views limited only by the surrounding peaks. Some of these peaks are jagged with rubbled gray rock, but for the most part they are low, rounded, and green. Despite the absence of any strong wind, the clouds continually move, casting the landscape into shadowed dull green and then picking out an emerald slope or a rocky, pale gray peak in a bright spotlight. The forms of the landscape dominate the scenery more than the subdued colors. The regular spacing of ridges and small ravines makes them resemble giant building blocks arranged for pleasing symmetry.

At my feet lies a landscape of miniatures. Everything is dwarfed, hugging the poorly drained soil that clumps into tussocks in the low spots, or the rocky slopes where frost boils churn up patches of bare soil. Thin, crusted lichens and thick-leaved cushion plants form little tufts among the angular blocks on rocky knobs. More sheltered spots hold a variety of berry plants among the birch and willow growing no more than a foot high. Matte-black crowberries, shining scarlet cranberries, pale yellow cloudberries, and plump blueberries hang from plants so tiny that they barely rise above the ground. Yellow Arctic poppies and potentillas, dark purple monkshoods, blue Arctic lupines, and other flowering plants are smaller versions of their lower-latitude cousins. The beautifully varied textures and colors of the lichens set off any fungi or vascular plants rising above them. The bonsai garden at my feet makes the tundra look as diverse up close as the canopy of a tropical forest.

These plants hug the ground to conserve warmth and moisture. They have small leaves that are hairy or waxy, and as much as 95 percent of the plant's living mass is below ground. The plants also grow slowly in the poor soil of the tundra. Almost none of them are annuals, because they cannot squeeze a whole life into the short growing season of the far north.

Only the mosses luxuriate in this climate. Thick sphagnum moss growing in vivid shades of green, yellow, orange, pink, and red cushions many of my footsteps. The mosses do well in this climate of little rain because the tundra excels in conserving moisture. All of this terrain is underlain by frozen ground that acts as an impermeable layer a few inches below the surface. When the sun reappears in springtime and air temperatures begin to rise, only the upper few inches of the soil thaw. The moisture released by thawing snow and ground ice cannot infiltrate the frozen ground beneath, so the upper soil becomes saturated and flows slowly across even gently sloping surfaces, leaving lumpy hills mantled with subtle green lobes. The wet soil also supports marsh plants such as cottongrass in the slightest hollows.

The puffy white seedheads of cottongrass provide warning flags ahead when I am enjoying the firmness of a well-drained patch of hill slope. I know that soon I'll have to endure the tussock two-step, a stumbling, lurching, tiring dance. Tussock grasses are like spring-loaded mushrooms; I never know which way they'll tilt me off when I step on them. Stepping between the tussocks makes for uneven strides and an awkward gait, and my feet sink into muddy water or sphagnum moss so thick that it throws me off balance. Scouting the terrain helps. After a little practice, I can interpret differences in soil moisture from a distance by observing the patches of different plant colors and textures.

I can scout the terrain for a long distance ahead. The air is so clear that it provides no sense of separation. Yet the light is also gentle, bright, and all-encompassing, warming but without the threat of intense heat and brightness present at lower latitudes. Even at midday the sun is never sufficiently high in the sky to bleach out colors and lose shadows.

Toward the end of the trip I have a morning of clear weather when I climb to a rocky knob with spacious views of the coastal plain. I sit on a cushion of lichens over boulders forming patterned ground. White lichens grow upward like miniature coral skeletons, with shiny red cranberries and pale olive-green pixie cups interspersed through the white. The tundra is an inversion of a tropical rain forest, where the broader landscape is closed but the immediate details are giant as plants spread huge leaves to compete for sunlight and nutrients. Here the broader landscape is immense but the immediate details are miniaturized as plants huddle into themselves to conserve warmth and moisture.

Below me the Wulik has left broad sweeps as it braids downstream toward the coastal plain, which is now shrouded in fog. I can see for tens of miles beyond the outlying hills to the plain. This landscape has the clean, spare lines of the desert, and its geographic distance from population centers has kept it as close as we can now come to a landscape untouched by humans. Looking at the vegetation succession along the river, I see a place where things still work themselves out naturally, although humans are changing the context of the landscape via global climate change. The Wulik is unfettered by flow regulation or bank protection. Its watershed has no grazing or crops, no timber harvest or roads.

The coastal fog gradually burns off, revealing distant asymmetrical peaks tilted westward. I gaze across a landscape apparently empty of animals, which likely reflects the relatively few species and low population densities, as well as migratory patterns. There is no sense here of ceaseless activity and change, despite the perpetual daylight, warmth, and moisture of summer. Again like the hot desert, this seems more like a timeless land, where change is less perceptible. Features such as the thick windrows of tiny birch leaves still piled in a ravine where winter winds left them belie the apparent low energy of the landscape. The tundra must experience fast changes like the hot desert's episodic flows of materials and energy; migration, spring meltout on the rivers, rainstorms that flood creeks, flowering and setting seed, and breeding and rearing all occur in shorter intervals than those that intervene. And so I sit on a ridge for hours and see very little move and then—a grizzly! My attention shifts to calculating distances. The bear is at a long distance and moves away from me, though, and inner quiet returns. The landscape once more looks deceptively immobile and inanimate. ◈

THE LITTLE ICE AGE, GLACIOLOGY, AND THE SUBLIME

Let's spend a little longer on what is the best-documented and most studied period of climate change in human history, not counting our own time: the Little Ice Age. This period is most evident in Europe, partly because that's where the most data have been collected, but also because some of its changes may have been concentrated around the North Atlantic. But it is also visible in other parts of the world. Many historians identify its beginning at roughly 1300 and its end sometime in the second half of the nineteenth century, though some focus mostly on the shorter period between about 1580 and 1850. The story is complex: we have a lot of information now about both climate and human events, and quite a bit of work has been done on this period, again especially for Europe (including notable books by anthropologist Brian Fagan, climatologists Jean Grove and Hubert Lamb, and, somewhat earlier, historian Emmanuel Le Roy Ladurie). So we'll note here only some of its major characteristics.

One was a sharp increase in climate volatility. Charts of temperatures and rainfall patterns are especially jagged, with conditions as likely as not to vary dramatically from month to month, year to year, decade to decade. A summer of heavy, persistent rains that drowned crops might be followed by another that was dry and hot, a terrible pattern for societies that still mostly depended on subsistence-level agriculture. Hunger and famine became more frequent, bringing with them social and political unrest, then, sooner or later, innovations in social order and agricultural practice. Storminess increased too, and the list of memorable and very deadly storms in Europe, especially around the North Sea, is long. In the second half of the sixteenth century, for instance, the incidence of severe storms quadrupled. In America, while the northeast was similarly stormy and cold, farther south a series of unusually intense droughts interfered with Spanish and English attempts to settle along the coasts of Virginia and the Carolinas. As Fagan points out, both the Jamestown and doomed Roanoke settlements coincided with the driest growing seasons and longest droughts in some eight hundred years.

In many places around the world, it grew colder. Although global temperature averages dropped just about a quarter of a degree Fahrenheit between the highest of the preceding warm period (about 1180) and the lowest of the Little Ice Age, local and seasonal changes were large enough to make real differences to life on the ground. That chart of temperatures in central Greenland shows a drop of just under 5°F between the highest and lowest points; in England and Switzerland, temperatures in the very cold 1690s were about 2.7°F lower than the 1920–1960 average. Among the possible partial causes for this especially cold period is the Maunder Minimum between 1645 and 1710, when an unusual absence of sun spots and auroras—phenomena long observed and recorded by amateur sky-watchers—suggests that the sun was emitting less energy. Perhaps the best-known effect is that the Norse settlements in Greenland collapsed, with regular ship traffic to Europe ceasing after 1369 and the last stray vessel arriving in 1406. Norse Iceland, similarly, began a decline that would last for centuries, pulled down into poverty and frequent famine by its isolation, increased sea ice and cold, and its ash- and gas-producing volcanoes; between 1100 and 1800, its population declined by half. And all the legendary explorers of the Arctic and Antarctic from the fourteenth century through the nineteenth, all those who risked and lost their lives looking for a Northwest Passage or trying to reach the poles—Barents, Frobisher, Davis, Hudson, Parry, Ross, Franklin—they, too, were in a sense victims of the Little Ice Age. Meanwhile, the Thule people consolidated into more scattered coastal villages from Alaska to Greenland and developed seafaring and whaling cultures.

Among other notable effects of the Little Ice Age were many paintings of snowy, icy scenes by such artists as Peter Brueghel the Elder. One striking example presented by climatologist Hubert Lamb, in his book *Climate, History and the Modern World,* is a contrasting pair of paintings of the three kings visiting the newborn Jesus, one done before and one after the first really cold winter of the time, 1565. The first is dominated by the human actors, the second by the snowy landscape surrounding them. In an idiosyncratic but intriguing effort published in the journal *Weather* (1970), meteorologist Hans Neuberger examined (first by eye and then statistically) more than twelve thousand European and American paintings done between 1400 and 1969 for their representations of sky color, cloud cover, and cloud type. He found, appropriately, that British skies were palest, haziest, and cloudiest in general, and also that for the total sample cloudiness and overall dimness increased during the cold centuries. At least thirty times between 1564 and 1831, the Thames froze solid through London (though as Lamb notes, this could happen more easily before modern water control allowed the tides to come farther inland) and became the scene of a series of frost fairs. Some four centuries later, Virginia Woolf used such a fair as the setting for a key part of her history-of-England novel *Orlando,* in which an intense but doomed love affair unfolds on ice skates, by bonfires on the ice, and under warming rugs. Ice appeared even in Marseille, Venice, and Spain.

All over the world, at different times during these colder centuries, glaciers grew: in the Andes, the Cascades, the northern Rockies, the

Caucasus, the Himalaya, in Alaska, in China, and along Canada's west coast. In the late sixteenth century (both before and after a large volcanic eruption in Peru in 1600 that dropped global temperatures), mountain glaciers in the Alps advanced with often astonishing speed, devouring villages sometimes in a single day's surge, blocking drainages and travel routes, driving farmers and vintners into poverty, causing massive icewater floods, and generally creating havoc. Those whose lives were disrupted petitioned the church and government to have their tithes and taxes reduced, built chapels, and asked bishops to lead processions of prayer and song, bless the glaciers, and conduct exorcisms. Sometimes the glaciers stopped moving. More often they did not.

Beginning in 1816, a new ice advance occurred in the Alps, adding to already extended glaciers. That was the "Year without a Summer" caused by the massive explosion of the Indonesian volcano Tambora the year before, and it is entirely possible that Tambora triggered the decade of advancing glaciers half a world away—along with significant famine, disease, and social disruption. This was also the year that Mary and Percy Shelley spent their summer in the Swiss Alps writing what might well be called glacier literature. Several key scenes of Mary's novel *Frankenstein* are set on the ice. Her hero describes the solace he gains from the "sublime and magnificent scenes" of the "icy wall of the glacier," "shattered pines," the "fall of some vast fragment, the thunder sound of the avalanche." But then, as he walks on the glacier (the Mer de Glace and Des Bois glacier on Le Montenvers, which she calls Montanvert), he meets his abandoned creation, the monster who bounds toward him over the ice, confronts him, demands a hearing, and soon, as they shelter together in a mountain hut, launches on the tale of his own troubled life. Similarly, the book ends with the creator tracking his monster across the frozen stretches of the Arctic, a reminder that the novel's frame narrative—an Arctic explorer who temporarily rescues Frankenstein and hears his story—points to the ongoing actual explorations and how they were hampered by the especially short months of open water. At the same time, Percy Shelley wrote his poem "Mont Blanc," a classic of sublime mountain literature, as his response to the same advancing glaciers and colder climate. Characteristically, he evinces at least as much interest in the importance of his own mind and imagination as he does in the scene around him, those

> frozen floods, unfathomable deeps,
> Blue as the overhanging heaven, that spread
> And wind among the accumulated steeps.

Frankenstein's monster and the glaciers are so closely associated because they come together in the same aesthetic, that of the sublime, a way of looking at certain overwhelming landscapes as places of mingled awe and fear. (This wasn't so much a single aesthetic as it was, and still is, a family or tradition of them.) This notion developed during the Little Ice Age, and many core descriptions of this new aesthetic, from Thomas Burnet (late seventeenth century) through the Romantic poets (throughout the first half of the nineteenth century), derive from journeys through the Alps and feature scenes of intense and ongoing glaciation. Like the Shelleys, Coleridge often saw the world through these lenses, as we can see in his 1802 poem about Mont Blanc, "Hymn before Sun-Rise, in the Vale of Chamouni," which reads in part:

> Ye Ice-falls! ye that from the mountain's brow
> Adown enormous ravines slope amain—
> Torrents, methinks, that heard a mighty voice,
> And stopped at once amid their maddest plunge!
> Motionless torrents! silent cataracts!
> Who made you glorious as the Gates of Heaven
> Beneath the keen full moon? Who bade the sun
> Clothe you with rainbows? Who, with living flowers
> Of loveliest blue, spread garlands at your feet?—
> GOD! let the torrents, like a shout of nations,
> Answer! and let the ice-plains echo, GOD!
> GOD! sing ye meadow-streams with gladsome voice!
> Ye pine-groves, with your soft and soul-like sounds!
> And they too have a voice, yon piles of snow,
> And in their perilous fall shall thunder, GOD!

The science of glaciology—like the science of geology in general—also developed during this period. When in 1818 a glacial dam broke in Switzerland, drowning many, Jean de Charpentier embarked on a study of glaciers, speaking to locals who told him stories of formerly greater ice coverage (perhaps from earlier in the Little Ice Age) and tracking the erratic rocks scattered around the countryside. His resulting theory of a much greater previous ice age convinced the Swiss American scientist Louis Agassiz, who became the first to propose such an ice age in print, and whose 1840 book *Études sur les glaciers* was the first significant book on the subject of glaciers. It is because so much early glaciology took form in the Alps that the vocabulary used by English speakers was quite often adopted directly from the French: glacier itself (from *glace*, "ice"; pronounced "glass-ier" in Britain, "glay-shur" in America), névé (a stage intermediate between snow and ice, also called firn, from German), crevasse (ultimately from the Latin *crepare*, "to crack"), moraine, roche moutonnée, moulin (a deep hole in a glacier, kept open by swirling water and rocks pouring in), serac or sérac (a cliff or pinnacle on a glacier surface, named for its resemblance to a kind of Alpine cheese), and so on.

These two strands—the development of glaciology and the aesthetic of the sublime—come together in the work of pioneering conservationist and writer John Muir. From his birthplace in Scotland, through his adolescence in Wisconsin, to his many decades exploring California's Sierra Nevada, Muir spent most of his life in landscapes shaped by ice. He found and assembled the evidence that ice had carved the high Sierra and Yosemite Valley (a new and controversial theory at the time), and when later in his life he visited Alaska's coastal glaciers, he noted living evidence of previously debated processes like kettle formation.

Muir described the perils of travel over ice—most memorably in his Alaskan story of the little dog Stickeen, who reluctantly but bravely followed him over a terrifying ice sliver to cross a wide and fathomless crevasse, then safe on the other side "ran and cried and barked and rolled about fairly hysterical in the sudden revulsion from the depth of despair to triumphant joy." He exulted in the beauty of glaciers, those "crystal prairies," with their "magnificent uproar of pinnacles and spires and up-heaving, splashing wave-shaped masses." Most often, though, he wrote about glaciers as agents of change and creation, imagining in California how they had sculpted the mountain landscape, and in Alaska how new country was taking form beneath the ice. Sometimes Muir speaks of them as being alive and active, like animals.

In this, he invites comparison with late Little Ice Age stories told by native inhabitants of the same glacier-rich Canadian and Alaskan coasts he visited. Although access to such stories is difficult for most of us—they are primarily oral rather than written, and so outsiders aren't likely to hear them—we can learn from the work of people like ethnologist Julie Cruikshank. Exploring some of the ways that the Tlingit and Athabascan people spoke of this time of simultaneous and interlinked changes in climate, landscape, and social life, she notes that while glaciers might have been animate for both Muir and these people, Muir lacked the indigenous sense that glaciers heard and responded to human actions, that rash or disrespectful behavior (such as cooking with grease on a glacier or speaking rudely of or to the ice) would provoke them into such dangerous responses as ice surges (this area has a very large number of surging glaciers) or ice-dam bursts and floods. According to

Muir's traveling companion, their Tlingit hosts and guides were puzzled by Muir's risk-taking behavior and wondered whether he was crazy or maybe a witch, so casually did he ignore both the physical dangers the ice presented and the etiquette with which the Tlingit had learned to moderate their own behavior in that environment.

Even for Muir, Cruikshank thinks, humans and nature were clearly divided, as they certainly were for the other Europeans who were beginning to penetrate this area for colonial, commercial, and scientific purposes. Muir's sense of the spiritual might have been tied to the natural world, but it was separate from the social. For the Tlingit and Athabascans, though, those divisions were fuzzy and permeable. Communication, transformations, and consequences linked people, animals, and features of the landscape. Their stories about glaciers were stories simultaneously about changing landscapes and social relations. These were people whose cultures had long known—or never forgotten—how to live with the ice.

ON THE SPOT: TOWARD A GLACIER'S EDGE
Ana Maria Spagna

We stand on the Railroad Grade trail looking down. The so-called grade, a lateral moraine on the southwestern flank of Washington's Mount Baker, has never seen a train, never even held a track, though it ascends as steadily, elegantly even, as if it had been surveyed by clinometer. The trail isn't a railroad grade at all, we explain to the nine high school students with us today, just earth shoved up by a passing glacier. A glacier that we can't see.

The ground drops off directly in front of us, a couple hundred feet of dirt, steep and sloughing, dark as the best topsoil and peppered with round rocks, freshly exposed. Across from us, perhaps half a mile, like a mirror image, another moraine extends.

The two run roughly parallel and then diverge, hemming a wide trapezoidal creekbed where small clumps of subalpine fir grow in a sea of cobble in shades of gray, variations on the color of today's sky.

Actually, we're *in* a cloud, one high schooler says. Right in it.

True enough. We can hear the creek water gurgling and hanks of boot-kicked dirt skittering down. We can hear the shrill whistle of a hoary marmot in the meadow below and the shouts of mountaineers on the glacier above. But we can't see much.

Jon, my geologist friend, passes around a photo of Easton Glacier when it reached this far, when it filled this trough less than one hundred years ago. In the photo, from above, the glacier looks long and side-turned like the lolling tongue of a maniac: smooth and muscular, elastic, in motion. Now it's gone—or mostly so. The glacier has retreated, lost half its area and more of its volume. What's left is on the mountain in front of us, a wide swath of brightness, visible in snatches when the cloud lifts briefly, swirls like smoke, and then descends again. Jon is here today to talk about melting glaciers, and his message is plain: climate change happens, and sometimes it happens fast. He passes around graphs of the earth's temperature for the past two million years that show clearly how the eccentricity of the earth's orbit, the tilt, and the precession have led to ice ages at regular intervals. The lines rise and drop as rhythmically as a heartbeat on a monitor in the ICU. Problem is, right now, when temperatures are supposed to be going down, they're going up. The kids know this; they've heard it. They've come from all over the country to spend a month learning about it. But they have never in their lives been any place remotely like where we are.

We hiked through an old-growth forest—hemlocks, cedars, silver fir—and meadows of heather and huckleberry. We slipped through the dregs of last winter's snow. In July! We watched mountain goats scramble up talus. Now, atop the Railroad Grade, we stand beside wide tufts of phlox growing lush pink and purple, jutting out over the creekbed like a bushy mustache over a gaping mouth. And just above us, when the clouds lift ever so slightly, we can see the glacier's edge: craggy high slab walls of blue-tinged ice. Blue like the Mariners, we decide, not the Dodgers. Like the Pacific, not the Caribbean. Like Redford's eyes, not Newman's, I say. They have no idea what I mean.

Glaciers move even as they melt, Jon explains. This one is creeping toward us five or six inches a day. The weight of the ice on top squeezes the ice underneath like toothpaste. Here in the Northwest, where glaciers exist even at relatively low elevation (the Easton comes down to five thousand feet) they crawl right over forests, leaving a helpful carbon map for geologists of the future.

What could a glacier *not* move? A high schooler asks.

Jon shrugs. A butterfly?

The kids turn to descend. This is as far as they'll go today, but I keep moving along the grade, along the moraine, as the tread changes from soil to snow, then angle outward onto the ice. I pass several camps, clusters of tents, green and orange, pups and domes. An autumn flood closed the roads to other routes, so climbers have been funneled to this side. Ahead of me, they creep upward in vertical columns, like crosshatches on the white. Below me, near the glacier's edge, debris collects in long, dirty stripes. If enough debris covered the ice, Jon says, it'd stop the melting.

The sun breaks through long enough for a glimpse of the summit. On another day, a hotter one, the glacier could seem as desolate as a desert, hot and empty and exposed, alive only in minutiae: ice worms, hair thin, coming out in afternoon shadows to feed on pink algae, watermelon snow. The rest is ice. Brittle, it shatters. Elastic, it squeezes out. Melted, it careens down down down.

You can see a glacier in its absence. I stand looking downslope toward the creekbed, and farther on, toward the wide, green Skagit Valley. Sometimes I have a vision of the Cordilleran ice sheet, seventeen thousand years ago, how it covered everything, pooling like water, filling the valleys. It must've looked, from above, the way clouds sometimes do when you stand on a high peak looking down at a blanket of white. What if these high clouds were ice now? They'd squeeze me out like toothpaste, a few inches a day, and spit me out over cobble. The thought, ridiculous as it is, tickles me no end.

Today we walked six miles. The glacier moved six inches. I stood in the ever-moving place where the ice meets the land and saw, for a moment, the smooth hump of whiteness against the blue: stark and blinding, cracked and inviting. Now I've got to turn around.

I'm late to catch the others, so I begin to half-jog downward along the Railroad Grade, past the phlox perched over the abyss, clinging hard to roots stiff and strong as guitar strings, blooming before the fast-approaching fall. ◇

THE STORY NOW

The Little Ice Age is over. But ice and its relation to global climate are still important shaping forces in our physical and cultural worlds.

For one thing, there is still quite a bit of ice and ice-affected land on Earth. A catalog of such places, of course, would begin with glaciers and ice sheets, but it would also have to list other elements in what scientists sometimes call the cryosphere or (overlapping but not identical categories) periglacial and subarctic landscapes. This larger realm includes enough long-term and seasonal sea ice to cover some six Mediterranean Seas at the end of the northern winter, ten at the end of the southern winter. It also includes places where processes of frost and snow are active, such as alpine and Arctic tundra and places underlain by frozen soil either permanently (about a quarter of the exposed land surface in the Northern Hemisphere) or seasonally (a little over half the land in the Northern Hemisphere). These permafrost lands overlap with the vast boreal forest, or taiga. The realm of ice includes high-altitude Southern Hemisphere ecosystems like the *páramo* of the Andes. And for some purposes it includes seasonal snow cover; in the northern winter, snow covers an area larger than North and South America combined.

For another thing, we are very interested in telling stories about ice and icy climates. Several temperate-land scholars have focused their recent attention on how important Arctic and Antarctic exploration was during the nineteenth and early twentieth centuries to the English and Americans: how new scientific information about ice fed the psychological and cosmological imaginations of writers; how narratives of polar exploration served as spaces in which to explore questions of national identity, empire, gender, and race; and how these tales entered the cultural background so thoroughly that they became a kind of "imaginative compost," to borrow the phrase of Francis Spufford, author of *I May Be Some Time: Ice and the English Imagination* (1997). Nor were these stories popular only during the explorers' own decades. As Eric Wilson notes in *The Spiritual History of Ice: Romanticism, Science, and the Imagination* (2003), the *New York Times* reviewed at least twenty-two books about the Arctic and Antarctic between April 1997 and July 2001 (compared to just four between April 1990 and June 1996), a phenomenon Wilson links with our own recent imaginings of a millennial apocalypse.

The tradition set by visitors to cold climates telling stories about their experiences continues to be robust. These writers include not just details of their own travels but also retellings of earlier journeys, current natural history, tales of ongoing scientific investigations, and stories about and from people whose cultures have been learning for thousands of years how to live with climates and ecosystems shaped by cold, snow, and ice. Often they write about their own curiosity about the world and their yearning for adventure, their desire to see new landscapes, new people, plants, animals, qualities of light. They tend to regard the high latitudes as worlds apart, or at least as worlds so different from their usual habitations that time spent there makes the ordinary seem strange, lets them understand that ordinary anew—typically in physical terms, but also, in some cases, in terms of cultural differences between themselves and their hosts. Like deserts and high mountain peaks, very cold places offer physical challenges that some writers particularly relish: are they tough enough, strong enough, to withstand the cold, the wind, the disorientation, the deep strangeness of their surroundings, without losing their lives, their health, or their sense of self? These writers are often fascinated by the seeming contradiction between barrenness and lushness, as the frozen desolation of winter gives way to the biological explosions of spring and summer, and in just a tiny shift of perspective, by the alterations of their own perceptions, at one moment of overwhelming emptiness and vastness, at the next of intricate detail and intimacy.

Quite often, too, they are also concerned with questions of meaning and imagination, emotion and spirit, and they hope to find answers by paying careful attention to these unfamiliar climes. As Gretel Ehrlich says in the preface to *This Cold Heaven: Seven Seasons in Greenland* (2001), in "trying to cut through to what was real," she "used the island as a looking glass: part window, part mirror." Barry Lopez's *Arctic Dreams: Imagination and Desire in a Northern Landscape* (1986) is particularly eloquent about these more philosophical questions. In one casual instance in a chapter about migrations, for instance, he notes that watching wild animals can lead us to "questions about the fundamental nature of life, about the relationships that bind forms of energy into recognizable patterns." Although it might seem that landscapes shaped by ice and cold are in their essence marginal to most of our lives and concerns, these writers find them instead to be central to some of the most important human questions.

Another kind of history is more literally, or at least more physically, emerging from the ice, in the form of natural and cultural artifacts and animal and human bodies whose frozen surroundings are melting. Whole mammoth and steppe bison bodies have emerged in Alaska and Siberia. A football-stadium-size area of caribou droppings some five feet deep (along with such things as rare wooden tools) emerged in the Saint Elias ice fields of Canada and Alaska, proof that caribou grazed in those alpine fields for at least eight thousand years: a bonanza for biologists, anthropologists, and First Nations people interested in their own history. In Wyoming's Wind River mountains, a glacier full of the bodies of the extinct Rocky Mountain locust helped solve the mystery of the notorious insect plagues suffered by pioneer European American settlers. Meteorites regularly appear on ice surfaces, to the pleasure of scientists and meteor enthusiasts. A very large meteorite is part of the mystery in Danish writer Peter Høeg's thriller *Smilla's Sense of Snow* (1992). This book's heroine has a Danish father and an Inuit mother and lives in a kind of cultural and environmental exile in Copenhagen; she embodies the tense colonial relationship between Denmark and Greenland, the effects of that history on Greenlandic and mixed-heritage people, and the kinds of intimate knowledge about snow and ice that are possible with long familiarity. The 1997 film version removes most of this complexity but adds some spectacular footage of ice.

Most dramatically and evocatively, melting ice and eroding coasts have begun to reveal human bodies emerging from their frozen preserves: A man who died crossing the Alps more than five thousand years ago—the Iceman, Ötzi—and now rests in a museum in Bolzano, Italy. Incan children sacrificed about five hundred years ago to the powers of high, icy volcanoes in the Andes, one of whom—Juanita, the Lady of Ampato, the Ice Maiden—now lies on display in a refrigerated glass box in a museum in Arequipa. Kwäday Dän Ts'ínchi, Long Ago Person Found, a young hunter who died in the mountains along Canada's Pacific coast about the same time, early in the fifteenth century. Agnaiyaaq, Little Girl, a child with a severe disability who had been lovingly buried on the north slope of Alaska about 1200 C.E., where she was preserved by permafrost. Thanks to the political power of Native Alaskans and Canadians, the last two were studied briefly and then returned to their rest. Each of these "ice mummies" (and there are others) suggests rich tales of the past, tales we can only work to decipher and imagine.

But the most urgent story today about ice landscapes is of course the story of climate change. Ice has taught us many important things about the past; now we are looking to it for clues about our future. As journalists say, this is breaking news. Books can hardly make it into print before some of their information has been superseded by new research or new events, and so newspapers, magazines, television videos, and Web sites have especially important roles to play, though many key stories are really open questions, likely to remain so for some time to come.

The basics are certain. Over the past 800,000 years, through the last eight major cycles between glacial and interglacial periods, carbon dioxide levels have fallen and risen to match global temperatures. During these eight cycles, atmospheric carbon dioxide has oscillated with remarkable consistency between some 180 parts per million during glacials and some 280 parts per million during interglacials. As recently as the early nineteenth century, while Wordsworth and Keats were writing poetry and Lewis and Clark were crossing the American West, we were at this higher 280. But by 2015, thanks to land clearing and fossil-fuel burning, our atmosphere will contain about 400 ppm, well higher than at any time for at least these 800,000 years. We know how such greenhouse gases act to warm the surface of the planet. We know that sometimes the climate has changed with shocking speed, pushed to some cliff edge and over. And we know that ice is a key player in global climate, intricately interwoven with the movements and temperatures of air and ocean, with the evolution, growth, and decay of organic life, and with human histories and actions.

The details of what will happen and when are complex and interlinked. Many surprises are probably in store. We still have much to learn about how ice behaves and just how thoroughly it can alter climate and weather conditions all around the world by amplifying changes caused by other factors. How quickly will the ice recede, and what are the mechanisms that control the speed of its melting? (The rapidity of the melting keeps surprising scientists.) How much will sea levels rise, swamping which fields and cities and islands, washing away which coastlines and villages, pressing which cultures into radical adaptations? (A moderate current prediction is for about three feet of rise this century.) Will melting ice slow or even stop the warm Gulf Stream and North Atlantic currents, and if so, how much colder will Europe become, how much hotter the tropics? (The Gulf Stream probably won't stop during this century, but the possibility figured in a report prepared by the U.S. Pentagon, "An Abrupt Climate Change Scenario and Its Implications for United States National Security.") As highly reflective ice gives way to energy-absorbing soil and plants, and as we keep pumping carbon dioxide into the atmosphere, how much warmer will the air become? One midrange projection for much of the American West is for a rise of 7–8°F by 2100, a projection based on much lower carbon dioxide emissions than now seem likely. Using the same data, one scientist has calculated that we have a one in six chance of seeing a 25°F rise in global temperature averages this century: that's the odds of Russian roulette.

If the winds change, where will it rain more, and at what seasons? What kinds of drought will ensue in which parts of the world, and how severe will be the famines in the poorest places, where millions of people still live at or near subsistence level? Both flooding and drought are expected to affect poor people most severely. Cities from Lima to Lhasa, Calgary to Calcutta, depend on glacier and snow melt for drinking water; and areas that are dry today are expected to get even drier in the future. How many plants and animals will be moved into extinction by climate-driven habitat changes and limits to their movements—ice-edge walruses, seals, polar bears, high-mountain pikas and fritillary butterflies? Some estimates by reputable ecologists are shockingly high: between 15 percent and 37 percent of terrestrial species. How much more carbon will melting permafrost add? What kinds of feedbacks will create what kinds of runaway effects, or, we must all hope, moderating countereffects? What will control our own behavior? Our future use of fossil fuels is the main uncertainty in all climate predictions: what will we do about it? Facing the stress of such fundamental changes in our world, will we respond mostly with competition, self-defense, and fear? Or will we create some new flowering of cooperation, innovation, and hope, producing another moment when—as in the crisis of the Younger Dryas—our adaptability and ingenuity help us to invent new ways of living?

Faced with such questions, such uncertainty, how do we determine how to think, how to feel? Changes this large and fundamental offer formidable challenges to our imaginations. Artists of many kinds are beginning to respond, on their own and in partnership with scientists. Sometimes their creations involve ice itself. The British Cape Farewell project, for instance, which takes ships full of scientists, educators, and artists into the Arctic, has produced, among many other things, photographs of ice sculptures that reflect and refract the light and images around them, then melt away. One photograph by director David Buckland shows the words "THE COLD LIBRARY OF ICE" projected in an eerie green light against the deep glowing blue of ice itself. (The blue of glacier ice, incidentally, results from the same process of Rayleigh scattering that makes the sky blue, though the two tints are not the same.) For a show at the Museum of Contemporary Art in Boulder, Colorado, curated by Lucy Lippard and titled Weather Report: Art and Climate Change, American artist Jane McMahan created a reliquary holding a cube of glacial ice (from the glacier whose water supplies the city of Boulder), kept frozen, though imperfectly, by solar panels and shading curtains. In a bit of pure artistic luck, as the ice slowly evaporated inside its container, some tiny, long-frozen bones gradually emerged. The possibility that those bones were a pika's added to the conceptual and emotional resonance of McMahan's piece, for these small tundra animals may vanish in a warmer world.

Such work is at its most compelling when it is filled with a passion for action, when it balances on a fine line between art and activism, when it brings together, as art sometimes can, our emotions, imaginations, ethics, and determination. Journalist, poet, and essayist Marybeth Holleman writes about her fear for polar bears, comparing our human efforts to grapple with climate change to the bears' searching for ice on which to rest and to hunt: "It's as if we, too, are swimming, swimming, toward a shore we cannot see but still believe is there, somewhere." And photographer James Balog has mounted an ambitious glacier project—the Extreme Ice Survey—involving several years of time-lapse photographs on Iceland and Greenland, in North and South America, and in the Alps. The project involves magazine photos, televised video footage, radio interviews, spectacular large-scale photographic prints in a traveling museum display, streaming videos and photos mounted at busy Denver International Airport, and an elaborate Web site.

Balog's team, which includes scientists, field managers, mountain climbers, engineers, photographers, and videographers, documents the rapid melting of glacial ice, collecting information and footage valuable to science. They are producing a set of dramatic, action-filled images that enact their belief and the project's motto that seeing is believing. At the same time, these images illuminate the astonishing beauty of ice landscapes. Melt ponds spread sapphire-blue across the black-and-white striations of old ice; white waterfalls plunge into moulins, deep holes that transfer melted water to the base of glaciers and speed their downhill flow. Ice cliffs and canyons mimic the sharp crags and rounded curves of rock. Layers of sediment mark ages long past. Crystalline ice fragments rest on black volcanic beaches or beneath violent ocean waves, there just for days or hours. Sometimes the sky's blues match the glacier's blues, and the white of the clouds the glacier's white; sometimes the ice blues are like nothing else on earth. Icebergs splash ostentatiously from glaciers, churn and flip, and float calmly in the sea. Tiny human figures stand at glacier edges, trail across vast white fields, and rappel down white walls, pointing their cameras at what they see so that we can see it too. Together, we watch and wonder at the world's iciest landscapes, their beauty and power, the fascination they hold for us. And maybe we hear the warnings they offer us.

3

Wet and Fluid

PROLOGUE

Watery creatures afloat on a watery planet, we are surrounded by oceans and seas, ponds and puddles, rivers and rivulets, melting snow, ice, rain, dew, fog, and humid air. Water keeps our eyes moist, our cells plumped up, and the blood running through our arteries and veins. It covers more than 70 percent of the globe and makes up well over half of an adult human body, some three-quarters of an infant's. And, together with the earth beneath and the air around us, it moves all the time, in ocean currents and tides, shifting sea levels, vapor rising until it falls again to cascade and trickle from high to low places.

Naturally, we see in water innumerable meanings, so many that we might better think of it as a medium for all meanings—a substance whose variations and contrasts can be used to represent virtually all others. Water has had immense religious significance. It falls from the heavens as a blessing or punishment: traditional Hindu culture holds that Ganga, daughter of the Himalaya, sheds her purifying waters on the sinful earth to bring salvation to humanity, while traditional Chinese culture depicts rivers as dragons that must be propitiated to prevent floods. Purity and clarity find their most obvious symbols in this element, the most basic cleansing substance, as do their opposites, taintedness and murkiness, and another pair of opposites, calmness and agitation. We use moving water to think and speak about the constant passing of time, its transcendence or essential illusoriness, and the way it accumulates things from the past and carries them into the future.

Water connects things and separates them. It flows in complex and often hidden networks; it takes one form, then moves or transforms itself into something different. Crossing the Ohio River, the Jordan, the Styx, the River of Lethe, bathing or having our bodies burned and carried away by the Ganges, we too are transformed, passing from slavery into freedom, life into death, one world or state of awareness into another. But we also journey along paths made by rivers, and the water we drink connects our cells with the most distant seas and the highest clouds. Water provides metaphors and symbols with which we express a wide range of attitudes, including complex ambivalences. "If I were called in/To construct a religion," wrote the poet Philip Larkin, "I should make use of water."

Imagine a natural landscape—any one you know—not as solid, stable, complete, dominated by the landmarks that are always there, but instead as a process of constant movement, as something that is not really a "thing" because it is constantly being produced by action. Martin Heidegger recommended this thought-experiment in his essay "The Thing," in which he turned the nouns *world* and *thing* into verbs: "the worlding world" and "things, each thinging." Similarly, in his book *Wholeness and the Implicate Order,* theoretical physicist David Bohm proposed what he called "the rheomode," an adaptation of language that would give verbs primacy over nouns. Our noun-dominated grammar, with its basic sentence formed by the separation of subject and object, seemed to Bohm seriously misleading. He wanted a language truer to the material reality revealed by quantum theory, one of indissoluble process and interaction. And he felt that such a shift in language would be a part of the fundamental change in culture and consciousness brought about by ecological awareness. Henri Bergson, another philosopher, said of all evolutionary life-forms in the animal world that "the very permanence of their form is only the outline of a movement." We know that we can say the same of landscapes. Perhaps only a Heideggerian philosopher or experimental poet would say "it's stoning" (never mind "it's continental crusting") or "it's atmosphering," but we do regularly say "it's snowing" or "it's raining"—or, as the nameless author of that fifteenth-century poem "Western Wind" wrote, "The small rain down can rain."

In this chapter, then, we'll take advantage of the literal and meta-phorical properties of watery landscapes to think also about certain elements of our fluid world: the processes of dissolution and becoming, the ubiquity of change, the connectivity of things, the permeability or illusoriness of boundaries. We'll focus, too, by looking mostly at places where dry meets wet, places where we humans can at once stand on earth, breathe the open air, and see, touch, or immerse ourselves in water. After a quick visit to an intertidal zone, we'll begin with a tour of the water cycle. Then we'll consider three kinds of places: rivers, where bounded by earth, water moves in ways we can't help but see; wet places in deserts, where islands of water enable life; and wetlands, where mixed with earth, water may seem totally motionless, though it is not.

ON THE SPOT: IN THE ROCKY INTERTIDAL ZONE

Kathleen Dean Moore

A low tide in southeastern Alaska this morning, -3.4 feet at 6:58 A.M. I've come out to see what I can find in the rocky intertidal zone on the seaward side of the island across the cove. There's dense fog this morning, so I can't see much of anything. But I pull on tall boots, grab a tin bucket, and head out across the cove, which is almost emptied now by the falling tide. At first, it's easy wading through shallow water that flows along a gravel course down the slope of the cove's floor. Then I'm at the end of the streambed, picking my way across the mudflats. This is tough walking. When the mud sucks at my boots, the smell of sulfur rises into the wet salt air. In front of me, clams squirt streams of water into the fog as they retract their siphons, leaving empty holes in the mud. But soon I am splashing across the pebbled channel where the last tidal water flows from the cove. Here the rocks are pink, painted with crustal algae. Blue mussels grow in this moving water, and here is the first of the bladder wrack, a seaweed with swollen, branched tips that function like air bladders. When I step on them, the bladders pop like firecrackers under my feet. In this green and gray foggy world, the bladder wrack is mustard yellow, a glowing band drawn around every island in Southeast, marking the mid-intertidal zone.

I can hear bald eagles chittering from the spars of the island's highest spruces, even though I can't see the trees for the fog. Ravens are talking somewhere up there, too. A bald eagle's cry is metallic—a dented winch or a steel pot dragged across sharp stones. But a raven's cry is all wood. I think of oaken bells clonked with wooden clappers. And now the smell of the island reaches me, the green dampness of Sitka spruce and rotting hemlock, and I call out to alert the bears. If there are bears on the island, they will be brown bears, uneasy in hard times before the salmon enter the streams. I'd like them to know I'm here, so we have no surprises. *Hey-up!* The cry echoes against rock piles I can't see.

Now it's up over the island's limestone spine and wildflower patches, down a crushed-shell beach past the windrow of flotsam at the high tide line. Here are logs polished silver, spruce cones chewed down to their cobs, shreds of kelp, a crab's lavender carapace, rope torn loose from some mooring. I pick my way down the steep slope of rocks on the seaward side, moving carefully because the rocks are slick with sea lettuce and bladder wrack and, closer to the water, slabbered with the brown, rubbery blades of the winged kelp. Some places, I use my hands and climb backward to keep from skidding. Under the fog, the smells are thick and close: iodine-salt air, the deep green smell of decaying algae, the white salt-smell of drying barnacles. And do I smell diesel in this damp air?

When I stand up, I can just make out the rumbling of seiners across the water, and the faraway shouts of young men setting nets for pink salmon. Water sloshes against shore and the broad kelp blades whisper as they lift and shift in the swell. A flock of peeping things approaches and veers away in the fog—western sandpipers, I would guess. When they have passed, the only sound I hear is the familiar confetti of the intertidal's small noises, the tick and pop of barnacles, bubbling of kelp crabs under the brown algae, tiny squeaks and sucks of rubbery flesh retracting, the anemones and geoducks, clicks of shell on stone. And now a loud huff, and *this* catches my attention. Two things huff like that around here—humpback whales and brown bears. I listen closely until I'm convinced this sound comes from the sea. How peaceful it is then to sit on a slippery rock and listen to the exhale, then silence, then another breath, finally another, until the whale rounds the island and heads out to sea, and I can't distinguish the sound of the whale's breathing from my own breath or from the rhythmic lifting and falling of the sea.

But now to business. I have hoped to find giant red sea cucumbers, *Parastichopus californicus,* and so far, I haven't seen a single one. So I begin a steady search through all these heaped-up, flopped-over creatures that can be seen only once a month—that particular time when the sun and moon line up on the same side of Earth at the new moon. Then their combined gravitational pull lifts the sea in a giant bulge, dragging it away from where I stand. If I stayed here until lunchtime, until the sun and moon dragged the bulge a quarter-turn around the earth, I'd be under eighteen feet of salt water.

On smooth basalt intrusions, I find mask limpets, one-shelled mollusks that the Tlingit people call "ravens' hats" because of their conical shapes. Limpets slide like snails along the rock, grazing on algae. But touch one, or even come close, and it sucks onto the rock and no amount of prying can dislodge it. Under limestone ledges are purple stars, five-rayed starfish humped over mussels, pulling them open and then sticking their stomachs inside the shells to digest the meat. I lift rocks and find, always, a blenny, a tiny fish shaped like an eel that flips and flips until it launches itself into a safer place, and dozens of porcelain crabs, speckled like sand, either scurrying away sideways or sinking into the sand until only their eye-stalks protrude. Grainyhand hermit crabs, their abdomens twisted into the spirals of periwinkle shells, tiptoe over sand. I find a sea lemon, a rarity, a mollusk that would look exactly like a lemon if lemons moved around on a single, slimy foot. The sea lemon is in the same tide pool as a leather star, brown, soft as wet suede, smelling, when I pick it up, like rotten garlic. And then, hidden under kelp, a sea cucumber.

It's a foot long, as thick as my wrist, maybe thicker. Bright orange-red, cylindrical, fleshy, with orange points that look like fat spines. When I poke at one of those points, it is soft and retreating. I nudge the cucumber onto its back. Its bottom surface is alive with white wiggling tube feet. Not that I want to pick it up. I've heard that it sometimes blows all its guts out its anus when it's disturbed, to tangle up potential predators. When I finally pick up the sea cucumber, it droops over my hand, as big and flaccid an orange sausage as I've ever held. I put it in my bucket. A sea cucumber is good to eat, I'm told. You poke it with a knife to let all the water drain out. Then you nail one end to a board. The whole animal stretches out, maybe to four feet, and you slice the length of it and open it like a narrow book. There are five strips of white muscle. These you strip out, roll in batter, and cook. Or so they say.

But as I look at this extraordinary creature in my bucket, its body shortening and lengthening, its fleshy points swaying, each of hundreds of tube feet struggling for a foothold, as I look at its color, which is as orange as a marigold in the fog, bright as a teenager's Camaro, I think, *There is nothing, no hunger, that would justify killing this wondrous thing.* So I carefully lift it from my bucket, put it back beside its rock, and draw a sheet of kelp up like a blanket to the place that would be its chin if it had a head, unless I have confused its two ends.

I hike the long way around the island on my way back, staying above the high tide mark, where the walking is easier. By the time I get to the cove, the tide is lapping at the sedges on the shore, too deep to wade. So I cross from the island to the mainland on a gravel isthmus, gathering a handful of the sea asparagus that grows there, and take the forest trail around the cove to the cabin, banging my bucket against trees now and then to tell the bears I'm coming. ◇

THE WATER CYCLE

For many of us, the enormous cycle through which water moves is part of our early and basic education about the workings of the planet. Parts of this process are subtle and hard to see, but much of it is visible and works on a time scale we can easily perceive. And so it provides us with a rich source of images, metaphors, idioms, and other ways of finding, making, or expressing meaning.

Let's begin with water moving from the sky to land—as rain, sleet, hail, and snow, as mist, or dew, or simple humidity that condenses or is absorbed directly by plants. It may come announced by lightning and thunder; it may fall silently in the night, turning a rough brown landscape white and soft by morning; it might, as American poet Carl Sandburg wrote about fog, "come on little cat feet," sit "looking over harbor and city on silent haunches" and then move away. It may bring a benediction to dry lands and thirsty plants and animals, or it may start floods that wash away whole worlds. It makes the world look and smell different, and it makes many plants and animals change what they're doing: worms rise toward the surface and robins stalk them, children dash for gutters to play in, the scent of shrubs like creosote bush and sagebrush intensifies, adults unfurl umbrellas, duck under broad trees and doorways, or turn their faces to the cool water. Its aftermath may be marked by rainbows, which we interpret as signs of luck or fortune or as a covenant between ourselves and God.

Once on the ground, fallen water trickles and flows as runoff into streams, rivers, and seas, sometimes after spending days or millennia

as snow or ice before it melts and begins to move again. Streams running down from high ground meet to form and feed rivers that find and
carve their way, making and maintaining valleys. Some rivers cut deep
canyons in rising ground and turn brown or red with sediment. Some
fall so abruptly they turn white with bubbles, while others meander at
their leisure across flat spaces, snaking sinuously through meadows and
making braids in gravel. Sometimes a river will disappear beneath the
surface and run underground before emptying into the sea or magically
reappearing on land.

We think of rivers as images of time's passage and of life's mutabil-
ity. We recall the comment attributed to the early philosopher Heraclitus
that we can't step into the same river twice: both we and the water will
be different with each passing moment. Or as Leonardo da Vinci wrote,
"In time and with water, everything changes." And we watch how rivers carry us away from things, and things away from us. Psalm 90 of the
Bible says that the Lord carries away a thousand years "as with a flood."
And the popular hymn "Oh God, Our Help in Ages Past," a version of
this psalm by Isaac Watts (1674–1748), includes these poignant lines:
"Time, like an ever rolling stream, bears all its sons away; they fly, forgotten, as a dream dies at the opening day."

Water that doesn't run off sinks through porous rock or the soil,
becoming groundwater. Some is held by the soil, in an upper "zone of
aeration," in which air and water fill the spaces between particles of soil,
and a lower "zone of saturation," where water fills all these spaces. The
top of this lower zone is the water table, and when the hole dug by a
spade begins to fill with water, the table has been reached. Groundwater
that is not held by the soil sinks until it reaches impermeable or saturated
rock. Then it begins to flow, often very slowly, in subterranean streams
and rivers called aquifers. These can be very large indeed, underlying
wide regions and providing water for drinking and agriculture. Often,
these days, thanks to technology like center-pivot irrigation, aquifer
water gets used much more quickly than it is replenished. Some aquifers
no longer refill because changes in surface geology direct "new" water
to other places.

When the slope of a mountainside or valley cuts through buried layers
of rock and brings an aquifer to the surface, water emerges as a spring
and runs down the slope, usually into a river on the valley floor. Or a
natural well or spring will occur when the water in an aquifer meets
an obstacle underground, is trapped, and rises; it "wells up," as we say
also about emotions, and about tears. Some springs rise from immense

depths, bringing up water that has been underground for hundreds or even thousands of years; such water is exceptionally rich in minerals. Springs and the rivers they create can be seasonal. In porous rock such as chalk, for example, in the dry summer months the water table may regularly fall so low that the flow ceases until the autumn and winter rains. Streams that dry in summer for this reason are in England called winterbournes. When we say that hope springs eternal, perhaps we might imagine this sort of spring, rather than the kind that runs all the time; hope subsides, our spirits sink, but then hope rises to the surface again.

Limestone landscapes in humid climates are good places to find such springs, underground rivers, and "sinks," where water disappears back beneath the surface. (The reason is that this rock dissolves relatively easily in water.) One important account of such a place was written by Quaker naturalist William Bartram, who traveled through southeastern America between 1773 and 1778. His book, which is usually called simply *Travels* (though its complete title is much longer), influenced Romantic writers like Wordsworth and Coleridge. Among Bartram's descriptions are several of remarkably crystalline springs, pools, streams, and sinks in northern Florida, where the clarity of the water, he thought, somehow, though temporarily, turned ordinarily voracious fish and insects into creatures of peace and friendship. Other passages seem to have provided inspiration and details for Coleridge's poem "Kubla Khan," where the fantastic, dreamlike landscape includes a powerful spring, a surface river (the sacred Alph), a sink, and an underground river that runs to the sea. Here is his description of the spring itself:

A mighty fountain momently was forced:
Amid whose swift half-intermitted burst
Huge fragments vaulted like rebounding hail,
Or chaffy grain beneath the thresher's flail:
And 'mid these dancing rocks at once and ever
It flung up momently the sacred river.

Although these lines are more exaggerated than the matching passage in Bartram's book, they do echo Bartram's frequent response of wonder. As he exclaims a little earlier in *Travels,* "how is the mind agitated and bewildered, at being thus, as it were, placed on the borders of a new world! On the first view of such an amazing display of the wisdom and power of the supreme author of nature, the mind for a moment seems suspended, and impressed with awe."

As streams and rivers flow, turbulence—from surface waves and currents, splashes, whirlpools, and waterfalls—traps oxygen in the water, where it dissolves. More oxygen is added by algae and larger aquatic plants, there because they receive enough sunlight through the water to conduct photosynthesis. Then this dissolved oxygen is consumed by the breathing of aquatic creatures and the rotting of organic matter, including sewage and farm waste discharged into rivers. If oxygen levels fall, aquatic creatures that require large amounts, such as fish, will be in danger. Moving waters also carry bits of soil, plants, decaying fish, dead insects, and other organic materials, nourishing plants on the banks and in this way fertilizing the soil. Where the banks lack plant cover, though, as in land that is intensively farmed, the opposite will occur and nutrients will leach from the banks into the rivers, impoverishing the soil and—if fertilizers, herbicides, or pesticides have been too intensively used—polluting the water.

When its flow is slowed, narrowed, or blocked, water deposits the particles it carries; this is how soil builds up in wetlands and how reservoirs silt up behind dams. When a river hits the ocean, the pressure from seawater and waves forces it to drop the sediment that forms deltas. There it slows and spreads out into lacy patterns of water and dry land. As ecologist Aldo Leopold wrote about a canoe trip he and his brother took in the Colorado River delta in 1922,

> On the map the Delta was bisected by the river, but in fact the river was nowhere and everywhere, for he could not decide which of a hundred green lagoons offered the most pleasant and least speedy path to the Gulf. So he travelled them all, and so did we. He divided and rejoined, he twisted and turned, he meandered in awesome jungles, he all but ran in circles, he dallied with lovely groves, he got lost and was glad of it, and so were we. For the last word in procrastination, go travel with a river reluctant to lose his freedom in the sea.

Quoting another Psalm, Leopold wrote, "'He leadeth me by still waters' was to us only a phrase in a book until we had nosed our canoe through the green lagoons." This account ends with the note that this delta now grows cantaloupes—and some of Leopold's most quoted words: "I am glad I shall never be young without wild country to be young in. Of what avail are forty freedoms without a blank spot on the map?"

Sometimes, when moving water meets some sort of blockage or basin, it pools and spills sideways, creating a pond or lake. (The usage of these two words varies; typically ponds are smaller than lakes, but other distinctions occur in some places.) If it is to persist, such a body of water must have an impermeable bottom or somehow reach down into the water table. Lakes may also be fed from springs—colder than the rest of the water in some places and seasons, warmer in others. Many of the largest are the result of glacial and tectonic events, both of which can create dams and basins. Others result from human activity. Dams built to fuel hydroelectric power, provide flood control, store water for irrigation, and so forth may be small or very large, as, for instance, the Three Gorges Dam along China's Yangtze River, or Glen Canyon Dam and Hoover Dam on the Colorado. Quarrying and warfare also create lakes. Ponds created by exploding bombs are still easy to see from an airplane flying over Cambodia and Vietnam, and the bombing of Britain during World War II produced ponds, many in cities, most of which have now been filled in naturally or artificially. One that survives is Bomb Crater Pond, in Walthamstow Marshes, East London, which was made by the explosion of a V-2 rocket.

Small chains of mud-bottomed lakes spill slowly down some valleys, held by dams built and maintained by beavers. In northern Wisconsin similar features, called flowages, were created by late-nineteenth-century loggers (sometimes in conjunction with beavers), who needed water on which to store cut logs before floating them out to mills, and sometimes also to help power that floating. The ones maintained today serve as valuable habitat for waterfowl, for other water-lovers like canoeists and anglers, and for the restoration of native wild rice (Zizania palustris), an ecologically and historically important food in this region. When the Ojibwa people moved into this part of the country from farther east, they settled where this "food that grows on water" fulfilled a prophecy about their future homeland.

Lake and pond waters are relatively still, though there are currents in places, caused by inflow and outflow. Where water flowing in from a river is denser than the lake water, it sinks, forming an underwater current that sometimes ploughs a trench along the bottom. Water coming in that is lighter than the lake water makes a surface current, which is more likely to be quickly dispersed by waves and wind. Complex circulations of water between surface and depth sometimes occur because of the combination of currents, winds, and differences of water density, temperature, and how far sunlight penetrates. Scientists used to think that large lakes had tides caused by the moon's gravitational pull, and indeed, for instance, the level on one side of Lake Superior (the world's largest in surface area) may be ten to fourteen feet higher than on the other. Now, though, it is thought that such massive movements are like water sloshing in a bathtub, created by wind, a difference in the air pressure above the surface, or sometimes other kinds of disturbances. These movements are called seiches.

Still water fascinates us. How many photos must there be of landscapes reflected in the mirror of water, doubled and glowing mountain ranges, trees colored by spring or autumn, or picturesque bridges? Such water serves us as mirrors for ourselves, too. Staring into the depths is a compulsion, a need, an excitement, and a willingness to search deeply, but many traditions and stories reveal it as a dangerous thing to do. One may be held there, unable to tear oneself away, like Narcissus in the Greek myth. Or one may be drawn into those depths, never to return. Ambivalent fears about love, sexual fascination, beauty, and self-absorption have often been represented by watery depths. Water nymphs—female spirits of the waters, usually pictured as beautiful young women with flowing hair, also from Greek mythology—were prone to fall in love with young men and tempt them into treacherous depths and currents. The famous pre-Raphaelite painting *Hylas and the Nymphs,* by John William Waterhouse, shows the adopted son and beloved of Heracles at the edge of a lake being tempted in this way, perhaps to drowning or perhaps to immortal bliss, by a throng of gazing nymphs amid irises and water lilies.

Narcissus had a famously unproductive fixation on his own watery image, but other more balanced gazers make good philosophical use of still water. When Thoreau came to write about the kettle pond near his cabin in the woods, he wished both to see the lake as a window upon the infinite and to measure it precisely, establishing its exact depth, proportions, and contents. "A lake," he says, "is the landscape's most beautiful and expressive feature. It is the earth's eye; looking into which the beholder measures the depth of his own nature." Then he takes systematic measurements and soundings—and draws a chart that marks the depth in feet at numerous points and pinpoints the deepest place. The results turn him to new wonder: "This is a remarkable depth for so small an area; yet not an inch of it can be spared by the imagination. What if all ponds were shallow? Would it not react on the minds of men? I am thankful that this pond was made deep and pure for a symbol. While men believe in the infinite, some ponds will be thought to be bottomless." Each time he moves between these perspectives—wonder and precise observation, imagination and science—Thoreau's understanding and appreciation deepen, and so, perhaps, do his reader's.

Especially since Freud, we've also used deep lakes to think about diving down to find something that has been concealed. This is the basic metaphor in Margaret Atwood's 1972 novel *Surfacing*, a book influenced by the Freudian psychoanalytical tradition and the feminist identification of feminine writing with fluidity and the dissolving of boundaries, as opposed to the hard, unmerging categories taken to be characteristic of masculine writing. Atwood's main character is a woman returning to her childhood home in a remote lakeland area of northern Canada. Searching for her missing father, she also wants to come to terms with traumatic events in her childhood and recent past, and one climactic event in the novel is her dive into a lake, where she encounters and confronts her greatest torments and fears—and then must surface to resume her life.

As Thoreau recognized, too, ponds and lakes have personalities of sorts, made up of such things as climate; their underlying bedrock; the depth, temperature, clarity, and chemistry of their water; and the varieties of plants and animals these things support. Scientists distinguish among oligotrophic, mesotrophic, and eutrophic lakes according to their richness in nutrients. The first is poorest in nutrients and thus in algae, provides the cleanest drinking water, and, as experienced anglers learn, is likely to support fish such as trout that like cold water that's high in oxygen. The second has midlevel nutrients. The last is too rich in plant nutrients and low in oxygen to support animal life. Like us, too, lakes age, and change their natures as they do so. Lake Superior is the youngest of the Great Lakes (because of the way the glaciers retreated), and it has not yet had the time to develop a high concentration of nutrients or accumulate much sediment, so it is close to barren. The shallower, older Lake Erie is much richer in nutrients, and thus in life. Many lakes, especially smaller ones, age themselves out of existence, filling in with sediment until they become meadows and then forest. When beavers move away from their ponds, their dams sometimes fail, the water drains out, and plants return; or even with the dams still in place, the ponds gradually fill with silt and meadows develop in their place. Under the right conditions, ponds may turn into bogs, then meadows, and then forests.

Some lakes have no outlets, and their levels rise and fall according to the balance between what water flows in and what leaves by seeping out, by being consumed by plants or animals (who make temporary use of it before returning it to the greater cycle) or by evaporation. Such closed lakes may become increasingly saline as minerals are left behind when water turns from liquid into vapor, and in arid places they may dry so far as to create salt flats. But most lakes drain into streams and rivers, and the waters they've collected and held for a while flow on toward the sea.

And so water that falls from the sky sometimes runs into the ocean, where it may lose itself for centuries in the long, slow dance of ocean currents. It might become trapped in rock that is pulled back into the mantle, slowing its dance until, far in the planet's future, it rises again through the throat of a volcano. It might be taken up and then released by living creatures. Or, caught by the bright, hot rays of the sun, lifted by soft breeze or strong gale, it might evaporate from the surface of the ocean or a lake, a puddle, a birdbath, a hot sidewalk: turn from liquid into vapor and once more become part of the atmosphere. There, in the air, it will cool and condense into the droplets that form clouds and move with the winds. These droplets may evaporate again, or when turbulence makes them collide and coalesce, or they collect around particles of dust or salt, they may become heavy enough to fall as rain, sleet, hail, or snow. And thus the cycle continues.

◇ ON THE SPOT: ALONG A RAIN FOREST STREAM

Ellen Wohl

The cool, clear water of the stream reflects a rapidly shifting mosaic of varying shades of green. The tallest trees rise two hundred feet above us, and a thousand intertwined layers of leaves and branches catch any sunlight passing beneath the highest canopy. We are working along headwater streams in Costa Rica, at La Selva Biological Station. Thirteen feet of rain fall here each year on average, and this wealth of moisture supports dense masses of greenery. Plants do not just grow from the soil at La Selva; they also grow on each other, sometimes stealing nutrients from their host, sometimes sharing them, sometimes just squatting where there's space. Many leaves of larger trees are covered with a green rust of mosses and fungi. The largest trees stand shaggy with dense coats of epiphytes, lianas, and bromeliads. The lianas in particular lace the forest together, and when a big giant comes down it takes many of its neighbors with it.

The stream froths downward over a series of short, bouldery steps, but we can't really hear the water for the louder background noise of the surrounding forest. Cicadas and other insects create a continual vibrating buzz that rises periodically to a crescendo. Shrieks, howls, and weird moans cut across this buzz at irregular intervals. Parrots passing overhead with short, swift wing beats scream as though they were being tortured. The gurgle of an oropendola is interrupted by the distant roar of a howler monkey or the sharp woof of a surprised peccary. The foot-long leaves and seedpods of the tropical trees do not sift gently down to the forest floor; they fall with a crash that makes us flinch and look upward nervously. And then a small troop of capuchin monkeys moving through the trees overhead grows agitated at our presence and begins to throw branches down at us. The rain forest is preeminently alive and changing, and when we pause even for a few moments we can discern details like the fist-size spider camouflaged as a knobby white fungus, or the tiny stream of leafcutter ants moving relentlessly over fallen tree and mossy boulder, each ant carrying aloft a miniature white flower.

We are here to study the logs that fall into streams. In temperate environments, these logs last for years or even centuries before they decay. The logs catch sediment and fine bits of leaves and twigs, helping to create a variety of habitats for fish and aquatic insects. We think that tropical streams might be different. Rain does not trickle down here in a gentle mist; it falls in torrents as though an enormous bucket were being emptied from the sky. Flow in the streams rises quickly in response, but then falls again rapidly once the rain ceases. The frequent intense rains of the wet season produce stream flow that goes up and down like a yo-yo, and logs lying in the streambed are lifted and carried downstream by the swiftly moving waters. The logs are equally vulnerable when they are stationary. There is no dormant period of freezing cold or exceptional dryness here, and microbes thrive. Rates of wood decay are much faster than in temperate climates, so the combined effects of flashy stream flows and rapid decay keep wood from remaining in the streams more than a few years, or in some cases days.

As we walk through the forest we occasionally encounter a pocket of air sweetly scented by flowers. Mostly the air smells like the freshly turned earth of a garden—the scents of fallen plant parts decaying. Rain falls at least briefly nearly every day, and the humidity remains so high that the air forms a palpable entity rather than an absence.

This is a climate for frogs. We continually flush them as we walk along. Those well camouflaged in browns and tans are visible only as sudden leaps among the ferns on the forest floor. With their vivid scarlet and azure hues, minuscule poison dart frogs appear like enameled miniatures in a jade-green setting.

This is also a climate for invertebrates. We constantly have to shift ourselves and our packs to avoid the truly remarkable variety of ants, spiders, and other insects that seek our flesh and blood. The rain forest is one of the few environments where I remember that invertebrates tremendously outnumber vertebrates and form a much greater proportion of the mass of living organisms in a landscape.

This is not a landscape for rocks. On the rare occasions when we encounter bedrock along the streambed, we can usually break off chunks with our hands. Relentless heat and humidity, combined with organic acids released by plants, accelerate the chemical reactions that change rock into soil, even as they ensure that these reactions proceed so thoroughly that all but the most resistant elements are leached from the soil and carried away in the groundwater. Aluminum and iron oxides are about all that remain in the clay soils that slime our boots and send us slipping down the steep trails. The dense vegetation that so effectively obscures the original geologic contours of lava flows underlying La Selva also fosters the decomposition of bedrock and the copious supply of silt and clay that threatens to suck off our boots when we wade into the streams to measure logs.

All streams change constantly. Flow rises and falls, earth moves from hill slopes into channels and downstream, fish migrate upstream, insects hatch and emerge into the air, and dissolved nutrients move from the subsurface into the stream. Tropical streams epitomize this continual change with their rapidly fluctuating hydrograph and fast decay of organic matter. In this, the streams reflect the continual cycling of matter and energy that characterize the surrounding forest. Dead plants and animals decay more or less slowly and incompletely in the temperate zones. The residue left behind becomes the thick mat of pine needles under a conifer forest or the half-rotten mass of fallen leaves in a deciduous forest. The lower portion of this layer adds to the organic-rich upper layer of the soil, which forms a reservoir of carbon, nitrogen, and other nutrients that grows incrementally over centuries. In a tropical rain forest this floor of leaves is only one leaf thick; it crackles loudly as we walk through it, but below it is pure clay containing very little organic matter. The thin skin of fallen leaves and branches is quickly decomposed by the countless insects, fungi, and microbes of the forest and reabsorbed as food by the living plants. The insects are eaten by frogs, lizards, and birds. Nutrients are stored in the tissues of living organisms rather than the soil humus in this most efficient of recycling systems. This landscape is alive in the truest sense. ❖

THE MOVING WATERS OF RIVERS

Rivers drain nearly every part of Earth's surface. The steepest slopes in the Himalayan mountains shed rock and snow in landslides and avalanches that provide meltwater and sediment to streams farther downslope, where the water flows in white cascades over huge boulders. Rivers wind sinuously across the world's flat coastal plains and extensive interior steppes. The great rain forests of Latin America and southern Asia shade the edges of huge rivers fed by the abundant rains of the tropical latitudes. Even in the driest deserts, rare storms create wide, shallow rivers whose channels shift rapidly back and forth across the broad valley bottoms.

Language reflects this ubiquity, and the long history we humans share with rivers is reflected in the names we give to them. In countries colonized by Europeans, rivers are more likely to reflect indigenous languages than are cities or other landscape features. Relatively few Native Americans remain in the eastern United States, for example, yet rivers retain the imprint of cultures long gone. The Tallapoosa, Chattahoochee, Yockanookany, and Yalobusha flow to the Gulf of Mexico; the Rappahannock, Chickahominy, Naugatuck, Quinebaug, Androscoggin, Mattawamkeag, and Aroostook flow to the Atlantic. The Murrumbidgee and the Warrego of Australia, the Waimakariri and Rangitata of New Zealand, the Yaqui and the Moctezuma of Mexico, and the Ubangi and Okavango of Africa record the presence of first peoples, as do the Quinault, the Kuskokwim, and the Futaleufu of the western Americas. English speakers also have a large vocabulary to refer to varieties of rivers and their basins: catchments, drainage basins, floodplains, watersheds; brooks, becks, creeks, burns, rills, washes, gullies, torrents, streams, floodouts, canyons, gorges, valleys, terraces, benches, bottomlands, banks, bars, freshets, floods, deluges, gully-washers, flash floods, and great floods. We say that rivers flow, gush, murmur, babble, pound, dance, whirl and swirl, rush, and surge.

The patterns of rivers reveal the structures of the land they cross. On gently sloping surfaces where the underlying rock doesn't vary, a network's channels can branch like veins in a plant's leaf. Where underlying fractures in the rock create zones of weakness, those channels turn in abrupt right-angle bends and make a trellis pattern. From a single high point such as a volcanic cone, channels radiate out and downward from the summit: a starburst. On a very steep slope, water might run in numerous regularly spaced, parallel channels with very few tributaries.

Geologists read these surface patterns on topographic maps or aerial photographs to discern the land's underlying rock structure and history, the forces that have made today's topography and drainage patterns. This is the bird's-eye view of a river across two-dimensional space.

The side view of a drainage basin also reveals the underlying geology and its history. A "longitudinal" profile plots elevation and distance downstream from the drainage divide. Many rivers drop steeply from their headwaters before gradually leveling out into a fairly flat profile in the lower reaches. Sometimes this pattern is so exaggerated that the profile looks like an L. One dramatic example is the Amazon, the world's largest river. It descends nearly twenty thousand feet from the Andes in waterfalls and whitewater cascades; but once it enters the vast rain forest below the mountains, it drops only another thousand feet as it flows more than two thousand miles to the Atlantic. This shape reflects the continuing tectonic uplift of the Andes and the contrasting tectonic stability of the lowlands east of the mountains, and it illustrates how rivers both shape the landscapes over which they flow and are shaped by them. Waters high in the mountains pour over cliffs in waterfalls because they are not forceful enough to erode more gradually sloping channels into the tough rock that tectonic forces are lifting upward beneath them. Lower down, the main stem of the river moves such vast quantities of sediment downstream and across its forty-mile-wide floodplains—delivering to the ocean thousands of tons every second—that it shapes the surrounding landscape.

A river's appearance reflects continual adjustments between geologic and climatic forces. Earth's internal fires push the surface up and around, bubbling it into volcanoes, folding and faulting mountain ranges, lifting whole plateaus without deforming them, and dropping the surface where tectonic plates thin, crack, sink, and tip as they pull apart. These forces determine the distribution of mass at the surface and thus the elevation. They also govern the types of rocks present and how easily water, fluctuating temperatures, and the organic acids produced by living organisms can break down those rocks into smaller and smaller bits. At the same time, solar energy moves around the globe in its great swirls of regular patterns and irregular eddies of heat and moisture, governing precipitation and how rock weathers into movable sediment. The Nile begins in high places that receive abundant seasonal rainfall, enough to support an annual flood along its four-thousand-mile path to the Mediterranean—even though almost no rain falls on the river itself, and no continually flowing tributary enters it along its lower half. China's Yangtze River, by contrast, begins with melting glaciers in the Himalaya, but it swells with rain that falls along its middle course.

Just as water flows downstream along a channel, so too the influences of geology and climate flow down to rivers through the intermediaries of topography, supplies of water and sediment, and the resistance of streambeds and banks to erosion. Geologists refer to a river in which these fundamental properties do not change with time as being in equilibrium, with the expectation that river form and process also will not change with time—a notion that largely reflects an ideal, not an attainment. Because they respond so slowly to changes in geology and climate, and because these changes occur so constantly, rivers are really more like figure skaters off balance, tottering this way and that without really achieving stability for more than a moment.

We're often caught off-guard by these continual adjustments in balance. We build cities next to rivers, expecting that the channel will not erode sideways or flood into our streets and basements. We build dams and bridges without thinking how they will push the river's processes to change. Months or years later, the flow will erode the streambed around the bridge piers and the bridge will collapse during a flood, or sediment will build up behind the dam and destroy the reservoir's storage capacity. Or we build a new subdivision without compensating for the way paved surfaces increase the flow of rainwater to the nearest river channel, abruptly widening its banks and eroding its bed. We have even more trouble thinking about the reverberations of changes we make much farther away in a watershed: cutting down forests or bulldozing new roads, thus freeing new soil from the hold of roots, piling fertilizers on fields far upstream, diverting or impounding water for irrigation and so reducing or changing the seasonal patterns of the flow.

But rivers integrate everything within their drainage basins. As anthropologist and writer Loren Eiseley put it in his description of floating on his back down a stretch of the Platte River as it crossed the Great Plains,

> I had the sensation of sliding down the vast tilted face of the continent. It was then that I felt the cold needles of the alpine springs at my fingertips, and the warmth of the Gulf pulling me southward. Moving with me, leaving its taste upon my mouth and spouting under me in dancing springs of sand, was the immense body of the continent itself, flowing like the river was flowing, grain by grain, mountain by mountain, down to the sea. I was streaming over ancient sea beds thrust aloft where giant reptiles had once sported; I was wearing down the face of time and trundling cloud-wreathed ranges into oblivion. I touched my margins with the delicacy of a crayfish's antennae, and felt great fishes glide about their work.

This connectivity is why we have so thoroughly altered rivers all around the world, often without meaning to do so.

Rivers are linked in many ways to the world around them. Along a single channel, water and sediment move downstream, while fish, mammals, birds, and humans travel both up and down. As flow rises and falls, water, sediment, and living creatures move back and forth between the main channel and the floodplain. Over longer distances, water, sediment, and organic matter, from bits of leaf to whole trees, move from hill slopes into rivers. Salmon migrate upstream to spawn and are eaten by bears, eagles, or humans, who turn them into fertilizer for the land; or they die and decay in the water, nourishing their own progeny and other aquatic creatures.

Rivers are also connected downward into the earth. At shallow levels, water, organisms, and nutrients move from surface flow to the hyporheic zone (*hypo* meaning "below," *rheos* meaning "flow"), where surface and groundwaters mix immediately below and alongside the channel. Layers of sediment beneath the channel and floodplain make all the difference here. Scientists working along the Flathead River in Montana first realized the extent and importance of the hyporheic zone when they found that a site right on the stream bank might be dry at shallow depths because lenses of clay impeded water movement underground, whereas a site a mile away might be connected by subsurface flows, thanks to porous and permeable ribbons of sand and gravel. At deeper levels, water moves from the river down to the water table, where all pore spaces are filled with water, or up from the water table to the river.

Above the river, aquatic insects hatch and emerge into the atmosphere for the final stage of their lives. Airborne dust, mercury, nitrogen, and other contaminants settle downward and enter the river. This form of connection, in particular, ensures that something occurring on the other side of the planet can influence a river. The rains that bring red dust from the Sahara to the rivers of England have increased within the past two decades, as oil-driven affluence has allowed people in the Middle East to replace soft-footed camels with four-wheel-drive vehicles that break up the surface crust of the desert soils and promote wind erosion.

Ecologists emphasize the importance of connectivity to river health. Flood control and bank stabilization along rivers such as the Rhine, Danube, Mississippi, and Murray-Darling have dramatically reduced the links between floodplain and main channel that once supported large populations of invertebrates, fish, and birds. But humans can also increase connectivity across natural drainage divides in ways that are bad for a river's health. Oceangoing ships dumping ballast water into the Great Lakes of North America also dumped larval zebra mussels. The invasive mussel, which is not native to North America, has proliferated to such a degree that it now clogs filters and structures along the Great Lakes and the upper Mississippi River drainage, requiring millions of dollars for control each year and threatening the existence of the mussel species native to the region. Well-meaning individuals and government agencies seeking to limit channel erosion introduced the riparian (streamside) tree tamarisk *(Tamarix)* to the southwestern United States in the early twentieth century. (Alternative stories include its introduction as an ornamental plant, or a much earlier arrival with Spanish explorers.) Native to the Old World, this tree, which is also called salt cedar, found its new home so congenial that it has spread across the western states, where it forms dense stands that crowd out native riparian plants like cottonwood and willow and shrink the habitat for other riverside species. Staggering sums are spent every year trying to control or locally eradicate this import. A recent ploy has been the introduction of a tamarisk-eating beetle that rafts from stand to stand piled up on bits of driftwood, landing here and there to transform a thick green stand into a dead one waiting to be cleared by fire or flood.

Archaeological sites around the world show that rivers likely shaped our earliest patterns of settlement and ways of using resources, but we humans have actively shaped them in turn for thousands of years. The earliest known dam was built about 2800 B.C.E. in Egypt. In 1590 C.E., the Japanese shifted the Tone River more than sixty miles to the east to prevent flooding in Tokyo. French settlers in New Orleans began to construct levees in 1717 to protect the city from flooding along the Mississippi River. Attempts to manipulate and engineer rivers have increased along with technology and population density, and in extreme cases rivers are essentially obliterated. Roughly a quarter of the channels throughout Switzerland have been moved into underground pipes to improve land access for growing crops. Some large cities have a long history of burying rivers: London, for example, sent the River Westbourne underground during the 1850s. Not surprisingly, some of these alterations backfire in later years, when the physics of water moving over land assert themselves strongly enough.

Some of the ways we have seen ourselves as connected with rivers in less physical terms are evident in literature and the other arts. Not surprisingly, flowing waters and the metaphors they offer pervade such imaginative works as novels and poems, where they serve sometimes simple and sometimes complicated functions. Often these instances tell us as much about writers' moments in literary history and their thinking about the human place in the world as they do about actual rivers, but sometimes they combine both kinds of information.

Relatively rainy and river-rich places like England and Ireland provide many interesting examples. The classic nineteenth-century realist novel *The Mill on the Floss* by George Eliot (Mary Ann Evans) is set in a town in the English Midlands where the fictional river Floss is the basis of the town's economy. The river also provides imagery for many of the emotions, ideas, and experiences of the characters. Thus the novel demonstrates how the community's enduring relationship with its river is both material and cultural. Rich and painful dilemmas about the idea of being "carried away," for example, are explored by means of river imagery. In a feeling that gradually builds to an almost irresistible current, Maggie, the main character, is tempted by the possibility of a treacherous elopement. In a chapter called "Borne along by the Tide," she and the man who attracts her glide down the river in a boat, "helped by the backward-flowing tide." As they go, she hears an occasional bird singing as if with "the overflowing of brim-full gladness, the sweet solitude of a twofold consciousness that was mingled into one." They are carried along in silence, "for what could words have been, but an inlet to thought?" Perceptions in the narrative and changes in the characters' feelings are registered by small developments of the river imagery—the brim-full gladness, the fluid mingling of consciousnesses, the inlet to thought.

Realizing that they have passed the place where they were supposed to end their boat trip, Maggie yearns to "glide along with the swift, silent stream and not struggle any more." Later, in a way, she does, for the novel ends with her drowning when the river floods:

> And now, for the last two days, the rains on this lower course of the river had been incessant, so that the old men had shaken their heads and talked of sixty years ago, when the same sort of weather, happening about the equinox, brought on the great floods, which swept the bridge away, and reduced the town to great misery. But the younger generation, who had seen several small floods, thought lightly of these sombre recollections and forebodings. . . . There was hope that the rain would abate by the morrow; threatenings of a worse kind, from sudden thaws after falls of snow, had often passed off, in the experience of the younger ones; and at the very worst, the banks would be sure to break lower down the river when the tide came in with violence, and so the waters would be carried off, without causing more than temporary inconvenience, and losses that would be felt only by the poorer sort, whom charity would relieve.

George Eliot's understanding here of the behavior of flowing water and the limitations of human understanding is precise, as is her recognition that sometimes the natural forces of running water are impossible to resist. "Nature repairs her ravages,—repairs them with her sunshine, and with human labor," she writes in the book's conclusion; "Nature repairs her ravages, but not all."

Two great literary works from the Modernist period of artistic experimentation early in the twentieth century have rivers winding and running through them. T.S. Eliot's long poem *The Waste Land* (1922), a complex portrayal of the fragmentation of the modern world, is full of rivers, seas, and the longing for rain. The rivers of Babylon are there, and Cleopatra's Nile, and the Ganges, but it is the Thames that appears repeatedly, giving the poem its London setting and atmosphere. The poem's book III, "The Fire Sermon," opens with the lament that "The river's tent is broken; the last fingers of leaf/Clutch and sink into the wet bank." From this river, "the nymphs are departed." In classical Greek mythology, the happy presence of these delicate and alluring spirits of place meant all was well, but their departure marked some sort of violation. So their absence by the Thames is a sign of disenchantment. Eliot uses "nymphs," here, to refer also to women, the ephemeral beauties of the London smart set with their summer parties and picnics on the Thames. The evidence that they are gone is the absence of litter.

Yet even this ironical survival of the classical figures carries the possibility of regeneration. Perhaps we can see the same possibility in the survival of the word *nymph* in biology. *Nymph* also names the juvenile forms of certain insects, including mayflies, damselflies, and dragonflies, who share with the classical goddesses a delicate, semitransparent, almost ethereal loveliness. And the genus name for water lilies is *Nymphaea;* the lotus flower, which is no longer considered part of this genus, was once called *Nymphaea nelumbo.* William Bartram speaks often of *Nymphaea* and also uses the word "nymphaeum"—a classical grotto dedicated to the nymphs—for Manate Spring (now Florida's Manatee Springs State Park), the one whose description impressed Coleridge.

Eliot's Thames is also a river of time bearing fragments from the cultural past. Odd things bob up, such as the refrain that follows his description of the departed nymphs: "Sweet Thames, run softly till I end my song,/Sweet Thames, run softly, for I speak not loud or long." Eliot found the first of these lines in Edmund Spenser's poem *Prothalamion,* written in honor of the weddings of the two daughters of the Earl of Worcester in 1596, and then added the second line, another hint of decline, this time in the waning strength of the poet's voice. These bits of cultural detritus float past, like physical debris on the water. In *The Waste Land*, they provide a subtle combination of wistful regret for a nearly lost wholeness and a pleasure in the mingling of high and low, sacred and profane, and past and present. The river, with its dissolving and merging action, makes this possible. Its flow is the poem's flow.

Another famous Modernist work, James Joyce's *Finnegans Wake* (1939), takes this literary imitation of a river's flow and merging even further. In this extraordinary book—written in a style unlike any other, full of puns that make different kinds of language overlap and merge—the river is Dublin's Liffey, which Joyce identifies with a great female principle encompassing lover and mother that he names Anna Livia Plurabelle. Suggesting with its initial lowercase letter that the story, like the water cycle, has no real beginning, the novel opens with "riverrun, past Eve and Adam's, from swerve of shore to bend of bay, brings us by a commodius vicus of recirculation back to Howth Castle and Environs." Joyce's Liffey merges everything—mythical, personal, intellectual, carnal, learned, colloquial, sacred, profane—in the constant punning and shifting of his sentences. It is the river of time and the river of language, image of the endlessly changing forms of language and the way words from the past are carried to us and into the future, with tributaries pouring in from all directions. The optimism, wit, and humor of the novel come from its joy in the vastness and endlessness of this great flux, and the river is the natural image for the whole process.

A present-day writer who uses the analogy between river and poem is English poet Alice Oswald. Her book-length poem *Dart* (2002) traces the course of a river and brings an influx of many different voices into a single flowing poem. Starting at one of the Dartmoor springs that merge to form the river Dart in Devon (in southwestern England), she follows the river to the sea, mixing her own descriptions with the imagined voice of the river and those of various people involved with it along its course: workers in farms, mills, and sewers; naturalists; walkers; boatbuilders; scientists; farmers; bailiffs; fishermen—and a nymph.

Our culture's responses to rivers can similarly be seen in less artistic descriptions, including those given by scientists, whose attitudes have evolved during the past two centuries. Early geologists took a mostly historical approach, interpreting rivers as records of geologic and climatic change over millennia. Physical and biological scientists focused on inventory and classification, considering such questions as what shapes rivers take, what organisms are present, and how these shapes and organisms can be categorized to reflect underlying geological or evolutionary processes. During the middle of the twentieth century, physical scientists shifted to a mechanistic approach that focused on quantifying the movement of water and sediment and the resulting geometry of channels. This attitude is epitomized in hydrologist Luna Leopold's partly facetious 1964 description of rivers as gutters down which flow the ruins of continents—a metaphor and view not all that different from Eliot's litter-bearing Thames, or, if we consider cultural ruins as well as physical ones, from Eliot's Thames and Joyce's Liffey.

More recently, ecologists have led a shift—in attitudes and corresponding imagery—toward a more integrative approach that emphasizes the way a river fits into the context of its surrounding landscape. In a 1970s film describing the system of locks and dams aiding commercial navigation on the Mississippi, the U.S. Army Corps of Engineers signaled its managerial view with its metaphorical title *Shackles for a Giant*. In sharp contrast, a 2000 publication of the Upper Mississippi River Conservation Committee describes a desirable balance between "a working river" used to transport grain and other commodities and "a river that works" by supplying wildlife habitat and drinking water. Citizens now join associations of "river keepers" dedicated to protecting and restoring watershed health, and scientists routinely refer to river ecosystems, riverscapes, and the integrity of rivers.

Concepts like river health and integrity imply that a river is analogous to a living organism with a right to some degree of autonomy, one that can function freely. River ecologists work with definitions of ecological integrity that emphasize ongoing changes in response to natural processes, including interactions among and between species and other factors like climate. A healthy river can erode and deposit sediment in response to changes in flow or sediment supply, for example; it can flood across a valley bottom when headwater snows melt or widespread rain falls. And its health can be restored.

But when we think about river restoration, we need to remember that we cannot return most rivers to an ideal, preindustrial state if key components of surrounding landscapes have vanished—or if that ideal state never really existed. Deforestation associated with the introduction of farming to Britain during the fourth millennium B.C.E., for example, resulted in changes in the amount of water and sediment entering rivers—and forests have not yet returned to many parts of Britain cleared during that era. Between about 400 and 1400 C.E., in the deserts of central Arizona, Hohokam farmers dug more than six hundred miles of irrigation canals—by hand. Although their fields have not been farmed for several hundred years, ecologists still find in them significant differences in soil texture, chemistry, and plant life, differences that affect the processes by which water and sediment enter nearby rivers.

It would also be arrogant to assume that our knowledge about rivers is complete. We can act only from what we know or suppose to be accurate. In the nineteenth century, we saw that extensive logjams hindered navigation on the Ohio River, as did the periods of low flow that inevitably followed the great floods that limited settlement along the river banks. We could fix all of these problems, we thought, if we removed the jams, dredged the riverbed, confined the river to a single channel, built levees to contain floodwaters, and stored water behind locks and dams for release during dry periods. Between 1825 and 1930, we worked very hard to improve this river—to make it closer to what we thought it should be like. Similarly, during the early twentieth century, some fisheries biologists and fishermen believed that rivers should be engineered toward maximal fish production and greatest ease of access for fishermen. They wanted to increase the frequency of pools along rivers such as the Blackledge in Connecticut so that fishing would be less like golf, with its long, onerous walks between holes. Fishermen ruthlessly killed every potential predator, trapping and shooting raccoons, otters, snakes, and even kingfishers. One writer described sleeping better on a pillow stuffed with kingfisher feathers, for he knew that the wily kingfishers would not be stealing his fish.

We now tend to focus on the unintended and unfortunate consequences of such attempts to engineer rivers, which we now consider misguided and tragically short-sighted. Modifications to the Blackledge, it turned out, actually reduced pools along the river, and overfishing reduced the trout population in a manner that natural predation never had. And in recent years we've spent millions of dollars re-creating valley-bottom wetlands and stream channels in the Ohio River drainage, trying to restore such river functions as water purification, flood attenuation, and wildlife habitat that we damaged with earlier engineering.

Even with these caveats—that rivers cannot be returned to an ideal or pristine state, and that there is much about rivers scientists don't know—the international scientific community agrees that the world's rivers are in a poor state of health, if not in crisis. Falling numbers of fish, we understand, signal the decline in a river's ability to support life, and in the past few decades, more than a third of all freshwater fish species have become extinct, threatened, or endangered as their habitat has vanished.

Scientists modeling losses in river flow associated with changing climate and water withdrawal estimate that by 2070 up to three-quarters of local fish diversity will be lost because of extinction.

Very little of Earth's water is fresh—something like 3 percent of the total water on the planet—and most of that is locked in glacier ice. More than a billion people today do not have access to clean drinking water, and partly as a result, more than five thousand people, mostly children, die each day from water-related diseases. An estimated forty thousand large dams over fifty feet high, and more than eight hundred thousand smaller dams, store more than two thousand cubic miles of water—an amount equal to five times the volume of water in all the rivers of the world—yet each year millions of people suffer and die from water shortages that destroy their crops or their supply of drinking water.

Precisely because they integrate and reflect what happens across their entire drainage basins, many of the world's rivers are polluted, impoverished in native species, altered in form, and reduced in flow. We have engineered these changes in our centuries or millennia of using the resources of nature, with more intensive uses leading to more significant alterations. Programs such as the European Union's Room for Rivers, which restores lateral connectivity by setting levees farther back from river channels, represent important measures to restore some level of physical and ecological integrity to our planet's blood vessels. As stories of nymphs and naiads remind us, the health of fresh water is vital to both natural and human communities. The word *naiad,* we might wish to remember, comes from the Greek for "flow"—a root shared also by the word *nurture.*

THE DREAM OF WATER IN DESERTS

Even in deserts there is water, and life tends to cluster around the few sources that are fresh rather than salty.

These include what geologists call exotic rivers, meaning that their sources are not in the desert they cross. The wetlands at their ends and along their banks are very rich locations of desert sustenance. Notable examples include the Nile, Colorado, Rio Grande, Tigris, Euphrates, and Niger, which all arise in highlands, cross deserts, and end in the sea. Other desert rivers never reach the ocean. Africa's Okavango ends in the vast wetlands in the north of the Kalahari. The Humboldt River, in Nevada's Great Basin Desert, ends its journey in the Humboldt Sink, an intermittent lake bed. Because most of Australia's desert rivers arise in the desert, they are dry most of the time for much of their length, though after rainstorms flood them, water can persist for years in intermittent pools. These rivers do not reach the ocean; they drain instead into the Lake Eyre Basin, where, normally, they simply peter out before reaching the usually dry bed of Lake Eyre. Distant or adjacent highlands also provide the water that creates oases, a distinctive desert wetland type. These occur when the underground water table rises toward the surface where the surface dips down. Sometimes oasis waters appear in a slow flow, sometimes in a vigorous spring.

Other types of water sources can also be found. In Australia, the Walmajarri Aborigines of the Great Sandy Desert make important and lifesaving distinctions among types of waterholes. A *jila* is a shallow well that, though small, has reliable water year-round. The water is attributed to the presence of a Kalpurtu, a magic water snake. During the dry season, large groups of people can live nearby. A *jumu* looks much like a *jila*—that is, a hole in the ground—but with the crucial difference that its water vanishes in the dry season. Both *jumu* and *jila* can fill in with mud and dirt unless kept open by regular attention. Anglo-settler Australians often call these water sources soaks. A *wirrkuja* is a deep natural pothole in a rock that catches rainwater and holds it for a long period of time. In hilly areas, with large rocks, these *wirrkuja* might be large enough to hold water year-round, but they cannot be relied on after long dry spells. A *jiwari* is a shallower version of a *wirrkuja*, from which water quickly evaporates. A shallow, broad claypan that holds rainwater is a *warla*. After heavy rain, the largest of these might hold water for several months. A brimming *warla* also attracts much wildlife, so hunting camps are often established nearby. Finally, *jujul* is low-lying land that becomes swampy for a short period after rain. With know-how and luck, holes dug in the damp soil fill with water.

However thirsty one might be in the arid lands, though, and however tempting the cool waters might look, a traveler had better think twice before drinking. Water found in some deserts is not safe to consume. Even if it is protected from human-caused pollution, it may be too contaminated by dissolved minerals for humans and most animals and plants to use. American desert writer Edward Abbey warns that it's wise to check for the presence of aquatic plants or insects in potential drinking water in order to avoid selenium or arsenic, though even this test might fail, since some aquatic plants, such as a Chihuahuan desert *Eleocharis* (spikerush), are tolerant of arsenic. Maps of deserts are dotted with names like Bitter Lake, Poison Creek, Badwater, Alkali Creek,

Lake Natron (Tanzania), and Wadi el Natrum (Egypt), and there are five different Bitter Water creeks in Western Australia. Then, of course, there are the famous saline lakes such as the Dead Sea, the Great Salt Lake in Utah, and China's Qinghai Lake.

Many, but not all, of the minerals responsible for desert badwaters are salts, itself a somewhat elastic term that includes chlorides like sodium chloride ($NaCl$) or table salt, often sulfates, and sometimes, more loosely, carbonates. These minerals arrive in deserts partly by wind and partly because they dissolve easily into water that falls elsewhere, runs into deserts, and evaporates, leaving the solids behind either as denser concentrates in water or as mineral crusts (sometimes called duricrusts) on rocks and soils; some accumulate enough to support mining industries. Concretelike calcretes (calcium carbonate), or caliches, are the most widespread. Silcretes, which include silica (SiO_2), are common in Australia, where they are the source of the famous opals of places like Coober Pedy and Lightning Ridge. Gypcretes such as gypsum, or calcium sulfate (which is used in plaster of Paris, cement, wallboard, and fertilizer), are abundant in the Namib and also form the white dunes of New Mexico's White Sands National Monument. Nitrates are abundant in the Atacama Desert in Chile and were extensively mined for use in fertilizers and explosives until artificial nitrate was developed in the 1940s. Now the Atacama is littered with more than 140 nitrate-mining ghost towns. Borax occurs in Death Valley and is widely used in cosmetics and detergents. Natron—a hydrated soda ash, which gave sodium its chemical symbol, Na—is found on the shores of Lake Chad and the salt lakes in Egypt's Wadi el Natrun. It was used by ancient Egyptians as an agent for mummifying the dead.

Strikingly, all of the places on Earth's land surface that lie well below sea level are in deserts. (In wetter climates, such low areas stay filled with water.) These include (in order from lowest to highest) the Dead Sea, which divides Israel and Palestine from Jordan; the Danakil, Turfan, and Qattara depressions (in Ethiopia, China, and Egypt); California's Death Valley; and Australia's Lake Eyre. Here, as well as in other higher, arid basins with no outlet for water except evaporation, salt lakes form. These lake waters have concentrations of salt much higher than in the oceans; the Dead Sea, for instance, is six times saltier. These salt lakes dry further into marshes, dry flats, and the depressions that go by many names: salt pans, salt and alkali flats, salars, playas, chotts, sebhkas, kevirs, takirs, and salines. The geology in such places is quite varied: some salty areas occur when groundwater rises high enough to dissolve the salts in the soils; some pans are clay rather than salt; some depressions are scoured by wind rather than water. Salt crystals may build up into fields of knee-high pillars like those in Death Valley's Devil's Golf Course, or into taller pillars like those around the Dead Sea that are identified with the biblical story of Lot's wife. One especially odd phenomenon occurs at Racetrack Playa in Death Valley, where, under circumstances not yet understood (though they probably include wind and water), rocks a foot in diameter are actually blown across the playa floor, leaving conspicuous tracks to perplex visitors in the days to come.

Because they're so scattered, desert waters create a pattern of isolated islands for living things. Such places, as Darwin and many after him have understood, encourage the development of new species. But they can also put those species at risk: smaller populations are more vulnerable to all kinds of disturbance, and most creatures can't move across long distances to find substitute homes. Additional fragmentation and habitat disturbances caused by human land-use changes may make matters worse.

One striking example is fish. For these creatures, of course, the problem of water is especially serious, yet they have evolved strategies that allow them to live in surprising numbers and variety in the world's deserts. They also possess a high degree of endemism—that is, they are native to a single place—and are now in many cases rare and threatened. Even the deserts of Australia, which offer very little surface water, support more than thirty species of fish, many of them endemic. While all of the rivers in these deserts dry up, they are all also susceptible to major flooding events that leave behind deep water pools, so fish must be able to perform two seemingly contradictory tricks: survive for long periods in ever-warming and shrinking pools in which salinity may also be rising, and withstand and take advantage of sudden torrential floods to disperse their populations throughout the watershed. All Australian desert fish species can tolerate temperature extremes ranging from 60°F to 95°F; a few can manage a low of 40°F or a high of 104°F for short periods. They can also survive rising levels of salinity, easily living in 50 percent seawater. Reports suggest that the spangled perch may even estivate in mud or damp leaf litter in order to escape drying conditions, though this behavior has not been confirmed.

Another intriguing example is in the Sahara, whose scattered water sources provide habitat for about forty fish species, about half endemic. One of the most widespread and successful of these species is the North African catfish, which is very tolerant of temperature variation and has an amazing adaptation for living in desert waters that are inclined to evaporate: an organ that allows it to breathe out of water. When its home has dried up, it can linger in the mud, gulping air. Even better, it can abandon its drying water hole to crawl across land on its sturdy fins in search of a deeper pool.

The desert pupfish is one of thirteen species of widely distributed fish living in springs, marshes, and oases in the deserts of the southwestern United States and northern Mexico. Once, it seems, they were a single species inhabiting the region's large Pleistocene lakes, but as their habitat dried up, they became isolated and so drifted apart into different species. One of the best known, the tiny devil's hole pupfish, inhabits a geothermal pool in a limestone cave near Death Valley, where its numbers fluctuate from about fifty to five hundred. To guard against a catastrophe befalling their only pool, populations of this federally listed endangered species have been established in a number of locations, including the Shark Reef Aquarium at the Mandalay Bay Casino in Las Vegas. This fish features in a recent environmental novel, *How the Dead Dream*, by Lydia Millett.

The most remarkable assemblage of desert fish in North America can be found at Cuatro Ciénegas, "four marshes," deep in the Chihuahuan Desert in the state of Coahuila, Mexico. This desert oasis contains two hundred ponds fed by thermal springs arising in the nearby Sierra San Marcos, each with its own distinctive suite of aquatic life. Nineteen species of native fish can be found in these waters, eight of them endemic. In fact, the valley surrounding the marshes has the highest level of endemic species of any place in North America, rivaling the Galápagos Islands. It is protected as a biological reserve by the Mexican government, and the Nature Conservancy has combined efforts with Mexico's Pronatura Noreste to purchase ranch land in the area. These groups are also working with local farmers to increase the efficiency of their irrigation systems, thereby lessening demand for water needed by the oasis. But threats from irrigation, mining, and other human activity continue to endanger this unique ecosystem.

Humans have also settled in deserts around significant sites of permanent water, whether large oases or major rivers, but with more robust results. Sedentary agricultural societies emerged in many deserts, and some of these settlements evolved into the first human cities. Roughly six thousand years ago in the desert riparian zone between the Tigris and Euphrates rivers, ancient Mesopotamians built what are considered to be the first human cities of Eridu, Uruk, and Ur. By 600 B.C.E. their city of Babylon had a population of two hundred thousand and was encircled by ten miles of defensive walls. One story archaeologists tell about this development begins with the construction of irrigation canals to distribute water from the Tigris and Euphrates, an undertaking that required large-scale organization. Irrigation then increased food supplies, which fostered a growing and sedentary population. Abundant food supplies were soon noticed by surrounding nomads, who raided the early settlements, and this threat spurred the city dwellers to coordinated defensive actions, including wall building. Similar urban development occurred along the Nile in Egypt and along rivers draining the Andes to the coasts of Peru. In North America, the Pueblos along the Rio Grande followed a comparable though smaller-scale pattern of development, constructing irrigation canals and permanent communities that they needed to defend against raiding nomads such as the Navajo, Utes, and Apaches.

In locations without oases or reliable rivers, cultures across North Africa, the Middle East, and into China developed a clever irrigation system called the *qanat* in Arabic, from a similar Persian term. In a *qanat*, a well-like shaft is dug at the edge of a mountain's alluvial fan, high above the agricultural fields but where the water table rises close to the surface. This shaft is then connected to a system of underground, gently sloping tunnels, which deliver the water downhill to nearby fertile soils, with little seepage or evaporation. In the mid-twentieth century, more than fifty thousand such *qanats* were in use in Iran. At the oasis of Turfan in the deserts of northwestern China, water is provided by a system of nearly a thousand *qanats*, with a total length of more than three thousand miles. They have been in use for about two millennia, since the Han Dynasty. Grapes and watermelons flourish in Turfan, thanks to the snow just visible high in the mountains on the horizon. On a scorching summer day a visitor can cool off in the shade after seeing—right outside town—the sand-buried remains of ancient settlements and human bodies mummified by desert heat and dryness.

In the past century, modern technology has made even larger water schemes possible. Much of Saudi Arabia's water (including most of that used by the 6.4 million residents of the capital, Riyadh) comes from desalination plants, a process that involves more than two thousand miles of pipelines. And many desert dwellers (including many Saudis) get their water by drilling into underground aquifers to tap what is called fossil water. Large portions of Australia's deserts, for instance, rely on the vast aquifer that underlies about a quarter of the country, a water source that allowed European-style agriculture and sedentary pastoralism to spread into the arid and semiarid portions of the country, though generally this water is too saline for irrigating crops and is being used more quickly than it is being replenished.

The dream of bringing water to the desert has also led to the development of enormous hydrologic schemes that are among the largest of human engineering endeavors. Egypt's massive Aswan High Dam provides electricity and water for irrigation. Both of the major river systems that cross North America's deserts, the Colorado and the Rio Grande, have been extensively dammed and connected to irrigation canals and municipal pipelines, and their waters back up in vast impoundments behind such facilities as Hoover Dam, Boulder Dam, and Glen Canyon Dam. Much of this water is lost to evaporation in the hot desert sun; if you try to eat a sandwich along the shore of one of these reservoirs, you'll need to chew quickly, as your bread will dry out in a minute or two. All of these major dam projects are relatively recent, and their long-term sustainability remains to be tested. Problems include silt accumulation, salinization, alterations to riverine ecosystems, ongoing legal conflicts caused by increasing demands for limited supplies, and predictions of changing precipitation patterns that are likely to shrink the resource still further.

THE SLOW WATER OF WETLANDS

The word explains itself: wetlands are watery lands. If your foot squeezes water out of the turf, and this is the permanent or seasonal state of the land rather than the temporary result of a heavy storm, you are treading on wetland. Mud too soft and deep to stand on is wetland, as is any soft, wet ground, or any place where the border between land and water is unclear and shifting. Wetlands are characterized by particular kinds of substrates and specific hydrological and chemical processes. They may be coastal or inland; their water may be fresh, salty, or brackish (a mixture of salt and fresh). Every continent but Antarctica has them, at nearly every latitude, longitude, and elevation, and in every kind of climate. Backwater, billabong, bog, bottomland hardwood forest, dambo, estuary, fen, gator hole, hanging garden, low moor, mangrove island, marsh, mire, mudflat, muskeg, pakihi, playa, pocosin, prairie pothole, reedbed, riparian corridor, salt marsh, slough, swamp, vlei, wet meadow, wet moss—the alphabet of wetland words and ecosystems both large and small is long and rich.

Wetland is formed wherever water is held still or in very slow movement, just above or just below the ground surface. The water may be direct rainfall that does not run off or evaporate but is held in place by impermeable ground, peat, and plants such as spongy sphagnum moss. Many wetlands are formed in this way, and with the growth of the moss the bog spreads, sometimes becoming blanket bog. Or the water may be runoff, trickling down slopes, on or below ground, until it reaches flat land and slows. As it trickles it takes up and carries soil and plant matter, until the flow is impeded and the sediment builds up as silt. Aquifers emerging at the base of hills or rising to the surface when obstructed by impermeable rock are another source of wetland. A more common source is the overflow of streams and rivers onto floodplains. This flooding itself is seasonal, but where the water is held in place by a high water table and impermeable rock below the soil, the result is wetland such as water meadow, marsh, and fen. Plants can live here only if their roots can survive being saturated at least for part of the time, so wetlands are populated by water-tolerant or water-dependent plants, insects, birds, and mammals. It's easy to kill an ordinary house or garden plant by overwatering it—that is to say, robbing its roots of the oxygen they need—or by giving it salt water instead of fresh. But wetland plants have evolved to thrive in such conditions: they "know" how to cope.

The word *wetland* is an umbrella term, encompassing all kinds of places that are both aquatic and terrestrial, and it is a word that only recently entered common use. Although the *Barnhart Dictionary of Etymology* cites an appearance in 1743, and the word did occur early in the twentieth century as one of a set of generic terms including *muck, marsh, marshlands, overflowed land, swamp,* and *swamplands,* it was rare enough that it did not merit inclusion in the *Oxford English Dictionary* as late as 1971. Not until the 1950s and early 1960s did it acquire its current centrality for American politics and science, and consequently its common usage. British scientists often use the somewhat narrower umbrella terms *mires* and *peatlands,* since northerly latitudes around the globe host more of the peat-forming wetlands that these two terms encompass, while lower latitudes often produce varieties that don't form peat. Although English, Irish, and Scandinavian scientists studied peatlands early in the twentieth century, full-fledged wetland sciences are mostly a phenomenon of the past several decades.

These categories can be hotly contested. This is partly because they often have legal weight, so their definitions are subject to political, economic, and managerial pressures, especially in the United States. (The same is true of related categories like floodplain and river corridor.) One

U.S. regulatory definition of 1985, for instance, excluded from the category of (protected) wetlands all permafrost lands in Alaska with high agricultural potential; one especially controversial definition proposed (but abandoned) in 1991 would have cut out as many as forty million acres of lands that had until then been included, thus removing them from any legal protection against outright destruction. For most landscape words, not much is at stake in a definition. But in cases like these, for wetlands and their inhabitants the meaning of a word, the details of definitions, can mean, quite literally, life or death.

Nor is it only politics that keeps wetlands from being a crystal-clear category. For words nearly always have fuzzy, complicated, changeable edges, and the material world does not come in neatly outlined and stable packages. At what point does a wet marsh become a shallow pond? At what precise line on the landscape does a (protected) wetland stop and a (plowable) grassland begin? This last question has spawned a new profession: it is the job of wetland delineators to draw such exact lines on inexact and changeable landscapes. And perhaps the most difficult problem of all, especially when controversies about restoring or constructing wetlands arise (as they do in "mitigation" cases, where permission to destroy one wetland entails promising to build or enhance another): what must a wet place do, how must it function, to qualify as a wetland?

As scientists have learned more, their answers and definitions have become more sophisticated. Still, they continue to focus on different criteria, and their classification systems offer what can sound like, and be, quite different categories. One international system includes as wetlands underground karst systems, coral reefs, and spring-fed oases; Canadians distinguish eighteen kinds of bogs and seventeen kinds of fens; and Americans say things like "palustrine emergent wetland, persistent," "lacustrine littoral intermittent, saline," and "estuarine intertidal unconsolidated bottom." Some of these sets of categories include the same places under different names. Others divide the world into different kinds of pieces, variously determined by shape, chemistry, cause, vegetation, cultural history, location, and so forth.

Similar complications occur in common usage as well. In England, *marsh* most often refers to estuarine and coastal salt marsh; in America, inland freshwater marshes are a common and specific wetland type characterized by mineral soils, through-flowing water, and particular plants. A "swamp" in England or Africa is a "marsh" in the United States; the large forested wetlands that Americans call swamps have long been gone from England but are common in North America. Often, too, a single place will encompass various and changing types of wetlands and drier uplands, as in Florida's Everglades or any glaciated landscape in which dry eskers and moraines separate wet fens and bogs. Perhaps largely because of such hybrid landscapes, common historical and linguistic divisions don't always match those drawn by today's scientists. Moors and heaths in England, for instance, are sometimes, or partly, wetlands, but not always.

Given the many layers of uncertainty about these categories and landscapes, estimates of the extent of current and former wetlands are necessarily inexact and vary according to the definition used and surveying technology. In former ages, of course, when the world's climate was much warmer and wetter, they covered enormous areas. During the Carboniferous era, for instance, lush swaths of tree-size ferns and horsetails, rushes and club mosses, made good homes for dragonflies with two-foot wingspans—just as today's wetlands welcome our smaller varieties of these insects with their transparent wings and brilliant blue, scarlet, and brown needle-shaped bodies. Now, some 300 million to 400 million years later, the heated and compressed remnants of these ancient, carbon-rich wetlands provide coal.

Today, worldwide, wetlands probably cover some 4 percent to 6 percent of the planet's surface. Just a few centuries ago, these numbers were likely twice as large: New Zealand has certainly lost more than 90 percent of its wetlands, Europe probably as much, Australia more than half, China perhaps 60 percent, parts of Canada between 65 percent and 80 percent, the contiguous part of the United States some 53 percent. Although protective measures have slowed the rates of loss, humans continue to destroy wetlands of all kinds. Why does this matter? What do wetlands do? For individual landowners, they are often useless or worse: to a farmer or a housing developer, land that can't be plowed or bulldozed is worthless, and land that subjects them to additional governmental regulations can feel like an intolerable burden, especially when those regulations are characterized by confusing details, conflicting demands, and erratic enforcement. But on a wider scale, for both humans and nonhumans, wetlands are immensely valuable. This mismatch between the interests of private landowners and those of the broader community is what makes them such difficult political places, and, in the United States, has made them a flashpoint for ideological conflicts about private property rights.

Individual wetlands in all their variety offer crucial temporary and full-time habitat to large numbers of plants, insects, birds, amphibians, and mammals: from dragonflies to cranberries, salamanders to moose. Partly because so many of them have disappeared along with their inhabitants, wetlands are home to as many as half of the animals and a third of the plants listed as endangered by the U.S. government. Some kinds, like both freshwater and saltwater marshes, are among the richest ecosystems on the planet. For reasons we've already noted, they operate as giant long-term carbon storage bins—a major value in a time of global warming. They also moderate extreme swings in hydrological cycles, slowing floodwaters and storing water for dry times.

And they improve water quality, cleaning out all kinds of pollutants both natural and human-made. As water passes through the accumulated mud of marshes and swamps and the roots of the plants that grow there, it is filtered. Fragments of mineral and vegetable matter that have been carried from upstream are stopped here, along with many microorganisms, so the emerging water is relatively clean. Specialist wetland plants then take up many of these sediment materials as nutrients, "fixing" them and preventing further dispersal. Other materials are broken up and converted by bacteria in the sediment, including anaerobic bacteria, which live in environments with little or no oxygen. Sunlight penetrates the cleaner water that emerges from the wetland, stimulating the growth of plants, phytoplankton, and the range of microorganisms fed on by aquatic insects, fish, and amphibians, which in turn become food for larger fish, birds, and aquatic mammals. Not only, therefore, are wetlands themselves exceptionally rich in diverse species of wildlife; the filtration they provide also creates the necessary conditions for the biodiversity of streams and lakes.

This filtration can be regarded as a natural cleansing or waste-disposal process, and in the industrial era human beings have given it plenty of additional work. Agricultural runoff from fertilizers and animal waste, for example, has increased the nitrate content of many streams and rivers. Nitrates are necessary to plant growth because of their role in photosynthesis, but large, abrupt increases in the nitrate content of water can lead to devastating "blooms" of algae—clouds of it that turn the water green or red. Discharge from sewage systems can have the same result. In daylight, when algal photosynthesis is producing oxygen, the amounts of dissolved oxygen in the water increase dramatically. At night, the oxygen decreases just as sharply, when the algae are absorbing it. Also, the clouds of algae keep sunlight from reaching the plants, animals, and microorganisms beneath. Mass fatalities of fish and other organisms sometimes result, either from lack of oxygen or because the plants and microorganisms on which they feed have disappeared. In some cases the algae themselves are toxic. Generally, the popular idea of the balance of nature is now regarded with some skepticism by ecologists, who have come to see ecosystems as more complex and subject to radical fluctuation and change than this model suggests. But in this case—even though eutrophication occurs naturally in some circumstances, such as the sudden floods that follow seasonal droughts—balance seems an apt idea: a sudden and massive stimulus prompts the explosive growth of some elements in the ecosystem, and a kind of balance ends in a way we're likely to see as negative.

Anaerobic bacteria in marshland sediments also pull nitrates out of water and convert them back into gaseous nitrogen. The conditions in which these bacteria thrive are produced by marsh plants—reeds, bulrushes, cattails, water hyacinths, and marsh pennywort—that also take up other organic and mineral pollutants. Peat soils are particularly good for this process. To drain wetlands—for agricultural or industrial purposes, to extract peat, or to "rationalize" waterways—is therefore to destroy a natural cleanup system. Increasingly, constructed wetland systems are being used deliberately to treat water pollution. A spectacular example of this is a series of ponds built high in the Peruvian Andes to remove nitrates and metal residues from runoff from a large copper-zinc mine. On a smaller scale, many farms now use wetland systems to dispose of their biological waste.

◇ **ON THE SPOT: AT THE BOG ON CÉIDE FIELDS**

Gerald Delahunty

I arrive at Céide Fields early on a clear July morning. Yesterday's drenching rains have vanished, and beyond Céide Cliffs the Atlantic reflects blue sky and willowy cloud. I park and walk into the bog on an asphalt path, a modern version of the oak-plank tracks, *toghers,* laid down over millennia in bogs throughout Europe. Irish bogs are typically *bogach*—soft, wet places; in Seamus Heaney's phrase, "melting and opening underfoot," intimate and introspective. But today it is dry. Where the turf, as the Irish call peat, pushes out from under the sedges, it is surprisingly dry, hard, brittle, black, even after the rain.

Céide Fields lie about five miles west of the small town of Ballycastle, County Mayo, in the west of Ireland. They are part of the four hundred square miles of blanket bog that stretches over most of northwestern Mayo. *Céide* (roughly, "cage-uh") means a hill with a flat top, and the Fields look straight north over the Atlantic, crowning shale and limestone cliffs that are four hundred feet high and three hundred million years old.

Across the Fields, the ling heather flowers are still the dark purple they put on just before coming into full bloom. I roll their rough leaves between my fingers, remembering days working on a midland bog as a twelve-year-old, the hardest work I've ever done. Other heathers reveal themselves—I spot a few of the pink bell-like flowers of cross-leaved heath and some bell heather with its needle-fine leaves. I don't see the sphagnum I have half expected, but the area is open and quite dry. I do find some lichens, perhaps cladonia, but I can't be sure. I'm hoping to find some orchids and eventually come across a single heath-spotted orchid, *Dactylorhiza maculata,* called the bog orchid in Irish. Its petals are a rosy white with pink spots. I wander on, feeling the hardy textures of the plants that crowd along the edge of the path. I hear a solitary skylark singing from its camouflage, perhaps just twenty feet away.

The bog that covers Céide Fields rests on soils of ancient woodlands, whose clearance by Neolithic farmers allowed the rain that had been absorbed by the trees to reach the ground. Over the next couple of thousand years, a layer of plant debris developed, and later an impermeable layer of iron pan, perfect for the development of blanket bog. Hummocks, lakes, and islands covered with grasses, mosses, sedges, heathers, heath, rushes, and cottongrasses create a muted brown-green-violet landscape whose horizon is indistinguishable from the clouds on wet days, though tinted here and there by the wind-whipped flags of hare's tail, bog asphodel, and the insect-ingesting sundews, butterworts, and bladderworts that lurk in bog pools.

Blanket bog shelters and sustains snails and slugs, overwintering Greenland white-fronted geese, red grouse, golden plovers, curlews, snipe, meadow pipits, skylarks, Irish and brown hares, and the frogs that otters come to feast on. Insects abound. Dragonflies, moths, and butterflies visit, though the heath butterfly and the marsh grasshopper are permanent residents, as is the notorious biting midge that can be a misery for anyone walking across bogland in summer. Miraculously, perhaps because it is sunny and dry, I'm not getting bitten today.

On my way back toward the car park, I sit awhile on a peat bank where it is severed by the path, recalling many other bogs I've walked or driven through and indulging the mixed feelings we Irish have about these landscapes. For many of us, bogs are alien, barren, inhospitable places that support few plants and animals, where tillage is almost impossible and grazing is rough, though it has been continuous for thousands of years and is now overdone on many Irish bogs. The tragedy of a bog-hole drowning is the stuff of Irish lore and literature. The uncouth, the uncitified, are "bogmen." But for thousands of years, bogs have provided antiseptic mosses for bandages and fuel for country and city households. For the past several generations, they have given us fuel to generate electricity and the technology to harvest that fuel, technology that is now an important export to other countries with substantial boglands.

It's hardly surprising that we have such feelings. About 17 percent of the island's surface is covered by peatlands—by fens, raised bogs, and blanket bogs—a greater percentage of its land than all but two other countries in the world, Canada and Finland. At various times, Ireland was covered with ice sheets, and glaciers scoured its mountain valleys. The drumlins left after their retreat trapped shallow lakes that morphed into the fens that covered much of Ireland about eight thousand years ago. Most of that fen evolved into raised bogs covered with hummocks of peat, enormous water bags that occasionally break, releasing flash floods that scour channels in the peat and beyond. Blanket bog, which has a somewhat different development, requires at least 250 rain days and fifty or more inches of rainfall a year—a lot even in Ireland.

Irish bogs are changing; many are disappearing. The smaller ones are harvested by hand and tractor. To feed the power plants, huge machines mill down the larger ones inch after inch. But every layer stripped away testifies to a long relationship between peat and people. Their anaerobic and acidic conditions preserve whatever bogs enshroud, as Heaney reminds us in his poems. What he calls "the trove of the turf-cutters" reveals both "beauty and atrocity"; "cobwebbed, black" bog oak; hoards of weapons; treasured and storied cauldrons; butter, still salty and white; and preserved bodies—victims of murder, of uncaring gods, or of "tribal, intimate revenge."

Just here, at Céide Fields, around and under me lie megalithic tombs, foundations of houses and pens, and about 2,500 acres of stone-girth fields, patchworked more than five thousand years ago, a thousand years after the first inhabitants cleared the pine, oak, birch, and yew. This is the most extensive Neolithic settlement yet discovered and the oldest enclosed landscape known in Europe, a model of land use that I can see in the small green fields beyond Céide. ◇

PEAT, MIRES, BOGS, FENS

Many of our names for wetlands may be used in ordinary conversation for all sorts of waterlogged ground. But sometimes we want more precise meanings.

We might begin with *peat,* a frequent but not constant ingredient of wetlands. This is an interesting word, apparently first meaning a brick of peat cut for fuel, used in Latin sentences written in Scotland and the north of England, then gradually shifting to mean the substance. Peat is vegetable matter that because of permanently wet conditions—and the lack of oxygen that accompanies them—has not rotted completely away. In more technical terms, peat forms when an ecosystem produces more organic matter (including small proportions of animal matter such as insect skeletons) than it can consume through the various digestive systems it hosts, especially the microbial ones. In other words, energy that comes into the system through plant photosynthesis is not entirely taken up by animals and plants or released by oxidation. The residue, or reservoir, is preserved as organic matter in the soils, which vary in color from dark brown to a characteristic black. This typically happens in places with high precipitation, humid air, poor drainage, and (often) cool temperatures. In northern lands, peat made of sphagnum mosses is typical, but other wetland plants also create peat. Peatlands are especially abundant north of the 45th parallel, across most of Canada and Alaska, the northern edge of the United States, Siberia, Scandinavia, Scotland, England, and Ireland. But they also occur across much of Africa, southern Asia, New Zealand, Australia's east coast, the Andes, the Amazon basin, Central America, and the coastal plain of the United States.

Several characteristics of peat make it especially useful to humans. Because its carbon content is quite high, in some places it has long served as a good source of fuel for household heat and cooking; more recently, it has been burned to generate electricity on an industrial scale. Because it is lightweight, absorbent, and organic, it improves garden and nursery soils. (Peat continues to be mined for these two uses, but because it forms so slowly, this mining is unsustainable.) Peat also interests many scientists and other scholars because it decays and accumulates so very slowly and preserves its contents so well: in the Northern Hemisphere, many peat deposits date from the end of the last ice age and a depth of forty feet is not unusual. Climate and environmental historians learn from pollens and other identifiable plant parts about changing landscapes, plant communities, precipitation, temperature, and so forth. Global warming researchers work with complex calculations about the carbon and methane absorbed and released by undisturbed and disturbed peatlands. Archaeologists and historians of human culture learn from the preserved artifacts of early human settlements and even human bodies. The preservative power of peat also inspired a powerfully evocative series of poems published in the 1970s by Irish poet Seamus Heaney. In such poems as "The Tollund Man," "The Grauballe Man," and "Punishment," Heaney considers the stories suggested by Iron Age bodies exhumed from bogs in Denmark. These poems are delicate meditations on immediate and deep history, victimhood, sympathy, and war.

As with all wetlands, words for types of peatland are not simple, and the common and scientific usages of terms don't always match. *Mire,* for instance, doesn't feature much in ordinary conversation, though it survives in newspaper phrases, as when politicians are "mired in corruption." The word has an old-fashioned ring. One might think of Grimpen Mire in Arthur Conan Doyle's most famous Sherlock Holmes story, "The Hound of the Baskervilles." But in ecological science it is a current term with a precise sense, though one used more by European and British scientists than by Americans: a mire is a peat-forming wetland.

Fens are one kind of mire. (*Fen* is a very old word in English, appearing in *Beowulf.*) These wide areas of low-lying flooded land are open to overflow from rivers or tidal floods from the sea; often coastal tidal salt marsh merges with inland peat fen. They are watered both by precipitation and by ground and surface water, so they have regular supplies of oxygen and minerals. Rich fens are low in acidity and high in minerals (these are close to what Americans call marshes); poor fens approach the conditions of bogs; and many places are intermediate. As their level of peat rises above water-carried minerals, fens may evolve from rich to poor and finally into the distinctively shaped domes called raised bogs.

In ordinary usage in the United Kingdom, the term is applied to any landscape with a certain characteristic appearance. Wild fen landscape is pools or lakes interspersed with thick, peaty reedbeds, sedge, and partially submerged trees, often willow, oak, or alder. There may be dry islands—some, such as the Isle of Ely in Cambridgeshire, large enough to support towns. Rivers from uplands bring silt that may form natural levees, embankments that tend to raise the rivers above the low-lying peat and then give way, causing further flooding and sometimes diverting the river. The traditional livelihoods of the undrained fen were peat-cutting for fuel, reed and sedge harvesting for thatch, fishing (especially for eels), and fowling. Most wild fenland in Britain has been drained, and one of very few surviving examples is Wicken Fen near Ely, run by the National Trust as a nature reserve and maintained in its fenland state by pumped water.

Draining these fens was begun by the Romans, but the most extensive projects were during the reign of Charles I under the supervision of a Dutch expert, Cornelius Vermuyden. The result was some of the best farmland in the country: livestock pasture in medieval times and high-quality arable land now. Britain's only large-scale wheat fields (nick-named "prairies") are to be found there today. This history explains why *fenland* also often means a cultivated landscape in which fields or wet paddies are criss-crossed with straight drainage dikes. The extensive paddy systems of much of Southeast Asia could be called fens, as could the reclaimed, windmill-dotted coastal plains of Holland. In Britain the region known as "the Fens" is the extensive area of ancient fenland, nearly all drained and cultivated, around the large inlet known as the Wash, on the East Coast where Cambridgeshire meets Lincolnshire. This flat, richly fertile country has black, peaty soil, big skies, and villages with names like Hundred Foot Drain.

Bogs are also a kind of mire. This word is from Irish or Gaelic *bogash,* with *bog* meaning "soft"; its first appearance in the *Oxford English Dictionary* is a late 1505; and it is widely used for any soft, spongy, waterlogged ground into which a person will sink a few inches or more. It may mean the small "boggy" part of a meadow or wood, where a stream has become silted or blocked, or a wide expanse of wet moorland. A bog deep enough to swallow you completely is sometimes called a morass or mire. A bog with an unstable layer of turf floating over water beneath may be called a quaking bog or quagmire (*quake* and *quag* do come from the same word root), another term with cultural resonance. Wars in which an army gets stuck or "bogged down" are often compared to quagmires, as was the American war in Vietnam.

Again, ecologists and geographers use the term in a more specialized sense, for mires whose wetness results entirely from direct precipitation. Bogs are likely to have no drainage and lose water only to evaporation. As a consequence, they are nutrient-poor, acidic, and low on oxygen.

With their impermeable peat bottoms, they support specifically adapted plants, especially the matted sphagnum mosses that form the blanket bogs that cover much of western Ireland, spreading in all directions over landscapes both flat and sloped. The walker on peat bog, or on the moor with which it is often interspersed, will need waterproof boots and a very exact knowledge of the paths but will find a landscape of space, solitude, and rich and subtle browns, blacks, purples, and greens.

We have often feared such landscapes. The ground in them, we have thought, refuses to support our feet. We might find ourselves sinking, and the mud might hold us with its suction, refusing to release us; thrashing to escape will only make us sink deeper, and we have to keep still, feeling ourselves sinking. We feel an elemental terror of being held in this way, an archetypal helplessness. Sometimes the unstable surface looks like the firm ground adjoining it—we might step onto the mud unawares—and so such landscapes have often been called treacherous, as if they were traps deliberately laid. Other threatening landscapes, such as forests and deserts, may be full of dangers, but at least they are places we can enter and leave on our feet. There is a forbiddingness about some wetlands that is eerier. These alien territories may seem to be beyond our known world. They are expanses of strange, impenetrable space into which we can only gaze, unable to venture in and put them behind us.

These are highly traditional attitudes in Western culture, where wetlands have often been represented as dark, sinister places. Canto 7 of Dante's *Inferno* has swamps in which the wrathful struggle and the sullen sigh. In *Paradise Lost,* Milton describes the terrain of Hell as one of "Rocks, caves, lakes, fens, bogs, dens, and shades of death." In John Bunyan's *The Pilgrim's Progress* (1678), the first hostile allegorical landscape encountered by Christian on his journey is the Slough of Despond, a "miry slough" representing the despair that sometimes falls upon sinners when they become aware of their sinfulness and the danger to their souls. This despair acts like mud and tangled roots, impeding the progress of the wading pilgrims and slowing them down until some give up the journey. Stevie Smith's poem "The Ass" reworks the tradition of this allegorical landscape, telling the story of Eugenia, a holy fool, who walks along a causeway across a "soppy morass" that "sopped and shuffled/Either side" and is tempted by a fiend to step off the path. (But "She was not such an ass to try the green,/It would deliver her below.") Wetlands, in this tradition, are the mires in which the pilgrim gets stuck and loses resolve. The right way becomes obscured by oozing, shifting, deceptive surfaces until it is lost and the pilgrim sinks to perdition.

Dark depths and marsh gases have also been associated with corruption. English nature writer Richard Jefferies's apocalyptic novel *After London* (1885) depicts an England suddenly depopulated, save for a few survivors. Most of the country reverts to something like its preindustrial condition, but the features that reveal this as a postindustrial rather than a recovered natural world are the huge, black, poisonous swamps that mark the sites of the great industrial cities and symbolize their filth and moral corruption. Rotting vegetation, too, suggests human decomposition and the grave, a melancholy connection reinforced by the flat "emptiness" of marshlands, places routinely described as "desolate." In *The Lord of the Rings*, J.R.R. Tolkien continues this tradition when he has Frodo, Sam, and Gollum pass through the "Dead Marshes" on their way to the dark land of Mordor:

> Hurrying forward again, Sam tripped, catching his foot in some old root or tussock. He fell and came heavily on his hands, which sank deep into sticky ooze, so that his face was brought close to the surface of the dark mere. There was a faint hiss, a noisome smell went up, the lights flickered and danced and swirled. . . . Wrenching his hands out of the bog, he sprang back with a cry. "There are dead things, dead faces in the water," he said with horror. "Dead faces!"

Tolkien later wrote that this passage owed something to the horrors of the muddy and pestilential trenches in France during World War I, where he served. But it is also packed with words, images, and accompanying attitudes that have emotional resonance in English and European cultural history.

◇ ON THE SPOT: AT WICKEN FEN

Richard Kerridge

Wicken Fen, Cambridgeshire, April 1. Stepping off the duckboard, I squish water out of the spongy turf, but the ground beneath feels hard and springy. The path runs alongside one of the straight drainage ditches, a canal of deep water, clear but revealing a dark brown world in which pale green plants stand out. Beside the path, mounds of black earth have been left by moles. Drain and path are surrounded by sedge, a straw-colored forest of reed sticks rising from dark marsh water and from a mulch of squashed stalks and leaves. At their heads the big tufts tremble. It is about seven in the morning: cold, but the sun is making the turf steam and beginning to filter through the reeds. Everything waits for its touch.

The ducks are warming up. Their conversation is all around me, starting, stopping, splashing, cackling, protesting, explaining; sometimes far off, sometimes only feet away beyond the reeds. They fly overhead, and across my short horizon. Geese fly, too, in unfolding skeins. Their honking out on the lakes and fields is a distant crowd murmur. A coot gives its abrupt, single brassy squeak. Clear water must be very close, but I can't get a glimpse of these birds, only the flicker of a wren or warbler in the reeds.

I come to a gap in the sedge and can look across the fen to the skyline. Few colors but subtle variations: everywhere the pale straw of the sedge, washed out but potentially golden. Above the straw is the darker brown of the dancing sedge-tufts. Oaks and alders rise from the reeds. Only the footpaths are green. Otherwise, this is a black and golden ecosystem, under bright blue sky: black earth, black water, yellow sedge. Another line of ducks goes over.

The hide is visible now, standing out above a patch of woodland, a shed on stilts with the look of a watchtower. Before I reach it, an angry chattering starts near my feet. A shrew the size of my thumb has stopped its scurrying forage to look up at me, rise on its haunches, and pour out its irritation, as if with hands on hips. Velvety brown shoulders and white tummy: the same color combination as the whitethroats and sedge warblers in the reeds. I am interrupting its precious hunting, in the last moments before the sun gets here.

The stair in the hide is knobbly with pigeon droppings. Downy feathers stick to the wood and float with the dust in shafts of light, as the sun reaches the hide and, flashing, enters through every thin gap, as if the world outside were in flame. Pigeon dust fills my nose, and the new warmth of sun on wood. Two trap doors let me up to the platform, and I'm above the reeds and the low, tangled trees, looking out over them. The whole expanse is bright. There are the fields of sedge, and beyond them the lake, and some birds on the lake, and the big fenland sky. Start the sweep again, from just below the hide, across the drain and over the sedge fields. I notice now the grid pattern of the partly hidden drainage ditches, like the layout of an American city. There by the lake are some cattle, one trying to catch up with the others, ambling with just a touch of anxiety until she can't bear it and runs to join them. In one corner, a group of horses stand letting the sun warm their backs; a snort, a tail flick, a leg scratch but otherwise still, their gray flanks catching the sun. They are the Konig ponies, a "bred-back" breed resembling the extinct Tarpan wild horses of Europe, brought here by the National Trust to graze the fen. Gray with a slight reddishness in the mane and on the flanks and rump, they have a zebralike look in the shape of their heads and backs. I walk back, in sunlight that is now an established fact that surprises nothing. Now the day has a mid-morning, busy feel. In the drain, where a clump of reeds rises from the water, I see a movement: a strange rolling, moving, dun-colored lump. It is a ball of toads, males trying to climb onto one female, and now I notice their little quacks and squeaks. At this breeding time the flesh of their backs seems to liquefy inside their skins, so that their backs shake like jelly. One, ejected from the ball, climbs back onto it, eagerly reaching, stretching his long legs. Farther on, I see a small pike waiting just below the surface, flanks glinting, striped with the green of the trailing weeds but with touches too of the sedge gold that enters everything here, against the black of the water. ◇

MARSHES AND SWAMPS

Few Americans know the difference between a bog and a fen. Except for the Atlantic Coast pocosins—a kind of shrub bog given an Algonquian name perhaps as early as the 1580s (maybe by Thomas Harriot, who accompanied Sir Walter Raleigh)—bogs are rare south of the last glacial advance; and in the United States, *fen* is a word used mainly in science and crossword puzzles. Although technically the Everglades are a fen, and scattered fens elsewhere are protected by the Nature Conservancy and other agencies, the word is rarely used. It does not appear, for instance, in the official brochure for Everglades National Park, which otherwise abounds with names of landscape types. Instead, the key distinction in America is between marshes and swamps. While these words are also used in Britain, they have more precise meanings in the United States, where they refer to two common kinds of wetland landscapes. *Marsh* is an old word in English (appearing as early as 725), but the earliest appearance of the word *swamp* listed in the *OED* is from Captain John Smith in 1624, making it (along with *pocosin*) among the earliest terms describing a specifically American landscape, one new to English explorers and settlers.

Typically, both marshes and swamps are fed by surface, ground, and rainwater, are well supplied with oxygen and minerals, and so are biologically productive. They may or may not form peat, and they are most often neutral rather than acidic. Thus some marshes are also rich fens, and the English Fens would be called marshes in the United States, as would European reedswamps. Both swamps and marshes may have waterlogged soils without surface water or be submerged by up to three feet of water, and their degree of wetness often varies by season, tides, and details of elevation. What distinguishes them from each other is vegetation: marshes are grassy, herbaceous, and shrubby, while swamps are forested.

Major types of marshes include coastal saltwater and freshwater marshes, many kinds of inland lakeshore and streamside marshes, the extensive networks of seasonal or vernal ponds in New England and California, and the prairie potholes, sloughs, and playa lakes of the High Plains. Some marshes are shrubby tangles of alders and willows that scientists may call carrs, a traditional word from northeastern England, while an American might use a regional term like willow flat or willow bottom. Sometimes marshes form narrow strips, but more often they cover broad swaths in greens, yellow-greens, purple-greens, russets, chestnut browns, and silver. They glisten and ripple with the wind into complex patterns of light and shadow. Marsh plants are typically long and slender—segmented grasses with lacy seed sprays, sharp-edged sedges, round rushes punctuated by spiky dark brown burrs, papyrus with its firework explosion tops, tall cattail blades whose long seedpods of chestnut velvet burst into blizzards of silky white seeds. These landscapes teem with life—not just plants, but also insects, fish, amphibians, reptiles, mammals, and birds.

Coastal marshes are among the most productive ecosystems in the world. They encircle much of Europe (including the United Kingdom and Ireland) and North America, where they are particularly important along the Atlantic and Gulf coasts. Significant coastal marshes also occur around the southern end of Hudson Bay, the coasts of Chile and Argentina, the southeastern edges of Australia, and New Zealand's South Island. With waters ranging from considerably saltier than the ocean to completely fresh, these transitions between land and sea form in protected intertidal zones where bays, barrier islands, and spits of higher land offer enough shelter that sediments can accumulate and support plants, which then hold more sediments in place and buffer wind-generated waves.

Some traditional human uses of these marshes are relatively benign, such as the harvest of "salt hay" for livestock feed, thatch, caning, packing material, and more. Many of the paintings of nineteenth-century Hudson River School artist Martin Johnson Heade depict this landscape's striking horizontality, complex skies (often dramatic with storm clouds and sunsets), variations of light and shadow, and scattered piles of cut hay, raised above the tides on wood staddles until oxen with specially fitted shoes would haul it off during the winter. More destructive agricultural uses include direct grazing (often found in Europe) and the replacement of native plants with crops. In the "tidewater" regions of South Carolina and Georgia, slaves with rice-growing experience from West African floodplains cleared native plants; built dikes, levees, and floodbanks; and established highly profitable rice fields. The city of Savannah, Georgia, has a name that is sometimes used for marshes, a word use that points out the continuity between marshes and grasslands.

Where fresh and salt water meet at the mouths of large rivers, there are often cities and thus even more problematic disturbances. Near and in cities like Boston, New York, and New Orleans, wetlands were drained and filled for building, and wet "waste" areas became dumping grounds. In Boston, for instance, a steady pattern of landfill (often involving sand and gravel excavated from the local drumlins) reduced its marshes to ever more pestilential fragments. (The fashionable Back Bay neighborhood was built on a filled-in tidal basin and marsh: hence its systematic, linear, symmetrical street pattern, which resembles no other part of the city.) By the late nineteenth century, Frederick Law Olmstead, designer of New York City's Central Park, was commissioned to deal with Boston's remaining polluted wetlands and chose to construct a new kind of park that offered both water-filtering marsh and restorative open greenery. Borrowing the word from England for its comforting pastoral associations, he called part of his creation the Fens. Thus we have the name of the nearby Red Sox baseball stadium, Fenway Park, and the current name for the area, the Fenway.

At the mouth of the Hudson River, a once vast and fertile mix of marshes and swamps has been reduced to the notoriously polluted and paradoxical urban wilderness celebrated by Robert Sullivan's witty book *The Meadowlands: Wilderness Adventures on the Edge of a City* (1998). And across the broad delta of the Mississippi, marshes have been damaged by such actions as channeling the silt necessary to maintain wetlands (and the continent itself) directly out into deep Gulf waters and dredging for ship channels and oil pipelines—changes that exacerbated the destruction caused by Hurricane Katrina in 2005. Similar damage continues to be done to coastal wetlands (including mangrove swamps) around the world, though there are also growing efforts to preserve and restore these protective margins.

Like other wetlands, coastal marshes are critical habitat for wading birds and shorebirds as well as for migratory waterfowl and songbirds. More than 280 bird species have been counted just in the freshwater coastal marshes of the United States, from the familiar and sturdy red-winged blackbird to the ethereal snowy egret, from little marsh wrens to giant pelicans and great blue herons, from shy rails to raucous gulls. These marshes provide essential food and shelter for fish and shellfish, often for spawning, nurseries, and juveniles: more than 90 percent of commercially important fisheries on the southeastern Atlantic and Gulf coasts depend on salt and brackish marshes. Species that rely on marshes include eels, striped and largemouth bass, the now-rare sturgeon, shad, catfish, shrimp, and oysters. They make good homes for river turtles, water snakes (including the notorious cottonmouth or water moccasin), the rare American crocodile (which prefers salt water), alligators (which prefer fresh water), river otters, muskrats, nutrias (first imported from South America for their fur, now an exotic species problem), and swimming marsh rabbits. Some of these animals help shape their habitat. Muskrats, nutrias, and snow geese sometimes destroy large amounts of vegetation. Alligators open up "gator holes"—pools that stay wet during the dry season and serve as oases for many other species.

For a visitor to the coastal marshes, the overwhelming experience is likely to be one of open space, of changing light and shadow, of moving air—and, it must be said, of mosquitoes. In the words of nineteenth-century poet Sidney Lanier, these landscapes, with their enormous skies and smooth expanses of green and gold, seem "candid and simple and nothing-withholding and free." Up close, though, the marsh suggests more that is hidden. On a day in March, say, with the dark clouds of a thunderstorm rolling in over a Gulf Coast marsh, the air will feel salty, thick, and heavy. The wind rustles and swishes through tall, slender leaves, and invisible migratory songbirds utter fragments of melody. A bittern booms. Ibises squawk as they rise from the water, startled by fishermen bringing their boats back to solid land. Nutrias and rabbits swim among the reeds, visible just for a moment, then hidden. An alligator hisses to warn walkers away from a scatter of babies, some so small they cling to a single narrow blade of green. The sound of large wings moving through the air announces a long, ragged line of roseate spoonbills, each one an odd, winged oval of pink following a long-handled spoon. They take one's breath away, glowing like firelight against the darkening sky.

Inland marshes are also very productive and have been similarly important to the growth of human cities and cultures. The civilization of ancient Egypt grew along the marshes of the Nile, and the infant Moses in his "ark of bulrushes," laid safely "in the flags by the river's brink," was a child of the same landscape. When the translators of the King James Bible chose their words, they were careful here to evoke traditional associations: not only *ark* rather than *basket*, but also *flag*, whose earliest meaning in the OED is rush or reed. Papyrus, the first paper, is made from marsh sedges. Babylon rose above the marshes of the Tigris and Euphrates rivers in modern-day Iraq. Across North America, many pre-Columbian towns and cities clustered along the marshes of the Mississippi and other rivers, and the European American migration routes westward across the Great Plains followed the wide riparian bands that often marked the only reliable water, forage, and wood.

Most inland marshes are freshwater. (Some are salty, as, for instance, around Great Salt Lake and in places on the Great Plains.) All are critical habitat for wildlife—natural homes for water-adapted species and refuges for creatures whose once-broader territories have been taken over by humans. During the spring, countless small vernal ponds in New England and elsewhere shelter frogs, turtles, and salamanders. If their ancient territories and migration paths have not been fatally altered, salamanders will travel overland on warm, rainy spring nights to breed in these oases; the young will race to transform themselves from creatures of water to creatures of land before their refuge dries up in the summer's heat. In willow-filled marshes (which often overlap with beaver ponds) such as in Yellowstone and Grand Teton national parks, moose may browse, hidden by summer's green and silver foliage, or winter's thickets of saffron, russet, and lavender branches, until they venture out into a patch of open water or grass. Strands of dripping vegetation hang from their pendulous dewlaps, their near-black coats seem to swallow the light, and the antlers of the males look like enormous branching plates.

One of the major inland marsh areas in North America—and one of the world's largest wetland areas—is the prairie pothole region, which covers parts of Alberta, Saskatchewan, Manitoba, Montana, the Dakotas, Minnesota, and Iowa. When the last glaciers pulled back some twelve thousand years ago, they left behind them a softly crumpled, poorly drained landscape of glacial till, a dimpled, pocked, and undulating land. This is essential territory for the continent's waterfowl and migratory birds: breeding ground for some half to three-quarters of all North American waterfowl and source of springtime food and rest for hundreds of thousands of other birds that migrate along the Central Flyway. In a May following a good snowy winter, an aerial photograph of an undrained reach of this vast landscape will show a thick splatter of water over land—dots and splashes, blobs shaped like tadpoles, genies rising from bottles, punctuation marks, maple leaves, mostly the color of the sky but sometimes the bright, rich green of marsh plants. Maybe three hundred little wetlands will be scattered across a square mile, a measurement that's easy to see in the plains, marked as it will likely be with the straight-line, dirt-road framework of surveyed "sections." Among these prairie potholes (or, as they are often called in this area, sloughs, pronounced to rhyme with shoes), soft mounds of wild prairie grasses wait for June's warmth to turn green. Here and there the crops in plowed fields just begin to emerge, some of them in lines curved to match the land, others in clean stripes and rectangles that stand out in sharp counterpoint to the sinuous water and hills.

Like marshes, swamps are ecologically varied by climates, water conditions, and dominant trees. Black spruce and tamarack (larch) grow on watery, acidic peatlands in northern latitudes: these are characteristic swamps of the taiga, the boreal forest that rings much of the Northern Hemisphere, most expansively in Canada and Siberia. Once there were significant cedar swamps in New England, one of which was the setting for Sarah Orne Jewett's 1886 story "The White Heron," in which the young heroine has to choose whether to hide her beloved wild bird from a charismatic visiting naturalist-hunter. New England's red maple swamps inspired Thoreau to exclaim, "When I would recreate myself, I seek the darkest wood, the thickest and most interminable and, to the citizen, most dismal, swamp. I enter a swamp as a sacred place, a *sanctum sanctorum*. There is the strength, the marrow, of Nature." *Dismal* is another early American word for swamplands (especially those near the Virginia–North Carolina border, including one named the Great Dismal), and as Thoreau notes, most of his fellow citizens would have thought of these landscapes in such negative terms. In his day, maple swamps were the wildest places left near his home, and it was in the same essay, "Walking," that he wrote, "In wildness is the preservation of the world."

Before they were cut and drained, several large swamps of mixed hardwoods (oak, sycamore, maple, ash, elm) lay south of the Great Lakes. One of these, Indiana's Limberlost, was home to the writer Gene Stratton Porter, whose best-selling children's books *Freckles* (1904) and

A Girl of the Limberlost (1909) captured the swamp's final years and a telling mixture of attitudes. Porter herself loved the swamp for its wildness and its rich birdlife, and so do her young hero and heroine. But both books also speak of its terrors—quaking bogs that devour humans (in fact, the swamp was named for a lost man, Limber), bands of violent outlaws, poisonous snakes—and her hero Freckles works for the logging company that is about to destroy the beloved place. In the words of the girl he loves, the Swamp Angel, "I feel dreadfully over having the swamp ruined, but isn't it a delight to hear the good, honest ring of those axes, instead of straining your ears for stealthy sounds?" Yet Freckles is nearly killed by the felling of the swamp's most valuable tree (a maple that will provide bird's-eye veneer for furniture), one of the many sharp and poignant ironies in these two novels.

In the South, along the Atlantic coastal plain, through much of Florida, along the Gulf Coast into Texas, and over the wide lowlands of the Mississippi River as far north as Illinois, swamps are variously dominated by bald cypress and tupelo, by pond cypress and black gum, and by Atlantic white cedar. Sometimes these are called deepwater swamps because they are nearly always flooded, or blackwater swamps because their tannin-rich water is the color of strong tea. In addition, along the wet-soil, seasonally flooded riparian plains and side channels of rivers throughout the same region, bottomland hardwood swamps or forests form, populated by a rich mix of trees including sycamore, sweet gum, elm, oak, hickory, cypress, and tupelo. (All these swamps are likely to adjoin marshes or be mixed with them.) Again, all have been much reduced from their former extent. William Bartram spent a good deal of time in cypress swamps, and his *Travels* is a good source of descriptions of them at the end of the eighteenth century. William Faulkner wrote about the Mississippi bottomland hardwood forests in *The Bear,* a coming-of-age story that turns on a boy's experience of vanishing wildness.

These are dramatic and evocative landscapes by any measure, dim, still haunts of alligators and cottonmouths, landscapes doubled by the dark mirror of the watery ground. In luminous black and white, and in the tradition of Ansel Adams, the (often large-scale) photographs of Florida artist Clyde Butcher capture the complexity, beauty, and mystery of these swamps and the big-sky open marshes with which they're mixed. Tall trees are wrapped in narrow, spreading buttresses and surrounded by the bulbous wooden stalagmites called knees. Long gray tangles of Spanish moss (not really a moss but a relative of the pineapple) hang from what seems like every branch, and along trunks and branches, resurrection ferns wait shriveled and dry for rain to restore them to green plumpness. On fallen trunks, turtles line up ready to topple into the water at the slightest alarm. Birds may be anywhere: bright yellow prothonotary warblers; tall pileated woodpeckers in red, white, and black; white ibis with their down-curved hot-pink bills; hunched night herons; and stately great white egrets.

Indeed, it is not too much to say that it is birds that best characterize North America's wetlands. Not surprisingly, for instance, many of the early nineteenth-century bird portraits by John James Audubon have wetland settings. His three marsh wrens perch on their clever nest of live rushes and cattails, dried grasses, and cattail fluff. His long-billed curlews stand against marsh plants across the water from the city of Charleston, South Carolina. His snowy egret (which he called the snowy heron or white egret) is framed in marsh plants with a plantation house in the background; a tiny human hunter is stalking the hunting bird. It was the impending extinction of such spectacular birds as the snowy and great egrets during the nineteenth century, hunted first for food but increasingly for plumes for women's hats, that led activists (many of them women) to found the National Audubon Society and take other protective measures.

Birds also stitch together the continent's wetlands, from the peatlands of the north, through the prairie potholes, to the coastal marshes and swamps: a map of current and former swamps and marshes in the United States closely matches both the ancient migration flyways and the current system of national wildlife refuges. The whooping crane once nested throughout the prairie marshes, and Audubon depicted it against a background of coastal swamp, curled tightly into the frame of the portrait despite its five-foot height, bending down to eat a pair of baby alligators. Now this endangered bird—North America's tallest, at about five feet—nests in the peaty muskeg of Wood Buffalo National Park in northern Alberta and winters in the salt marshes of the Aransas National Wildlife Refuge along the Texas Gulf Coast. Along with other migratory birds, the other fifteen remaining species of crane make the same kinds of connections among wetlands in Europe, Asia, and Africa, weaving together distant parts of the world.

WET/DRY

As we promised at the beginning of this chapter, we've looked at how water embodies the permeability or illusoriness of boundaries: in terms of word meanings, cultural categories, the ways water distributes itself across landscapes, and the ways our attitudes are often mixed rather than straightforward, changing rather than permanent. But we ought to think a little more about how it is perhaps particularly in the nature of wetlands to complicate, confuse, or blur lines we'd prefer to think of as simple, clear, and stable, and how these complications can be both deeply disconcerting and also extremely productive. For wetlands are—as the very word tells us—both water and land, wet and dry, neither clearly one nor clearly the other, or at least never so for long.

Perhaps it is this quality that leads us to associate them with the primitive. *Backwater* is a word that, strictly, means a stretch of water that has spilled out of the mainstream or backed up when it reached the mainstream and so sits trapped and stagnant. (It is also the name of the watery coastal complex of India's southwestern coastline in Kerala State.) Tellingly, though, the term is mainly used as an image for rural places that are remote and overlooked, places whose inhabitants fail or refuse to move with the times. When evolutionary ideas began to enter the popular and political imagination in the second half of the nineteenth century, the prehistoric landscape was commonly imagined and depicted as swampy, a habit that reinforced the general association of wetlands with primitive ways and forms of life. Later the idea of primordial ooze as the very earliest generator of life came along, adding to the notion of primitiveness a more positive sense of creativity.

Or perhaps the indeterminacy of wetlands makes us uncomfortable in some way psychoanalysis would recognize. Because they have surfaces into which things can sink and be preserved out of sight, they make a ready symbol for the Freudian idea of the unconscious. This is one idea explored by Australian cultural theorist Rod Giblett in his book *Postmodern Wetlands* (1996). Assembling various cultural significations of wetlands—aesthetic, symbolic, literary, political, military, and psychoanalytical—Giblett analyzes the combination they make. Feminist analysis has frequently identified traditional images that demarcate masculine and feminine identity. Important among these are the associations of hardness and rigidity with masculinity, softness and fluidity with femininity. But wetlands are neither masculine nor feminine in these terms. Neither solid nor fluid, they merge the two identities, and Giblett sees them as especially threatening—symbolically—to the dominant masculine identity: they look solid but are soft and engulfing.

It is this indeterminacy as a gender symbol, he argues, that makes wetlands so valuable. They have a subversive cultural or countercultural value to go with their ecological value. "Wetlands have by and large been the locus of the horrifically uncanny to be shunned, and destroyed," he writes, using the Freudian concept of the uncanny or unhomely— the known place suddenly become strange, nightmarish, and uneasy, disturbing in its very familiarity. Now, instead, he says, "they need to become a place of the fascinatingly uncanny, their sights and sounds and smells, even their tastes and textures appreciated and conserved." So a culture that begins to emerge from its repressive fears of the female, the body, and the unconscious is a culture that will delight in wetlands. A culture may emerge that relishes the symbolic indeterminacy of wetlands, instead of finding them scary. This is Giblett's intriguing implication and the way he brings the cultural and the ecological together.

There's certainly no disputing the physical difficulties that wetlands can offer to human bodies. Fear of these places fitted naturally with the assumption that wetlands were unhealthy places, where poisonous gases bubbled up and diseases were bred. For a long time, stagnant water was thought to produce "miasma," unhealthy air that caused such illnesses as malaria and yellow fever, two diseases that traveled to the Americas in the seventeenth century from Europe and Africa; traditionally, malaria (from Italian *mala aria*, "bad air") has been called marsh fever or swamp fever. This assumption is to some extent accurate. Mosquitoes continue to be major carriers of diseases like encephalitis, West Nile virus, and malaria, which is still a deadly force over much of the world. When, at the beginning of the twentieth century, mosquitoes were found to be the carriers of these diseases, rather than the air itself, increasingly aggressive drainage campaigns were waged, and with considerable success. According to Ann Vileisis's thorough environmental history of American wetlands, *Discovering the Unknown Landscape*, between 1900 and 1914 the malaria mortality in New York City dropped by more than 86 percent, and nobody then much cared about the loss of the wetlands with their rich repertoire of what ecologists today call "ecosystem services." We know now that malaria can often be prevented by sleeping under treated mosquito nets, and that disease-carrying mosquitoes are also likely to breed in abandoned tires and muddy tire tracks, and so we're less likely to think that destroying these landscapes is the best action to take. (We also don't try to eradicate cars.) But we may well still feel especially uneasy about seeing a mosquito land on our arm if we're walking in a bog or swamp.

Given their traditional array of negative cultural associations, wetlands were above all obstacles to rational order for Enlightenmentminded technologists and administrators, from the eighteenth century until very recently. Indeed, in their physical complexity—their innumerable pools and channels, serpentine coils, and congested waters—they were more than obstacles; they were the very antithesis of order, the epitome of disorder. (It would take a later ecological perspective to discover order and function in the wetlands themselves.) They signified the brute, random, recalcitrant messiness of nature. To the men of the Enlightenment, backwater superstitions seemed to be products of their natural home, bogs and marshlands. The ideology that set out to eliminate superstition also set out to eliminate the marshlands themselves. The two aims were seen as parts of the same project.

In *The Conquest of Nature,* his study of German attitudes and policies regarding the natural world, historian David Blackbourn describes this official point of view toward marshlands in the mid to late eighteenth century, a period when huge areas were drained and reclaimed. Hard to map and classify, hiding places for unruly resisters of rationalization, wetlands were an affront to the absolutist Enlightenment state, with its determination to order, measure, and discipline the world. "Every association of marshland and swamp was negative," Blackbourn writes. "Their inhabitants were considered taciturn, clannish, and superstitious, people who saw marsh gases and believed them to be will-o'-the-wisps." The sequence of engineering projects throughout the nineteenth century to straighten (or "correct") the path of the Rhine and drain numerous areas of wetland around it was hailed as one of the triumphs that defined the emerging German nation. As Blackbourn points out, those in charge regarded the people who inhabited and understood wetlands in the same way as they regarded the lands themselves: they had no more compunction about ending these cultures than they did about destroying the wetlands that sustained them. Similarly, in Italy in the 1920s, the draining of the Pontine marshes was Mussolini's propaganda centerpiece, defining him as a modernizer of the country.

During World War II, the great redoubt of partisan fighters against German occupation in eastern Europe was the Pripet Marshes, a vast wetland area—still the largest in Europe—that in 1939 covered extensive parts of eastern Poland and the Soviet Union. Even before the war, Blackbourn recounts, German geographers and engineers had plans for draining them, while Nazi race-theorists liked to contrast the ordered landscapes of modern Germany with these marshes, which they claimed were the birthplace of the Slav races. Draining the marshes was a major project envisaged by the Nazis when they invaded Russia in 1941. Right away, the Nazis took advantage of the concealment the marshes offered, using these places for the mass murder of Jews. Later in the war the drainage plan had become a military rather than civil objective, but by this time it was beyond Germany's means.

Resistance based in marshlands and reprisal campaigns directed against them have been features of many wars from antiquity to the present. In the Vietnam War, the immense Mekong Delta was a vital stronghold for the Communist guerrillas, and therefore the object of numerous attacks and drainage plans, first by the French and then by the Americans. A more recent example still is Saddam Hussein's assaults on the Shi'ite Marsh Arabs of southern Iraq in the 1980s and 1990s, which combined military attack with the draining of extensive areas of the Tigris-Euphrates marshes, the largest wetland in western Asia.

But as these examples also indicate, the difficulty wetlands present to travel has also meant that they have long offered refuge for marginalized peoples and hunted fugitives. How to think about this is entirely a matter of perspective. For those with military and colonial minds, these landscapes have seemed primitive and savage, homes to rebels, bandits, and outlaws. For the hunted, though, and for those resisting invasion, government control, or coercive modernization, they have meant safety. If their human inhabitants have been feared and denigrated for their presumed backwardness and brutality, they have also been celebrated for their fierce resistance to mainstream culture.

It was from an impenetrable hideout in the East Anglian fens that legendary Saxon leader Hereward the Wake resisted the Norman conquerors in eleventh-century England. Especially during the early years of European settlement in North America, native people often retreated for safety and strategy into marshes and swamps, where they could maneuver much more comfortably than could the English. Well into the twentieth century, the Seminole and Miccosukee people of Florida maintained their traditional culture by living in the Everglades. In turn, settlers associated the swamps with their negative attitudes about the native people: in the popular seventeenth-century "captivity narratives," the swamps of New England often serve both as literal landscapes and as evidence of the devil's powers as manifested in the Indians. Similarly, the resistance of the Irish to English colonial rule led to the English term of contempt "bogtrotter." In the American Revolutionary War, the "Swamp Fox," Francis Marion, earned fame by leading a band of guerrilla warriors based in the swamps of the southeastern coastal plains. Later, escaped slaves often found refuge in swamps, and the

Underground Railroad featured safe routes through these forbidding landscapes.

A recent fictional character in this tradition is the DC Comics super-hero Swamp Thing, a virtuous monster, part human and part plant, who defends his wilderness home and fights for environmental causes. Another is Clinton Tyree, or "Skink," the renegade former governor of Florida, living wild in the Everglades and wreaking revenge on corporate corruption, who features in several of Carl Hiaasen's comic environmentalist novels. Although Skink hides out in the swamps and doesn't always follow the law, he and the Swamp Thing are both clearly in the business of protection. They embody and defend the health of ecosystems, a rich kind of order that needn't follow straight lines or eliminate natural complexities.

Something like this development of attitudes can be found in the best-known recent English novel about wetlands, Graham Swift's *Waterland* (1983). This book brings together many of the themes we have identified in this chapter—mythical, ecological, magical, scientific, technological, indigenous, historical, political, military, and psychoanalytical—in its calamitous story of two young people growing up in the East Anglian fens during World War II and consequences that befell them forty years later. The fens are a physical presence explored in ecological and zoological terms, a landscape to be understood historically, and a symbolic landscape that demonstrates, among other things, complex relationships between small, accumulating events and large explosions: "So forget, indeed, your revolutions, your turning-points, your grand metamorphoses of history. Consider, instead, the slow and arduous process, the interminable and ambiguous process—the process of human siltation—of land reclamation." They are also a tangible landscape, encountered by individuals at particular dramatic moments that give that landscape particular, momentary meanings.

Terry Tempest Williams's memoir *Refuge: An Unnatural History of Family and Place* (1991) contributes to a similar literary revision of our culture's responses to wetlands, this time centered in Utah's Great Salt Lake and the Bear River Migratory Bird Refuge just to the north. Like Giblett, though in a different mode, she associates the desire to control the movements of water (though not to eliminate wetlands) with a patriarchal system, one whose complicated strands include the Mormon Church, the builders of highways, and the military, which carried out atomic bomb tests in the area. And she associates the wildness of birds and women with a resistance to that system. Williams, too, whose career has also encompassed work as a natural historian, knows well the many benefits of such unruly places to the land's ecological health and to its inhabitants, nonhuman and human.

And we do now increasingly value wetlands. We value them as reserves of biodiversity, as refuges for creatures driven out of terrain where farming and building are easier: the Florida panther, the starling flocks of the Somerset Levels, and—a rather more difficult case—the man-eating tigers of the Sundarbans. We value them as places of beauty and as sites for recreation. Perhaps most important, we value them as vital components of larger ecosystems on which we all depend.

We might see these emerging attitudes as a shift in the place watery places hold in our imaginations. We might see them as deriving from a more knowledgeable self-interest: we need the continued carbon storage wetlands provide, their water filtration, the food they help nurture; and we need these things even more than we need the convenience that comes from draining, plowing, harvesting, and paving such places. We might see them as coming from the increasing willingness of the dominant Western culture to listen to other cultures, to tap into the wisdom of what anthropologists and land managers now have an acronym for, TEK, traditional ecological knowledge. And we might see them as results of a perspective that is shifting, perhaps fitfully and only in places, from a strictly human-centered view of what matters to a more ecological understanding of how our planet works and what truly matters to the health of all its creatures.

◇ **ON THE SPOT: AT THE BILLABONG**

Deborah Bird Rose

A billabong in the monsoon tropics of North Australia is always in motion. Rain expands it, the dry season shrinks it, and no two years are ever the same. The 2005–2006 wet season (December through April, approximately) was one of the best on record, and I was keen to see Yinawurru billabong. I had mostly been there during drought years, and I knew it best as a struggling little body of water surrounded by desiccated savanna woodland and mesas.

Yinawurru is only a mile or so from the Aboriginal community of Yarralin, where I lived for several years and where I continue to visit, learning all that I can of the people's philosophy and ecology. Jessie Wirrpa became one of my best teachers, and she first brought me to Yinawurru. I was new to the place then, and I watched with awe as the women and girls waded out past the tangle of algae and grasses to cast their hand lines.

White settlers named it Crocodile Billabong, and with good reason. The crocodiles here are endemic freshwater crocs, not to be confused with the larger and more aggressive saltwater crocodiles, and they love this billabong. "Freshies" are opportunistic predators, eating fish, insects, crayfish, frogs, lizards, and other small creatures, but they are not interested in people and there are no reports of people ever being taken by them. Nevertheless, on my first trip to the billabong I was wary, and stayed back among the coolabah trees. These large, twisty, spreading trees like to have their toes in water, and they offer pretty good shade. As a nervous newcomer I felt protected by them, and my appreciation increased again when Jessie told one of the young women to take an ax and climb a tree to cut out the native honey she had spotted.

Later we cooked and ate at an open fire. Billabong water has a curiously muddy sweetness that intensifies as the dry season rolls on. Billabong fish have that flavor too, especially the bottom-feeding catfish, whose delicate white meat offers up a sweet billabong smell as it cooks on the coals. I came to love that taste of sweet mud. It is like eating the primal moment when earth, water, and sunshine come together to make life.

Technically, a billabong is an old river channel that has been cut off from the main river. Yinawurru is related to the nearby Wickham River, which is a major tributary of the mighty Victoria River in the Northern Territory. Yarralin people told me that the billabong and river are connected by an underground channel, and there were strange stories of horses that died in the billabong and were found floating in the river. While it is hard to know exactly what to make of such stories, they offer eloquent testimony to the tricky ways of water, how it connects sky and earth, how it travels in subterranean waterways, how it floods and then disappears, how it is dangerous and life-giving all at once.

In August 2006 Yinawurru was big, about a mile and a half long and nearly sixty yards wide. There had been a big freeze in the south, but by the time the Antarctic wind reached North Australia it had mellowed, and offered respite from the intense heat. The wind rippled up the water, and Yinawurru glittered.

Around the edges horses grazed in the shallow water, and brahma cattle moved slowly among the coolabahs, their silver hides looking spectral in the shadows. These nonnative animals had already eaten out all the lilies around the edges, and only a few hardy individuals survived in the middle, their tender white and purple flowers standing up toward the sun.

The large water birds were there. Brolgas are blue-gray cranes with red head markings; they ambled along pecking the soft ground and looking, as always, entirely elegant. Some of the egrets rode on the backs of horses, and some walked around in the shallow water. The fish-eating birds were there too, a blue-winged kookaburra diving around, and a pelican munching its way slowly and methodically along the edge. Plovers waded along grabbing anything that moved, and the small insect eaters were extremely busy. Rainbow birds were doing aerial acrobatics between the trees and the billabong, their iridescent feathers flashing blue and green and orange in the sun. Willy wagtails are not so glamorous, but they too were working nonstop catching food on the wing. Dragonflies, butterflies, and water skippers were all on the move, and so were the rifle fish. These small striped fish spit a jet of water at an insect that is hovering or resting on overhanging vegetation and knock it into the water so they can grab it. Clever fish! They are accurate to a distance of about one yard, and are able to calculate the way light bends when it hits the water.

"Every place is a story," my Yarralin teachers told me. Yinawurru is no exception. In the creation period, when the great ancestors known as Dreamings walked the earth, the Sun Dreaming created this billabong. Nearby in the gorges and mesas are Rainbow Dreaming sites. Water and sun are the two big driving forces of the monsoonal tropics, and Dreaming stories locate both of them right here. The locations emphasize complementarity: the sun is in the billabong and the rain is in the mesas.

I encounter many stories here—Yinawurru is a great place for watching emplaced philosophy in action. Yarralin people's philosophical ecology works back and forth between that which endures and that which is ephemeral. Here at the billabong we see the major powers—sun, water, earth, and wind. We see how their interactions are always changing and always returning. We see the flux and the stability, the brief and vivid transience of individuals, and the enduring relationships that sustain life.

To sit and look at the billabong is to see that creation is ongoing.

We can see it with our eyes, and taste it and breathe it; sometimes creation smells of native honey, sometimes it tastes like mud. It glitters quick and bright, or gleams slow and stately, and is both deeply enduring and immensely fragile. A few years ago there was a proposal to pump out Yinawurru to irrigate a mango farm, the logic being that the water was wasted just lying there in the billabong. Yarralin people didn't accept that proposal; they know there is no waste here—there is food for everyone, year after year.

Yinawurru is Aboriginal "country": this billabong, this dry land, these plants and animals, these people; this holding together of that which passes and that which keeps going; this home for all who participate in its life. ◇

4

Desert Places, Desert Lives

PROLOGUE

Life on Earth began in the watery realm of the seas, moved to the shallows and marshes, and then crept onto increasingly drier land. But no matter how far we have evolved from that primal ocean, all living things need at least some moisture to survive, traces of that brine from which we were born. Life in very arid conditions, then, poses special challenges. Indeed, a major task of evolution has been solving the problem of how organisms can cope in ever-drier places. But for those creatures that have successfully adapted to aridity, the plants and animals as well as the human cultures, such conditions are not dire. To life forms adapted to it, the desert is not harsh but life-giving and, we might say, even pleasant. Desert plants and animals don't eke out a bare existence in their hot and dry environment, they thrive there. Transport them to a wetter and more temperate climate and you do them no favor; they will perish.

As we consider deserts, we must remember our biases. Some of the most enduring impressions of deserts for Europeans, Americans, and Australians have come from nineteenth- and early twentieth-century artists, adventurers, and explorers in North Africa and the Middle East. As compelling and powerful as these impressions are, we should recognize that they come from an era of European colonialism and have more than a little of the tinge of what cultural historians call orientalism, the depiction of Asia and the Middle East in ways that have more to do with satisfying psychological and political desires in European audiences than with the actual cultural and natural environments of those places.

Language itself can complicate our understanding of deserts. Cultures, their languages and their landscape aesthetics, evolve in particular places. And the English language is no exception. Evolving on a small, wet, hilly, foggy, and very green island, for a long time it had no need to render vast expanses of exceedingly dry, clear, rocky, rugged, and multihued desert terrain. To take an obvious example, consider what is implied by the use of *green* to signal ecological health. Implicitly, deserts, with their brown, tan, red, and yellow landscapes, seem ecologically injured or destitute, even when they may be vibrant with healthy biodiversity. On the other hand, English is also a famously omnivorous language, and as it has spread around the world, it has absorbed many words from more desert-oriented languages. Indeed, by now, English speakers can draw from a real wealth of words to speak of such places.

There's no denying that compared to many other kinds of environments, deserts *are* landscapes of exposure and extremes: heat and cold; wind, space, and bare rock; rain that vanishes long before it reaches the ground and canyon-carving flash floods; starlight bright enough to read by and blinding sunlight; nuclear bomb tests and spiritual quests; proximity to divinity and a version of hell. If wetlands are dominated by their abundance of water, deserts are ruled by its rarity, a fact that allows us to see especially clearly those other classical elements: earth itself, whose attribute for the Greeks was dryness; air, whose quality and motion are so important to any desert experience, and which, in the guise of climate, controls what lives there or dies; fire, sometimes the planet's interior heat but more obviously the sun's bright burning; and, as always, change.

Given their aridity, deserts would not appear to be particularly hospitable to life. Yet most deserts are far from lifeless, and desert plants and animals display a considerable array of clever adaptations to the difficult and varied conditions of climate and terrain they face. The same is true in other landscapes, too, with amazing adaptations everywhere. Still, deserts are excellent places in which to consider the way life evolves to fill nearly any possible niche, and so we'll linger later in the chapter with some examples of this marvelous process. We have separated plants, animals, and human cultures; but as you will see, this separation is artificial, for all desert life-forms—including humans—depend on one another in a myriad of complex ways and at times offer some quite remarkable examples of symbiotic mutualism.

For living things in a desert, the margin between life and death is narrow. Each tiny shift of conditions—the shade of a few leaves, a sliver of shelter from the wind, a trickle of water—creates a stage for some clever strategy, some pocket ecosystem where a desert plant or animal can spend its nights and days. Although our attitudes about deserts have sometimes been as extreme as their temperatures, we humans, too, have adapted to these difficult environments through cultural innovations as diverse as the delicate but enduring tracery of Aboriginal Australian songlines and the more imposing but perhaps less durable edifice of Nevada's Hoover Dam.

While the word *desert* is derived from the notion of a "deserted" place (another linguistic bias), most deserts aren't actually deserted at all. They're just dry. And because of that dryness, living things are usually spread out, giving one another a bit more elbow room. Deserts often seem empty, lifeless, or forbidding to visitors. The name Taklamakan, for example, means in Uigur "go in and never come out." Still, to many people whose cultures have inhabited such terrain for long spans of time, deserts are filled with life and stories. As Navajo author Luci Tapahonso writes, "This land that may seem arid and forlorn to the newcomer is full of stories which hold the spirits of the people, those who live here today and those who lived centuries and other worlds ago." And Australian Aboriginal poet Jack Davis notes that, though European settlers call the arid portions of Australia a "desert," the word is inaccurate because, as his people know, the landscape is full of life. You just have to know where to look for it.

So what are deserts? We'll consider this question first through the layered lenses of geology and climate. Then we'll look at some of what we can see in the light of natural history, the second ancient branch of science, kin to natural philosophy but focused on living things. Like earlier natural historians, many of us remain fascinated by the characteristics and behavior of plants and animals—and by their connections with ourselves, both how we have made ingenious, practical, material use of them and other, less concrete ways in which our lives and theirs are intertwined, through metaphor, religion, storytelling, and art. So we will focus here on matters of biology, and without separating ourselves from this realm: we will consider the characteristics and behavior of human desert dwellers and visitors, too. In this focus, we can add to the interests of men like Pliny the Elder and Aristotle the insights of Charles Darwin's successors, men and women who have continued to find in the intricacies of the biological world a repository of wonders and marvels. We might say that we have expanded the classical list of elements by one—or perhaps just that we have teased out that ubiquitous "change" to highlight one of its ingredients in particular: adaptation. This is another of Earth's great shaping forces, one that is at work everywhere but is easy for us to see in the clarity offered by desert places.

◇ ON THE SPOT: DOWN A DESERT RIVER CANYON

SueEllen Campbell

It's the last week of May and the river in this desert canyon is running full, fast, and muddy. The weather is perfect—cool, clear air, intensely blue skies, laser-sharp sunlight, shadows, and reflections. With only one small dam far upstream, the Yampa is as close to wild as rivers around here get, and the landscape we'll travel on this five-day float from Colorado into Utah is further protected by being inside Dinosaur National Monument.

What do we do all day? Well, we drift. We look through binoculars, daydream, chat, take a hand at the oars, try to read the river as the guides do and learn bits of their lingo: we ride the wave train; avoid sleepers, keepers, and other holes, haystacks, exploding haystacks, and whirlpools; cross a fence into an eddy by a cut bank opposite a point bar; watch for domes and boils. We meander along goosenecks, watch the canyon walls narrow and widen again. Moving from riffle to rapid and back into still water, we slide past side canyons, slot canyons, box canyons, amphitheaters, hanging gardens, parks, holes, and bottoms. The rafts spin slowly, and we gaze at the river's endless variations of gloss and gleam, at the banks, at the high rock walls.

Now and then we stop to hike. In the pale limestone we find fossil crinoids, corals, brachiopods, and dark, shiny knobs of jasper. We ascend a series of small ledges that a few days ago sparkled with waterfalls. We follow a chain of desert potholes full of still water and water striders, caddis fly larvae in their pebble and stick cases, horsehair worms so attenuated it seems impossible they could be alive, tiny ostracods straight out of the Cambrian. In a hanging garden of ferns, monkey flowers, and columbines, we marvel at a helleborine orchid being pollinated by a fly disguised as a bee carrying a yellow backpack of pollen.

The native-fish biologist who's with us catches, shows us, and then releases a Colorado pikeminnow (much smaller than the five-footers that used to live here), flannelmouth and razorback suckers, and a humpback chub—fish made to live in the tough and change-able conditions of desert rivers whose current threats include dams and introduced fish species. I'm appalled by the nightmarish looks of the insect he uses as bait, the finger-long hellgrammite that pro-vokes another biologist to speak of the beauty of the bizarre. But with their delicate, watery hues of silver, gold, yellow, and pink, the fish are beautiful.

Above us, countless swallows swing and dip. Desert bighorn sheep admire the view and lizards with beaded turquoise bellies skitter and bob. Far below a speeding peregrine falcon, great blue herons flap their prehistoric wings among clicking cicadas. We study a scorpion's ominous curl, trace the wandering doodlebug tracks and miniature craters made by ant lions, examine a Mormon cricket whose swarming kin almost ruined the crops of the earliest Mormon settlements in Utah until large flocks of gulls came to the rescue— the same insect (not really a cricket, but a katydid) that, ground into flour, helped nourish local Native Americans.

Flowering plants are everywhere: Mormon tea and fragrant three-leaf sumac bushes, flaming claret-cup and pastel prickly-pear cac-tus, desert wildflowers in white, cream, scarlet, orange, blue. Box elders, cottonwoods, and willows cling to the river, while higher up, native grasses frame silvery sagebrush, solitary piñon pines, and Utah junipers with their fibrous, peeling bark. Even the ground is alive with its lumpy gray cryptobiotic crust.

Floating down a river in the desert springtime: I don't know about everyone else, but I'm also feeling afloat in the present moment. In charge of nothing, wearing no watch, following no map, nowhere near my car, phone, or computer, I'm free to step away from the march of my regular life and into older rhythms: hunger and food, wakefulness and sleep, the day's brightness and evening's cool breezes, the arcs of sun, moon, and stars, the easy lilt of bodies drifting on moving water. The days are long and leisurely, both lazy and stimulating. I don't know, don't care, what time or day it is. It is simply *now* and I am simply *here*.

And yet *now* and *here* are not really so simple, for a canyon like this is also about deep time and some of the most powerful trans-formative forces on Earth. We're moving down and up through lay-ers of time—from surface materials just now laid down to rock that has been in place for more than a billion years. Much of our journey is through petrified sand dunes, seashores and bottoms, and coral reefs created more than three hundred million years ago when this part of Earth's crust was south of the equator, buried for eons, finally exposed by moving water and rising land to become these canyon walls. We float, too, through one of the richest known beds of dinosaur bones, the reason this area was protected as a national monument.

We see folds from major fault lines and places where the land domed up and plunged back down, crossbedded sand dunes that look like hundreds of fingerprints piled together, layers of contrasting grays, purples, creams, tans, pinks, and the reds made by iron oxides brought to the surface by percolating rainwater. Enormous slabs hang on the cliff walls waiting to fall into the boulder piles below; some will leave behind them hanging alcoves with arched roofs. Brown streaks on the walls mark recent moisture; white streaks mark salts. Most dramatic of all are the black streaks and washes of desert varnish—manganese oxide pulled out of dripping water and fixed, I've read, onto the rock surface by metallogenic bacteria that may be as old as cyanobacteria. To a knowing eye, these tiger walls tell stories about which surfaces are fresher, which older; where the water most often flows; even the records of climate change. On one high ledge, where we climb to look at pictographs painted by the Fremont people more than a millennium ago, we find chunks of a pack-rat midden, its contents rich with stories about past ecosystems.

Which moments best capture my sense of this place? One is the night I sleep tentless at the base of a smooth sandstone amphitheater whose ivory walls soar some thousand feet over my head. In the great stone-ringed arc of sky, swallows give way to bats, the stars emerge, and the gibbous moon lights the cliffs, one white disk encircling another with the depths of the universe between.

In the other, I lie on the raft and watch the canyon rims swing past as the oarsman spins us slowly along the flow. The lacy lavender fronds of invasive tamarisk sweep by, the glossy hearts of cottonwood leaves, the box elder's new emerald, the soft red cliffs, all mirrored in the river's gleaming planes, fractures, and swirls. I hear the riffle of water, the slap of oars, the thin thread of a hawk's scream, and then—just now, just here—the heart-stopping waterfall song of a canyon wren. ❖

DRY, HOT, WINDY, AND DUSTY

Most deserts occur in the subtropical zone of Hadley Cells we considered in chapter 2, along stretches of land centered at roughly 30 degrees of latitude below and above the equator, where the global air circulation patterns produce dry air: the Sahara is the largest instance. Many occur in the rain shadows of mountains, where warm, moist air forced upward to pass over mountains cools rapidly, releases its moisture to those higher elevations as rain or snow, and then, once it is dry, descends: the Gobi, for example, or the Mojave, which is in the shadow of the Sierra Nevada, and Death Valley, which is also downwind from the Panamint Range. Some deserts are dry in part because they lie far from moisture-producing oceans: Asia's Taklamakan, Australia's Gibson and Simpson. Finally, in two important cases, Africa's Namib and the western edge of South America, they adjoin cold ocean currents that don't allow the air above them to carry much moisture; these deserts collect nearly all of their moisture from fog. (In El Niño years, this pattern changes, and the coast of Peru can receive significant rain.) Several of these factors—subtropical latitude, the shadow of the Andes, and a cold ocean—combine to make the driest place in the world, the Atacama Desert, which receives on average only a tiny fraction of an inch—a millimeter—of moisture per year. Parts of it have gone decades with no measurable rainfall at all.

Altogether, roughly a third of the planet's land surface is desert to some degree. Some 4 percent is extremely arid, with less than four inches of precipitation a year. (Polar deserts are sometimes included here; much of Antarctica, for instance, receives just two inches in the form of snow.) Arid land accounts for another 15 percent, with up to ten inches a year. And nearly 15 percent is semiarid, with up to twice this much moisture, a category that merges with steppes, thirstlands, bushlands, and so forth.

But precipitation amounts don't tell the whole story. Some deserts receive their water in the winter, when it evaporates less quickly but may also be less useful to then-dormant plants; others experience summer monsoon rains so heavy that most of the water runs off in flash floods. Crucially, rainfall measures must be balanced against evaporation rates, which can be extremely high in hot, dry, sun-drenched, windswept lands. How big an effect can evaporation have? In England, rainfall is double the evaporation rate, so forty inches of annual rainfall is offset by only twenty inches of evaporation, for a net gain of water of nearly two feet per year. In North America's Chihuahuan Desert, however, the evaporation rate is not half of the rainfall rate, but thirteen times greater. That is, the eight or ten inches of annual rainfall in this desert are countered by a whopping ten feet of potential evaporation. Calculated in this way, the driest deserts are the Atacama and Sahara; the wettest are those in North America and India's Thar.

Rainfall regularity is also a key issue. In many parts of the world, plants, animals, and cultures are adjusted to a rainfall pattern that, with minor variability, follows a predictable and reliable annual cycle. The rhythms of agriculture follow those of rainfall and temperatures. This pattern is deeply ingrained in European and American cultures as the familiar four seasons, but it is often misleading when applied to desert climates in other parts of the world, where thinking in terms of two seasons, or six, or none at all might be more sensible.

In many deserts, rainfall is highly variable not only throughout a single year but also (and this is more important) from year to year. We call extended dry spells *droughts,* but this term is inaccurate for deserts, where they are not unpredictable disruptions of a normal pattern; they *are* the normal pattern. This long-term variability has been especially challenging for settlers from landscapes with more regular climate cycles. For example, when settlers from England or Ireland (that *Emerald* Isle) moved into the arid portions of Australia (which make up more than 70 percent of that country), they had to adapt not just to the overall general dryness but also to rainfall patterns that (in response to the El Niño–Southern Oscillation) cycle between several years of little or no rain and periods of torrential downpours. Average precipitation numbers can be worse than meaningless in such a climate; they can be fatally deceptive. In such variable climates, native plants, and especially their seeds, may stay dormant for many years (sometimes for decades or even centuries), suddenly bursting into life when sufficient rain falls, but imported plants and animals—wheat, corn, sheep or cattle, and sometimes the settlers who depend on them—quickly succumb during extended dry spells.

Another variable is how much moisture is retained in the soil. In some deserts, rain soaks readily into the soil where roots can easily absorb it. But in many deserts, soils are low in nutrients, sparsely vegetated, sandy, rocky, and steeply sloped, and so rain quickly runs off before plants can capture it. In short, in many deserts the amount of effective rainfall is significantly less than even the meager average totals might suggest.

Other desert conditions can seem equally extreme and challenging to living things—and often exacerbate the effects of aridity. Although not all deserts are hot, in many of them summer temperatures often reach well over 100°F. The hottest temperature ever recorded on Earth, in Al 'Azīzīyah, Libya, in the Sahara Desert, was 136°F. And surfaces can be far hotter. In Turkmenistan's Kara-Kum, for instance, when at midday the air is 113°F, the ground might be a searing 176°F. However, since dry air does not hold heat, nights may be much cooler, enough so that humans may need to add a layer of clothing. At the other extreme are winter temperatures that can fall well below freezing: −20°F is not unusual in interior Asian and North American deserts.

Unimpeded by vegetation, desert winds can desiccate seedlings in minutes and carry enough sand and dust to cut visibility to zero, make bare skin bleed, bury human settlements, and drop dust thousands of miles from its source. These winds, which typically come at regular times of the year, have such strong personalities that they have names: khamsin, sharav, simoom, shamal, Santa Ana, chergui, ghibli, haboob, harmattan, bad kessif, brickfielder, sirocco.

Sand and dust storms come in three basic sizes. The smallest are the whirlwinds called dust devils (originally an Anglo-Indian term) or, in Australia, willy-willies, a term probably derived from a Yindjibarndi Aboriginal word. (As slang, perhaps in ironic understatement, willy-willy can also refer to a tropical cyclone, a vastly larger phenomenon.) These whirlwinds are just a few yards across and normally last only a few minutes. For desert-dwelling people, they often appear as spirits or ghosts. Among many Arab countries they are thought to harbor jinns. In Egypt they are referred to as *fasset el 'afreet,* "ghost's wind." Among the Kikuyu of Kenya the dust devil is called *ngoma cia aka,* meaning "women's demon." In Aboriginal stories willy-willies also represent spirit forms, and in central and northwestern Australia there is a sacred Willy-Willy Dreaming, the Winpirlirri Jukurrpa, which is frequently represented in contemporary artwork. Among the Navajo, Whirlwind-Boy is a familiar character in mythology, and his image appears regularly in sand paintings.

Stronger storms (like the haboob of Sudan) pick up bigger particles of sand to a height of three to nine feet and may last an hour or two. In these sandstorms, a thick, low-lying cloud of driving sand has a clear upper limit, and people's heads and shoulders may sometimes be seen rising from this layer into clear air beneath a blue sky.

The biggest storms have sustained winds averaging thirty miles an hour and gusts twice that strong, involve more dust than sand, may last for days, cut ground visibility to zero for several hours, leave behind piles up to ten feet deep, send thick clouds high enough into the atmosphere to endanger airplanes, and raise the finer dust even higher, where it gets transported around much of the globe. These massive dust storms are especially common in the Sahara, the world's largest source of atmospheric dust (perhaps half the total). Dust from Saharan simooms reaches Europe, North America, and the Near East, and climate historians find it useful as it appears in cores from ice sheets and the bottom of the Atlantic. Early in his voyage on the *Beagle,* Charles Darwin noticed Saharan dust out at sea and speculated about its role in transporting microorganisms and plant spores to remote islands. Dust from the red Australian deserts regularly coats the snowfields of New Zealand, and climate records can be traced back ten thousand years by drilling cores in the glacial ice and peat bogs in New Zealand's mountains, where levels of dust correspond to drier and wetter conditions fifteen hundred miles away in Australia. Chinese desert dust reaches North America, too.

Sometimes this windblown dust accumulates over the centuries. Like the finest glacial debris, desert dust is a source of loess soils, which may cover large areas to considerable depths, a hundred yards or more, and offer very rich farming soils. China's loess plateau, the Huangtu, a gift of the central Asian deserts, covers nearly 250,000 square miles and has historically provided some of the richest farming soils in that country, though it is now badly eroded because of poor agricultural and grazing practices; its dust is washing ever more rapidly into the Yellow River, the source of the coloration that gives the river its name. Elsewhere, too, dust storms have been made worse by human activities such as overgrazing, plowing, construction, and, most recently, the growing use of off-road vehicles. Increasing dust blown east from Utah's deserts is contributing to earlier snowmelt in the Colorado Rockies by absorbing more solar heat than clean snow would do.

◇ **ON THE SPOT: IN JABAL AJĀ**

Othman Llewellyn and Aishah Abdallah

A dry spring in the north of Arabia. It is the last week of March and we are waiting for the demoiselle cranes. We are here to discover the boundaries of their flyway, to estimate their numbers, and to find the staging posts where they alight to feed and drink, gain strength for their onward flight. Each day we arise in the dark to take our morning meal with the workers at the farm, pray the dawn prayer at the first streak of light, and drive out to watch for the cranes. The lonesome plains of Hā'il, dusty rose grit stippled with yellowish saltbush and the burrows of the spiny-tailed lizard, slant up to the Ajā mountains, which rise abruptly, rank on rank, across the southwestern horizon: a maze of granite domes and spires, pinnacles and boulders. Each day we climb up among these pink and lavender rocks to the high, windy summits, where we sit, scanning the sky with binoculars and telescopes. We do not eat or drink from the streak of dawn until the setting of the sun, for this is the month of Ramadan.

Today we drive into Al-Asmar, a knot of knobs and domes on the northeastern edge of Jabal Ajā. It has drizzled off and on throughout the night and there is a misty haze throughout the day. We park and climb by narrow rifts floored with pools of rainwater that mirror the white-barked boles of little Punjab fig trees, bright green cushions of moss, and mats of liverwort in shady clefts. As we creep through caverns under tumbled boulders we hear the tinkling of drops that drip into the pools, sending ripples of reflected light across dark walls.

We come out in the pass of Al-'Ajwah, a pair of meadows touched with a flush of green from the small but recent rains. All around, the rugged Ajā mountains rise in dappled splendor. The meadows are bright with wildflowers into which native honeybees are shoving their furry bodies. A spring runs silently down the lower meadow, reflecting the blue of the sky. It issues from a low wall of stone at the base of the upper meadow. In this wall the larger boulders are outcrops of the natural landscape, but look closely—some stones were placed here long ago by human hands. With recurring floods, soil settled behind this wall until it could hold sufficient water for what crops they planted—most likely dates, for that is what Al-'Ajwah means. Perhaps they grew grains as well, for here grow wild barley, wall barley, and bearded oats. After all, away to the north, beyond the mazy red sands of the Nafūd, beyond the tilted black lava flows of Al-Harrah, lie the ancient fields of the Fertile Crescent, where humans first learned to farm.

The soil stores sufficient water for this spring to run for several months after a good rain. No crops are cultivated now, but acacias and a rich herbaceous layer thrive in the accumulated soil. These mountain terraces represent the labor of hundreds of years; they now provide rich habitats for wildlife. This is the noblest kind of garden we can imagine: enhancement of the natural landscape to bring new life to the land. The Prophet Muhammad declared that "Whoever brings lifeless land to life, for him is a reward in it, and whatever any creature seeking food eats from it shall be reckoned as charity from him." The roots of a land ethic were put thus by his son-in-law, the caliph 'Alī, to a man who had reclaimed abandoned land: "Partake of it gladly, so long as you are a benefactor, not a despoiler; a cultivator, not a destroyer."

A pair of russet kestrels is nesting above us in the crags. We startle a covey of pinkish-brown sand partridges, which flee running and hopping up the bouldery slope, their chicks struggling behind. Palm doves flutter two by two among the trees; their heads are lilac with dark eyes and their throats speckled copper and black, while the feathers covering their backs and wings are edged in reddish-gold. A brawny star lizard—the painted rock agama—regards us from the edge of the terrace wall. We edge close to admire the pink and gray bands of rearward-pointing scales across its back and tail, its spiky head and jowls, its fine dark eyes with delicate scales like lashes.

The Ajā massif is crisscrossed by a lattice of straight, narrow ravines, which follow the fault lines and major jointing planes of the granite. The rains run rapidly off the bare, impermeable rock and flood the basins, ledges, clefts, and canyons, which receive far more water as runoff than directly from the clouds. The waters seep into jointing planes and fissures and emerge as seeps and springs, ephemeral streams and rock pools that may last into the summer; a few, festooned with maidenhair fern, flow all year round, for the Ajā massif is cooler than most parts of Arabia, receiving regular frosts in winter and, rarely, snow.

In spring, such upland meadows are yellow with Asteraceae, yellow or mauve with mustards, or white with chamomile. They harbor plants and animals that were widespread during the Pleistocene but vanished from Arabia as the land dried out, withdrawing to the cooler, moister climes of Syria, Turkey, Iran, and Turkistan. Among the rare species found in Jabal Ajā are the painted rock agama, the alpine Punjab skink, the earthworm (nowhere else in the central deserts of Arabia have we seen earthworms), the fragrant Judean wormwood, a gladiola, a mullein, and balangu mint. The relict populations that cling to life in this mountain refuge may become new endemic strains and species as they adapt to their surroundings over time. A grass and a wallflower, a globe thistle and a milk vetch are known only from Jabal Ajā and the nearby sands—they have been found nowhere else on Earth.

We descend to the plain and wend toward Al-Maslūkhah—*stripped bare* as its name tells—a lofty, massive granite dome that rises in sweeping curves above the pinnacles of the adjoining ridge. As we draw near, it arcs above us like an enormous wave about to break. At its foot, we turn westward into a narrow canyon where the granite, vertically jointed, is weathered into serried rows of rounded pinnacles like twisted fingers pressed together. Spheroidal boulders are emerging from the rock as the outer layers of the stone spall from their surfaces in thin, curved flakes, hammered and chiseled by the sun, the frost, the dew. The cluster of rounded ribs rears skyward in rhythmic lobes. The rocks have a deep blue sheen from the rusting of iron and manganese, long exposed to the air: this desert varnish is dark and shiny; and if you see it from certain angles it reflects the blue of the sky. Here a stunted palm tree ekes out its living where a damp seep oozes from the rock. And over there a little hide of stones was placed by hunters, here to shoot the partridges that come to drink.

A fan-tailed raven is pirouetting in the air against the sun—wheeling up from a rib of rock and back in a cartwheel, the sunlight flashing off the planes of its glossy black feathers as it turns. Nine griffon vultures used to roost upon those bluffs stained white with their droppings and soar in circles over the canyon, flying single file. Their flight expresses not so much the grace of other vultures as enormous power, as they carve short arcs in the air, the wind whistling keen in their pinions. But we haven't seen them this year. We pray they have not been poisoned, shot, or hit by speeding cars when feeding at road kills. It may be that they have migrated elsewhere, since this place has been dry for several years. They may return with the rains.

We scramble up a ravine among dry rock hollows and pools half filled by last night's showers. There are surely toads here in the streambed, buried deep in the soil. When strong rains fall, perhaps another spring, the hollows in the rock will be brimming over, dripping from pool to pool, and all night long the toads will be singing. Here is an overhang—when the rains come it will be a waterfall. Although the fall is dry now, the basin beneath it is nearly full. The waters have left their image on the rock, a pale violet veil that marks their path: smooth upper surfaces with ripples of rougher, rusty rock below. They look like sand ripples on a dune.

We double back to scale the backslope of Al-Maslūkhah; the aromas of lavender, mountain germander, and sweetrush fill the air as the leaves are crushed beneath our boots. The sweetrush, a lemongrass with dense domed tussocks and seedheads like clustered bees, is used to treat fever, repel insects, and perfume homes and graves. The germander heals wounds and cures intestinal ills. We pass a pergularia, a spangle of azure leaves like disks of sky—its caustic sap is used to remove the hair from hides—and a group of silkweed shrubs with long, curved, whiplike twigs, a heart tonic. Touch the hornlike seedpod and a cluster of seeds with sails of silk drifts away on the breeze. At the saddle where the ridge of pinnacles joins the dome, a cave with the cliff stained white beneath it is the aerie of a griffon vulture. It is empty.

Scaling the topmost granite sheets, we dance about the summit of the dome; our exhilaration has overcome our thirst. We examine the blossoms of an asphodel, like small white bells or dewdrops. We take out our binoculars and scan the southern sky. Fan-tailed ravens soar around us, playing with the winds, gliding, tumbling—masters of the air. But we see no griffons, and no cranes.

We look out over the plains beneath us, gaze south and west to the high crests of the Ajā range. These ramparts rise from the desert plains and sands like islands from the sea. Like islands, they are landmarks—a bottleneck for migratory birds. These rugged peaks have preserved populations of the Nubian ibex with its great curved horns, the bold black ratel with its gray, white-edged cape, the tawny caracal, striped hyena, Arabian wolf, rock hyrax, and crested porcupine, the griffon vulture, and the opalescent lilac-brown cat snake. Their crags have kept their flora safe from the goats, the sheep, and the camels that have laid waste the lowlands. They harbor the greatest richness of plant and animal diversity in the arid heart of Arabia. They are a bioclimatic refuge for species that have nearly vanished from this desiccating land. It is from these mountains that such species may one day disperse to reoccupy the land, if favorable conditions return. They are a natural seedbank from which the degraded rangelands that surround them might be restored. But will the global climate changes unleashed by burning fossil fuels be greater than these species can endure? Will this ecosystem with its unique diversity of life, its endangered, endemic, and relict species, its medicinal herbs, its local strains of dates and figs, barleys and oats, be lost forever?

As we descend, the cold shadows lengthen, engulf us, and rise to line the sculpture of the highest peaks. At the setting of the sun, the mountains blush in crimson alpenglow. The plains of Hā'il are an intense purple, the twilight sky suffused with a glow of rose. A far-off flash of lightning under lowering clouds inspires both fear and hope, followed by a rumble of thunder more felt than heard.

The next day is cool and cloudy. Exhausted from our exertions, we spend the early morning searching around the tilted sandstone mesas near the farm, but we see no cranes. We see sandgrouse, wheatears, blackstarts, babblers, bulbuls, shrikes, larks, swifts, and coursers . . . but no cranes. At midmorning, we park and sit to watch the sky. We wait, wait . . . it seems the cranes will never come. We begin to doze.

In our dreams we hear a distant clamor, a far trumpeting . . . startled, we are wide awake. Coming toward us from the south is a skein of cranes, flying from the direction of Al-Asmar, calling with staccato cries, *krrooah, krrruorrr, garr-oo-oo-oo.* We run up the little ridge to observe them. They fly directly overhead, resolving into a double V-formation, and their cries are louder. Forty-five, we count, and estimate their height. This is the first small flock—in the coming days more flocks will follow—more than ten thousand birds in all. It is ten o'clock. The cranes pass over and their cries fade, but their formation can long be seen dissolving and re-forming, growing fainter in the sky. We take the bearing of their flight. We see each other's eyes are shining. ◇

WHAT WE SEE

Deserts are topographically quite varied, and their many shapes and colors are undisguised by vegetation. Indeed, many people find the stunning and colorful topography deeply appealing. As Georgia O'Keeffe remarked, "All the earth colors of the painter's palette are out there in the many miles of badlands."

Some deserts are exceedingly level, including most of those in Australia. Many contain salt flats or playas (Spanish for "beaches"), such as Utah's Bonneville Salt Flats and Australia's Lake Eyre basin, both so flat they have been used as tracks upon which to set land speed records. And the official flattest place in the world is a desert playa, Bolivia's Salar de

Uyuni, whose 3,800-square-mile surface varies by less than two feet. Usually the sense of sublimity is associated with vertical topography such as high mountains or deep canyons, but a similar emotion of horizontal sublimity can be evoked in viewers surrounded by a flat, arid terrain that extends with no seeming variation into a vast distance across 360 degrees of horizon.

Other deserts contain mountain ranges, sometimes with significant moisture in their upper reaches and dry lands lower down, including the spreading aprons of rock debris called bajadas (Spanish for "slope" or "descent") or alluvial fans. Mountains cover nearly half the deserts of Arabia and roughly 40 percent of those in the Sahara and the United States; the Great Basin Desert alone contains some 160 ranges. The most famous desert mountain may be the biblical Mount Sinai, where Moses is said to have received the Ten Commandments. The traditional candidate for this peak is Gebel Musa (*jabal* is Arabic for "mountain"), which stands about 7,500 feet high in the old granite ranges in the southern Sinai Peninsula. Much higher and more dramatic are the arid mountains of the Hindu Kush and the Karakoram, where glaciers and precipitous gray rock walls descend into desert valleys. In Death Valley, elevations range from 282 feet below sea level to 11,049 feet above. Most such areas are still undergoing active tectonic movement and mountain building, and the resulting high relief controls the behavior of water and erosion as well as contributing to the rain shadow effect. In the American West and northern Mexico, some mountain ranges rise more than seven thousand feet above the surrounding desert, creating "sky islands," isolated landscapes of much wetter and cooler climate surrounded by a sea of desert.

Some deserts contain other large bodies of prominent rock, which, though smaller than mountains, are no less striking. Relatively unusual but visually notable are inselbergs (German for "island mountains"), isolated, usually rounded rock knobs or hills left standing when softer rock around them has eroded. Australia's Uluru (Ayers Rock) is perhaps the best known of these. Its unusually resistant red sandstone is very ancient, laid down some six hundred million years ago, before there was any multicellular life on land, and when the southern supercontinent Gondwana was still forming. Over many millennia, compression folded and tipped the stone until its layers were nearly vertical. Sediments covered it and then eroded away, revealing the present spectacle. It is an ancient landform, first appearing in the Cretaceous and visible to dinosaurs and early mammals as it rose ever so slowly above the surrounding landscape.

Because they are so striking to human observers, inselbergs can take on a powerful cultural significance. Uluru, for example, is the location of important sites sacred to Aboriginal peoples. Narratives of the Pitjantjatajara and Yankunytjatjara peoples occur there, sacred sites continue to be off-limits to tourists, and songlines tracing the journeys of Dreamtime ancestors connect Uluru to other locations throughout the central deserts in a complex mythology. More recently Uluru has become a defining icon of Australian national identity, competing with Sydney's Opera House as the most recognizable image of the country. Each year as many as a half-million tourists from around the world make the long trip to Central Australia to witness it. Sunrise and sunset light on the red rock draws thousands of spectators daily to designated viewing areas.

Another major category of desert landscape is the wind-scoured, stone-covered plains whose names include *reg* (from Hamitic) and *serir* (Arabic for "dry") in the Sahara, where it covers roughly a third of the surface, desert pavement in the United States, and gibber plains (after a Dharuk Aboriginal word for "stone") in Australia. Gibber stones vary in color. In Australia's Sturt Stony Desert, the landscape is covered with weathered pebbles of black and red ironstone intermingled with white quartz, many coated with a bronze layer of desert varnish. These multi-hued pebbles reflect the low sun at dawn and dusk with a dazzling sheen, resulting in the designation "adamantine plain." Little seems to grow in these plains, where often the wind has taken away the smaller soil particles, leaving a surface of pebbles and cobble that look as though they have been cemented in place. Following rains, however, Australia's gibber plains can be altered by rapidly growing Mitchell grass and suddenly blooming ephemeral wildflowers, which spring to life and transform a seemingly lifeless, rocky terrain into a magnificent if short-lived garden.

Rocky plateaus are also widespread, as, for instance, in the Sahara, where they are called hamadas (from an Arabic word). They may be flat and smooth, cut by deep watercourses, or covered with good-size rock rubble. Sometimes they erode to create the layered landscapes that cover much of the American Southwest: the flat-topped mesas (Spanish for "tables") with caprock layers of more resistant stone, cliffs and steep walls just below the caprock, and gentler slopes farther down; the smaller buttes (French for "small hills" or "rising ground"); and the still smaller pinnacles. Historical factors and geographical accident loosely link the appearance in American place-names of mesas and buttes—there are more buttes in those parts first explored by the French, including the northern High Plains, and more mesas in those areas where the Spanish did the first European naming. But the main distinction between these two features, the one that characterizes their widespread use by geologists, is size. All of these forms can be seen in the spectacular display of Monument Valley on the Arizona-Utah border.

This type of terrain is widely familiar because of its presence in so many Hollywood Westerns and in the paintings of Georgia O'Keeffe and other artists. Certainly it is visually arresting with its combinations of red, buff, and cream-colored rock faces, its striking silhouettes along the horizon, its polka-dot vegetation patterns of dark green juniper and piñon pine, green and yellow rabbitbrush and silvery gray-green sagebrush, and its intense blue daytime skies and sunsets whose lurid colors—orange, peach, scarlet, green—seem fake until they're witnessed. It is also the territory of several contemporary Native American cultures. In Arizona, for instance, Monument Valley itself is a Navajo Tribal Park, and the Hopi reservation is dominated by First, Second, and Third Mesas. In New Mexico, the old Sky City of Acoma Pueblo sits on the top of an especially striking small mesa, a location that provided excellent but ultimately not impregnable protection from the Spanish conquistadors. And the Early Puebloan (or Anasazi) ruins of Mesa Verde (in Mesa Verde National Park and the Ute Mountain Tribal Park) are cut into the high cliffsides of a large mesa in southwestern Colorado. Such landforms are not limited to the United States. Jordan's Wādī Rum, a similar landscape, was an important site for T.E. Lawrence's work with Arab fighters in World War I and one of the locations where the movie *Lawrence of Arabia* was filmed. In Australia, features like these are usually called jump-ups or breakaways.

As these examples suggest, erosion is one of the key shapers of desert landscapes. Without regular precipitation, vegetation does not develop in sufficient quantities to hold the earth in place against such sculpting forces as strong rains, sand-bearing and thus scouring flash floods and winds, the expansion and contraction caused by temperature swings and freeze-thaw alternations, or certain kinds of chemical weathering. The dominant agents vary from place to place. In many deserts, for instance, the sculpting effects of winds show in the long, streamlined, wind-carved rocks that look like whales just surfacing from beneath the sand. First described by the Swedish explorer Sven Hedin in China's Taklamakan Desert, these are called yardangs, from the Turkestani word *yar,* meaning "ridge" or "steep bank."

The occasional presence of water is most evident in the many varieties of creekbeds and riverbeds, ranging from the deepest of canyons or barrancas cut by rivers through slowly rising land over the course of millennia to ephemeral beds so wide and shallow that it's hard to believe the warning signs on roads—"stay out when water is present, flood danger." Sheets of floodwater thirty miles wide sometimes sweep across the channel country of western Queensland, stranding motorists for days and even weeks. The distinctive ephemeral creeks of deserts, which are wide-bottomed and steep-walled since floods erode mostly laterally, go by many names. In the United States, they are arroyos (Spanish; most often used in the Southwest), washes (short for dry washes), and draws (Blackwater Draw in New Mexico is where evidence of the Clovis people was first discovered). In Egypt and parts of the Middle East, they are wadis (Arabic *wādī*; or in French, *oued*). Because of the long Arab influence in Spain, the term also appears in many Spanish place-names as *guada*, meaning "river," such as in Guadalajara, "river of stones," or Guadalupe, blending Arabic and Latin into "river of wolves." In Australia, ephemeral watercourses are often just labeled rivers, even when dry for years at a time. These names, and the blue tracing that designates them on a map, can deceive unsuspecting travelers. During World War II, captured German U-boat submariners were interned at Camp Papago in Phoenix. Although they were treated very well, twenty-five of them tried to escape on December 24, 1944. They had looked at a map, noted the nearby Salt River, and decided to boat downstream and escape. But they failed to realize that the Salt is now dry most of the time (though it did flow historically) and were immediately recaptured.

Erosive forces can create some spectacularly intricate desert landscapes. The numerous national parks and monuments of the Colorado Plateau contain graceful sandstone arches and bridges; human-looking hoodoos; tall thin pillars, some of them in extensive gatherings; mushroom and balanced rocks; amphitheaters; gently rounded stone hills; slot and box canyons; fins; and more. These are the forms of what is called canyon country and the redrock or slickrock desert. And then, of course, there is the Grand Canyon itself, an astounding landscape that offers countless side canyons, buttes, mesas, cliffs, scarps, benches, walls, waterfalls, pillars, towers, hoodoos, and potholes (*tinajas* and *huecos* in Spanish) that collect rainwater, all in multiple colors of rock—an endlessly varied array of forms cutting through two billion years of geological history in its winding 277-mile length.

Often these complexes of carved rock are decorated by dark red, brown, and black streaks of desert varnish. Although this varnish appears on horizontal surfaces as well, it is most spectacular on vertical ones, where, its pattern modified by erosion and water seepage, it may cover cliffs hundreds of feet tall with erratic dark veils that may resemble the stripes on tigers or zebras. In many locations in the American Southwest (as in other parts of the world), the surface of dark varnish overlaying a lighter rock material has produced a perfect surface for the carving of petroglyphs, tens of thousands of which adorn the landscape.

These rugged landscapes are also home to many pockets of vegetation. Single trees and small clumps of bushes grow on ledges, in rock crevices, and along the bottoms of arroyos and canyons—in places sheltered from sun and wind, and, more important, where a bit of extra water lingers after rain or snow. Along the tops of rock plateaus, water accumulates in small potholes along with blown-in dirt, and miniature gardens appear. Sometimes, high on cliff walls, hanging gardens grow as small amounts of water seep out where soft, porous rock layers intersect with harder, more impervious layers; there, water-loving plants like delicate orchids, red and yellow columbines, and ferns flourish. While deserts are dominated by earth colors, then, in all their surprising variety, they are also punctuated by the colors we more readily associate with life.

The one landform most people imagine when they hear the word desert is the sand dune. Although they cover only 20 percent to 30 percent of the world's deserts, these are the landscapes that provide the iconic romantic images of robed and turbaned men riding camels in caravans across vast stretches of golden dunes; in North America, they also provide the setting for endless cartoons featuring gaunt, ragged, thirsty men (almost never women) crawling across the barren sand. Dunes may be many colors—bright red in the Namib; red and orange in parts of Australia; as white as fresh snow in one basin in New Mexico—and they may be more or less mobile, depending on current and past climate and whether any vegetation helps hold them in place. All dunes tend to form where something on the ground arrests the movement of sand grains, and small dunelike features called nebkhas accumulate around plants, where they also gather plant detritus and create fertile and protected microhabitats for plants and animals.

Based on wind speed and direction, the availability of loose sand, and the underlying ground shapes, larger desert dunes take several forms. Linear or seif (Arabic for "sword") dunes are long, narrow, straight or slightly wavering, and typically found in parallel sets; they match the direction of the prevailing winds. In Australia, where nearly half the deserts are covered with relatively stable dunes, these forms dominate. Desert travel along the troughs between and parallel to the dunes is relatively easy, but crossing them can be exasperatingly difficult. In 1845 explorer Charles Sturt finally turned back from his efforts to cross the Australian continent when he saw ahead of him wave after interminable wave of the red sand dunes of the Simpson Desert—the world's longest parallel dunes—and cursed the place as "the entrance into Hell." Crescent-shaped or barchan dunes are the most common dune type. (The word *barchan* comes from Russian, which adopted it from Kirghiz, which borrowed it from Turkish.) These lie perpendicular to the prevailing winds and form where the wind blows mainly from one direction all year long. The narrow dips of the crescent face downwind, and they may join together into long, scalloped ridges, at which point they are called transverse dunes. Barchan dunes migrate across the landscape; astonishing rates of a hundred yards per year have been recorded. The largest are in China's Taklamakan Desert. A third distinctive type is called the star (or radial) dune. These look rather like irregular starfish with varying numbers of arms, and since they form in locations that lack a prevailing wind direction, they also tend to stay put. Radial dunes dominate the Sahara's Great Eastern Erg (or Grand Erg Oriental), and they can be particularly tall, with some in China reaching 1,640 feet, the tallest sand dunes in the world.

Much of what is known about the subjects of blowing sand and dune formation and movement is thanks to a man named Ralph Bagnold. Sent to Egypt in 1926 as part of the British military presence there, Bagnold watched the excavation of the lion body of the Great Sphinx by workmen carrying sand away in baskets on their heads. When the excavations reached down to the front paws, the workers uncovered a tablet announcing that Pharaoh Thutmose IV of the Seventeenth Dynasty had performed a similar excavation three thousand years earlier. At that time, the statue was already ancient, having been built in the Fourth Dynasty (2575–2467 B.C.E.) by Khafre, who also built the nearby second pyramid of Giza.

Intrigued by the power of sand to bury and rebury such an imposing structure as the Sphinx, Bagnold and a few friends began to explore the surrounding deserts and to experiment with strategies for motorized desert travel. They customized vehicles, soldered lids onto gas cans to keep water fresh for years, and devised a simple, reliable solar compass. Then they figured out routes through Egypt's Selima Sand Sheet and Great Sand Sea. In the former, the only breaks to flat monotony were one rock just a foot high and one barchan dune. They camped in the dune's lee and left their rubbish to be buried by the sand. Fifty years later, that trash was found by archaeologist Vance Haynes a quarter-mile on the other side of the solitary dune, which had crossed entirely over it, allowing Haynes to calculate the dune's rate of movement at 24.6 feet a year. In the Great Sand Sea, Bagnold and his friends discovered that some sand surfaces were firm enough to drive on while others were "dry quicksand" in which their cars would immediately sink to the axle. Realizing that the largest dunes were also the firmest, on top of which softer small dunes had formed, they found that with care they could wind around and between these softer areas and find sinuous routes to penetrate these vast dune fields.

During these explorations, Bagnold became fascinated by how "the dunes seemed to behave like living things," though with a formal simplicity and order he had not expected in nature. One night, he notes in his 1990 autobiography, *Sand, Wind, and War,* he heard the dunes booming and "had the eerie notion that these great beings were talking to one another in the stillness of the night." Back home in England from 1935 to 1939, he built himself a wind-tunnel laboratory and experimented with such things as the way sand grains move by saltation (from the Latin for "jumping"), bouncing like Ping-Pong balls or hailstones, setting other grains in motion as they go. This, he found, was the main method of sand motion, aided, in the case of dunes, by the formation of a small cornice over the dune's leading edge or slip face; grains accumulate on this cornice until they reach the maximum angle of repose (34 degrees), at which point a small avalanche occurs and the dune has moved forward a tiny distance. He also investigated such intriguing tangents as the relation between sand ripples and dunes (they look alike but have different causes and structures; he compared ripples with the clumps that form in traffic, interspersed with emptier stretches of road); the reason that larger particles like pebbles or archaeological objects become scattered widely over sand surfaces rather than staying together; and the possible reasons for the various singing sounds dunes can make—the whistle of beach dunes (which are more humid, more vegetated, and often more disturbed than desert dunes) and the low-pitched booming or rumbling of the desert. The resulting book, *The Physics of Blown Sand and Desert Dunes* (1941), remains the classic in the field.

One final desert phenomenon is the mirage, especially the false appearance of water in a bone-dry landscape. The word comes into English through French, derived from *mirer,* "to look at oneself in a mirror," and is based—once again—on the Latin *mirari,* "to gaze at in wonder." Mirages are not optical illusions, as often claimed, but true images that can be photographed. Nor, though the word is related to the word *mirror,* are they in fact caused by reflection, as many sources incorrectly explain.

Humans have long noticed mirages, but it wasn't until the end of the eighteenth century that a Frenchman, Gaspard Monge, a mathematician on Napoleon's famous Egyptian expedition, turned his mind to explaining the phenomenon in scientific terms. French soldiers and explorers in North Africa had often been baffled by the appearance of inviting lakes that mysteriously vanished as they trudged toward them. Monge demonstrated that these visions were phenomena of light refraction caused by differing densities of air. In desert conditions, the layer of air closest to the ground heats up well beyond the temperature of the air only a few feet higher. This differential creates a difference in density, and the boundary between the two layers acts like a refracting lens, bending light. As the American Meteorological Society's *Glossary of Weather and Climate* explains, "an image of some distant object is made to appear displaced from its true position because of large vertical density variations near the surface; the image may appear distorted, inverted, or wavering." In desert mirages, that distant object is often the sky, whose wavering image is displaced by refraction so that pieces of the sky appear to lie on the ground, creating a remarkably realistic image of water. Many of us have seen such images on highway surfaces, especially on hot days: the road ahead looks wet, but the shiny pools keep receding. Anyone who has seen the utterly convincing image of crystalline pools of water shimmering in the near distance on a hot desert day can sympathize with desperately thirsty travelers who followed such visions to their doom. Deserts are the most likely sites for these "inferior mirages," in which the refracted image appears below the line of the horizon.

The second principal type of mirage, "superior," in which the image appears above the horizon, occurs most often in polar regions or over cool-water oceans, though it, too, can appear in deserts. In superior mirages, the light from objects that are beyond the curve of the horizon, and so should be invisible, is refracted by the boundary between the cool temperature of the air near the surface and the warmer air at a slightly higher altitude, a weather pattern known as a temperature inversion. These distant objects—ships, islands, cliffs, even cities—appear to float in the sky above the horizon. Especially striking versions of these superior mirages have come to be known as *fata morgana,* an Italian term evocatively referring to King Arthur's half-sister, the enchantress Morgan le Fay, who lured sailors to her castle that rose from the sea. When these superior mirages occur in the desert, it is usually in the morning following a very cold night. Commonly, distant mesas or mountain peaks are seen to be stacked in mirror image on top of themselves. When distant light sources are affected by superior mirages at night, they can appear as mysterious ghost lights seeming to hover above the horizon. The famous Min Min lights of Australia's Channel Country and the Marfa lights of West Texas, though fabulous supernatural and extraterrestrial explanations have been given for them, are no doubt cases of this effect.

Mirages of all sorts have had a strong pull on the human imagination. Navajo stories include an important character called Mirage Man and also refer to a group of people living far to the east called the *Hadahuneya'nigi,* the Mirage or Agate People. In the Qur'an, infidels are compared to people deluded by mirages: "As for those who disbelieve, their deeds are like a mirage in the desert which the thirsty takes for water till he reaches it to find that there was nothing, and finds God with him who settles his account, for God is swift at the reckoning." In a quite different cultural register, mirages are a staple of films set in desert locations, often luring travelers to their deaths. German film director Werner Herzog produced a hallucinatory film in 1971 titled *Fata Morgana* that was shot in the Sahara Desert and includes many scenes of mirages, his symbol for the insubstantiality and delusions of human understandings of nature. Perhaps surprisingly, given these associations, the word *mirage* has great marketing cachet, being used for everything from fighter aircraft to casinos.

◇ **ON THE SPOT: IN THE CHIHUAHUAN DESERT**

Tom Lynch

We've arrived. I lift Riley, age two, out of his car seat, plop him in the backpack, and swing him up behind. His brother, Cody, age five, has hopped out of the car and is tossing parking-lot gravel down into the arroyo.

On this humid, partly overcast July morning, with the temperature lingering in the mid-eighties, we head up the trail. We're at the Dripping Springs Natural Area, a Bureau of Land Management site tucked into a spectacular canyon that slices deep into the Organ Mountains, just a few miles east of Las Cruces, New Mexico. Though far from pristine, it's a wild and secluded refuge amid the growing frenzy of nearby Las Cruces and El Paso. And it's only a twenty-minute drive from our home, so it serves as a frequent location for family outings.

I tote Riley while Cody, still intent on tossing stones, trots ahead. Our trail skirts a hillside above an arroyo that flows through what's called Ice Cañon. The cool upper reaches of the canyon, shaded by massive purple rhyolite cliffs, must have inspired the wistful name. Desert willows line the arroyo bottom, their pink blossoms wilting a bit but still attracting plenty of native bees.

After a fifteen-minute stroll, we pause to rest beneath an alligator bark juniper. The boys clamber into the green shade of the lower branches, the scaly tree bark rough and richly textured beneath their small hands. On one side of the trail stands a male juniper; on the other side, a female. In spring, the male spills clouds of pollen into the air and gusts of wind set golden clouds adrift; now, in summer, the female grows small greenish-brown berries flecked with a pale white coating. The boys find my lesson on tree gender plenty dull, but the berries are still fun to throw.

Although they'd rather linger in the tree, I cajole them up the trail, and we stroll through a varied landscape. Spiny ocotillo wands dangle scarlet blossoms where hummingbirds and carpenter bees hover. And there are lots of prickly pear cacti. Beneath one expansive prickly pear bush, a pack rat has stacked sticks and cactus pads into a massive midden, a well-fortified home. The boys try to peer inside to see if anyone is at home, but the thorns keep them too far away to get a good look, just as Mr. Pack Rat had planned.

Trudging on, we round a bend and arrive at the livery stable of the old Van Patten Dripping Springs resort. We inspect the wreckage of a corral fence, and I note the various grasses that sprout in the clearing. Since Dripping Springs has been closed to grazing, many grasses have returned: side oats and black grama, bluestem, sprangletop, Arizona cottontop, lots of others I don't identify and the boys are uninterested in hearing about anyway.

This stable and corral are where the horses and carriages stopped as guests arrived at the swanky resort, built here in the 1870s by Colonel Eugene Van Patten. The spacious, sixteen-room structure included a dining room and a concert hall. In the days before air-conditioning, the 6,200-foot elevation provided a cool respite from the desert heat below. It still does. During its heyday, anyone who was anyone in this part of the world stayed here, including Pat Garrett and Pancho Villa . . . even, some claim, Billy the Kid.

Despite the resort's popularity, Van Patten went bankrupt in 1917 and sold out to a Dr. Nathan Boyd, whose wife suffered from tuberculosis. Boyd converted the resort into a sanatorium, and farther up the canyon he built additional airy screened cabins to house patients. Before antibiotics, dry high-desert air was considered the most effective treatment for TB. I always think as I wander through here that if I were dying of TB, this wouldn't be a bad place to spend my last days: sheer reddish granite and rhyolite walls flecked with green and yellow lichen, yuccas and agaves that spring from towering pinnacles and render dashing silhouettes against the bluest of skies, rock wrens that twitter and sing among rugged boulders, canyon wrens whose song cascades from the cliffs, silky black phainopeplas fluttering atop cholla, crissal thrashers that screech in the whitethorn acacias, ladder-backed woodpeckers clambering on the hackberries, and canyons that recede into the dim recesses of the mountains.

Yes, one could spend one's last days in worse circumstances than these, I think, and, sadly, most of us will.

We stop above the reservoir Van Patten had built to supply the resort and picnic on chips, oranges, cheese. I keep an eye on the clouds building gray overhead, as thunderstorms can arise quickly. Above us, water runnels down a natural trough in the rhyolite, into a splash pool, and then through and under gravel down to the reservoir. Oh, and what's this? An Organ Mountain evening primrose growing right here beside the creek, a large one, with some wilted orange flowers lingering from last night's blossoming. This is a rare and endangered species that grows only in these mountains. I've searched for it on other hikes, and here it is, unbidden and unexpected.

The boys toss stones in the splash pool, build dams in the creek. I worry about the little damage they might do, their effects on erosion. While they play, I explore a short distance up the narrow drainage. On the ledge above the trough, a sign reads, "Danger, keep out, live fire range. *Peligro, se prohibe la entrada, campo de tiro, armas de fuego en uso*." This is the boundary of Fort Bliss. Hah, while I worry about the boys tossing a few stones, the army is blasting away on the other side of the mountain with howitzers.

Many of the desert narratives one reads stress adventure and conquest, the daring exploits of intrepid adventurers. Well yes, I've done a few vigorous and perhaps foolhardy outings myself. But, truth be told, most of my desert journeys have been like this, casual strolls with my kids. Sure, there are scorpions, cactus thorns, spiny acacias, biting ants, spiders, and snakes to fret about. We've had a few close calls, and I could highlight those. But to do so would falsify my experience. I don't really worry about the dangers. It's rugged here, to be sure. But my overwhelming experience is of natural abundance steeped in serenity. ◇

CLEVER PLANTS

Desert plant communities range from nearly barren to fairly lush. In the hottest, driest, least stable dune fields of the Sahara or the Arabian Peninsula's Empty Quarter (Rub' al-Khali), or on some of the rocky plains of the Gobi, there are no plants. Portions of the Atacama Desert seem so dead that NASA used them to test the ability of the Viking Mars lander to detect life. But such extreme conditions are far from the norm, and only a tiny amount of rainfall can support a surprisingly rich and diverse flora.

Many deserts are characterized by open ground with more or less densely scattered bunch grasses, annual plants, small shrubs, and small trees: the spinifex on Australian dunes, the sagebrush in the Great Basin and other temperate-zone deserts, the piñon and juniper trees on American mesas, the acacias in subtropical deserts. Wetter areas have their own communities of plants that range from the reeds and rushes of salty and freshwater marshes, through the gallery forests that line wet arroyos and wadis and the date palms that cluster in oases, to the seep-watered hanging gardens high on canyon walls.

Although it conflicts with their popular image as barren wastelands, deserts, because of their great variety of habitats, are second only to tropical rain forests in the biodiversity of plant and animal species they produce. By the standards of European and American agricultural expectations, the desert soils of Western Australia are among the most infertile in the world, extremely low in organic material, nitrogen, and phosphorus. Nevertheless, they produce the planet's greatest diversity of flowering plants, with up to twelve thousand species, about half of which are endemic, and with new species discovered each year. Wildflower viewing has become a major tourist industry in the state; local wildflower festivals and backcountry tours are a thriving business. Similarly, part of the Namib desert, the Succulent Karoo, is among the most biodiverse arid regions on the planet, with more than five thousand plant species, a high percentage of them endemic. And it contains the richest succulent flora in the world, with a third of the world's ten thousand succulent species growing there.

To survive, desert plants need to do one or more of these things: persist when it is dry, act quickly when it is wet, collect as much available water as they can, and then safely store it. Many flowering plants excel at the first, biding their time out of sight and remaining inactive in the form of root systems, bulbs, or tough-shelled seeds, then rapidly sprouting, bursting into bloom, and setting seed when there is enough moisture after rainfall events, even when these events occur years apart. Travelers to the Kara-Kum desert may be startled to see vast displays of tulips, a species that evolved there. Altogether there are more than a hundred species of tulips in arid Asia, where their underground bulbs are well adapted to survive long periods of aridity.

Many other desert plants are annuals, or perhaps more appropriately ephemerals, since they may not grow every year. This is an effective survival strategy: in the most extreme arid regions, up to 90 percent of the plants are ephemerals. Desert ephemerals grow and flower quickly when conditions are right, lay seed in often vast quantities, and then die off. Even barren-seeming desert soils can harbor these seeds as they await the right opportunity to sprout. An astonishing five thousand to ten thousand seeds per square yard are common in the Sonoran Desert. In Australia, the seeds of Sturt's desert pea can lie dormant for several years and then, after rains, quickly produce extensive nets of low, trailing stems, hairy leaves, and intense scarlet blooms with nearly black centers. Other ephemeral flowers abound even in hyperarid deserts like the Atacama or Death Valley, where the infrequent blooms are spectacular and beautiful, carpeting normally bare ground with colorful, shapely blossoms of many species. Such flowers can complete their whole growing cycle in just a few weeks, before favorable conditions change. Many of the seeds they produce during this time can survive for years before sprouting; indeed, some desert seeds can survive much, much longer. A cache of date palm seeds was found by Israeli archaeologists in 1965 during the excavation of King Herod's first-century fortress at Masada. In 2005 scientists successfully germinated one of these palm seeds, which had been carbon-dated at two thousand years old.

For gathering water, large root systems are key, and a very high proportion of plant mass in deserts (sometimes as much as 90 percent) is underground. Some plants, especially shrubs and small trees like mesquite, palms, and some acacias, have particularly deep taproots; mesquite may draw water from fifty feet straight down. Other plants, including many cacti and euphorbias, have shallow root systems that spread very widely; pistachio trees may drink from a hundred-foot circle. This structure also ensures that other plants do not manage to sprout nearby and so compete for water. Other water-collecting tactics occur as well. Some cactus and euphorbia spines point downward so that water condensing on them drips into the root zone, and a few plants can survive by directly absorbing dew or mist.

What makes many plants *look* like desert dwellers is that they store water in their trunks, branches, or stems. The quiver tree or kokerboom of southern Africa and baobab trees of Madagascar, Africa, and Australia use their thick trunks; the elephant's foot yam of southern Africa seems to grow in a clay pot that is actually its own stem. Pleated trunks and stems allow many cacti to swell to hold water—an organ pipe cactus, for instance, can absorb enough in a few wet days to last for several months—and the pleats help the plants cool off during the day by casting shadows on themselves. Many other desert plants, including agaves and euphorbias, are succulents whose water is held by their thick leaves. Tough, waxy, hairy, and pale surfaces reduce water loss from evaporation and transpiration. Succulents such as cactus and agave don't require a lot of rain, but they do require regular rainfall, so they are largely absent from deserts with highly variable cycles, such as those of Australia. Also, because they retain water in their fleshy cells, most succulents cannot survive long periods when the temperature is below freezing. Hence, they are largely missing from cold deserts such as the Great Basin and the deserts of central Asia. The cacti that do live in the Great Basin are extremely low-growing. By contrast, the warm Sonoran Desert produces the tallest cacti, with saguaro and cardón reaching higher than fifty feet. Such warm deserts with reliable cycles of rainfall produce a great variety of cacti, agaves, and other succulents. Visitors to the Baja California peninsula may be surprised to find themselves in what seem like lush, tangled forests of desert plants, many of them quite tall.

Another common water-conserving strategy is to grow small leaves. Some plants have leaves that fall off when it is too dry and hot, as do the creosote bush and ocotillo. Some leaves hang vertically to reduce sun exposure, as on the jojoba, while other plants tilt their leaves away from the sun during the hottest part of the day, as do many eucalypts. Some leaves are reduced almost to invisibility, with stems taking over the job of photosynthesis, as in the American palo verde (Spanish for "green wood") and Australian mulga tree. Some leaves take the form of thorns and spines, as on cacti, many euphorbias, and acacias, which are commonly called thorn trees. Indeed, spiky, succulent old-world euphorbias and new-world cacti often look alike, not because they are related—they aren't—but because of "convergent evolution": living in similar conditions, they have independently evolved similar survival characteristics. Such features as sharp, serrated leaf edges (agaves and pampas grasses), sharp and sometimes fish-hooked spines and thorns, pungent scents (creosote bushes), and unpalatable tastes (wild onions) discourage herbivores as well, another indirect method of conserving water.

The least conspicuous desert plants are probably the most ubiquitous, the lichens and other biological soil crusts that occur in undisturbed and exposed sections of arid and semiarid areas around the world, where they may, in some places, constitute up to 70 percent of the living cover. Sometimes called cryptogamic (from root words meaning "hidden marriage"), cryptobiotic ("hidden life"), or microbiotic ("very small life") soils, these are complex mixes and mosaics of cyanobacteria (also called blue-green algae), green algae, lichens, mosses, microfungi, and other microorganisms. Producing filaments that glue together loose soil particles, they create a matrix that stabilizes and protects the surface from erosion; collects airborne dust; responds to water in just minutes with respiration, photosynthesis, and the creation of oxygen; and favors native over exotic plants, since natives are more likely to have self-burial mechanisms (such as spiral seed cases, which can work themselves into the soil) or to be collected and buried by rodents. Depending on climate details, such soils can be flat, rolling, wrinkled, pinnacled, or covered with pebbles. On the Colorado Plateau, where freezing temperatures are common and cyanobacteria dominate, frost heaves combine with erosion and varying growth rates to produce a miniature version of the landscape's extraordinary mesas, pillars, and canyons, but in shades of charcoal rather than the more familiar pink and tan. These living soil crusts are easily destroyed by foot, hoof, or tire and can take decades to recover. Without them, erosion—especially by wind—becomes much more of a problem, plant communities deteriorate in diversity and health, and once-productive ground may decline into barren sand and dust.

Grasses grow in many deserts, and their density is one indication of an area's position on the spectrum from desert to dry grassland. In response to aridity, desert grasses are typically bunchgrasses rather than turf grasses. Some grow in salty areas, some on sand dunes; some spring from their roots within hours after rains; some have sticky roots that help hold soil together. Tall pampas and plume grasses cover large areas of Patagonia and the Kalahari; camel grass feeds those animals in the Sahara. Larger clumps of grass, such as those formed by the spinifex hummock grasses that dominate much of Australia (genus *Triodia*, not the same as beach dune spinifex), can be more than a yard tall and many yards across. Spinifex provides a home to whole communities of animals, including some (like the rare sandhill dunnart and the spinifex hopping mouse) that are found almost exclusively in this habitat. Grasses once dominated portions of the Chihuahuan Desert; black and side-oats grama, tobosa, sacaton, bush muhly, fluffgrass, and burrograss were abundant, until overgrazing led to their being largely replaced by creosote bush and mesquite scrublands, species cattle avoid eating. Still, the remaining grasses are lovely. Isolated stalks etch faint circles in the surrounding sand when their branches are swept by the wind, and the low light at the ends of the day can transform their seedheads into delicate glowing filaments, a nostalgic trace of what once was.

Many native desert grasses provide food for birds, animals, and traditional human cultures. Very old caches of seed from panic grass (*Panicum*, a genus that includes millet) have been found in both Australia and the Sonoran Desert. In the latter case, people had clearly selected and saved the seeds for flood-recession cultivation, instead of harvesting them only in the wild. As with biological soil crusts, and often in tandem with them, the destruction of native grasses from grazing, changes in fire regimes, off-road vehicle traffic, and invasive species (like cheatgrass) contributes to the biological impoverishment of desert places.

Although cacti grow naturally only in the Western Hemisphere, for most people they are the iconic desert plants. In all, there are about twenty-five hundred species of cacti; some of the rarer ones are now nearly extinct in the wild because of collecting by overly avid gardeners. In size they vary from quite small—pincushions an inch or so across— to quite large, such as the fifty-foot saguaro. Cacti have short-lived but large and vivid flowers. Yellow, magenta, purple, and scarlet in every imaginable shade and combination make cactus blossoms among the most beautiful flowers in the world. These flowers then produce edible fruit. Tohono O'odham people of the Sonoran Desert use long poles to knock the fruits off the very tall columns of saguaro and organ pipe cacti and then use these fruits to produce a frothy pink alcoholic beverage consumed in annual rainmaking rituals.

Among the cactus in the deserts of North America are the various *Opuntia* species known as prickly pears for their purple or yellow pear-shaped, spine-laden fruits. When handled judiciously, these fruits are delicious and nutritious, though full of tiny seeds, and may be used to make jelly. Prickly pears are a common sight in the North American deserts (though they're not limited to these ecosystems), both in the wild and as plantings around homes, where their stunning multihued spring-time flowers are a welcome sight.

It's no wonder someone had the idea to introduce such an attractive and useful species to Australia, where no cacti are native. Sometime in the early 1880s, they were imported. At first, prickly pears *(Opuntia stricta)* were grown mostly as an especially effective hedge, though they could also serve as emergency stock feed during prolonged dry spells, as is done in the American West. (The nutritious pads are first burned to eliminate the spines and then fed to cattle.) Finding the climate much to their liking and relieved of any of the controls provided by their natural enemies in the Americas, prickly pears suddenly exploded into one of the most destructive cases of invasive-plant damage ever recorded. By 1925 some hundred thousand square miles of land in Queensland and New South Wales were overrun, rendering these vast areas unsuitable for grazing and forcing many settlers off their properties. Enormous efforts were expended to eradicate the plant: vast amounts of herbicides were sprayed and bounties placed on the birds (including emus, crows, and magpies) that ate the fruits and thereby spread the seeds. All to no avail. The spiny forces continued to spread at a rate of nearly two thousand square miles a year, and it seemed as though all of Australia might soon become one vast prickly pear forest. Finally, when hope seemed lost, a tiny caterpillar came to the rescue. Researchers discovered that the cactoblastis caterpillar had a voracious appetite for prickly pears but would not eat any other plants. These caterpillars were released into the environment in 1926, and within six years, in a case that still ranks as the world's most effective example of biological pest control, the prickly pears were all but gone.

Shrubs are another very common form of desert plant, one whose importance is often overlooked. Indeed, shrub-dominated landscapes are frequently derided, and when humans disturb desert vegetation they typically remove most or all of the shrubs in the area, causing considerable ecological disturbance. This misperception is partly a matter of scale. To humans, shrubs seem unimpressively short. To smaller creatures, however, shrubs loom large and have great utility, performing key roles in desert ecosystems and creating fertile microclimates that support a wide variety of other plants and animals. The soil beneath desert shrubs is often more fertile than it is in the surrounding open ground because shrub roots draw nutrients from deep in the soil into their woody tissues and leaves. Then, as their leaves fall, they deposit these nutrients beneath their canopy. Similarly, shrubs act as windbreaks in open country; organic debris accumulates around them, decays, and releases nutrients into the soil. Shrubs also offer protection from harsh sun and wind that could desiccate young plants, thereby sheltering the seedlings of other plants. Many animals, especially reptiles, small mammals, and birds, thrive in the cover of shrubs, using them to hide from predators and hot sun and eating their seeds, sap, and leaves.

Because of their enormous numbers as well as their ecological importance, particular shrub species are the definitive plant in many arid and semiarid regions, including the sagebrush flats of the American West, the chenopod shrublands of central Australia, and the creosote scrublands of the Chihuahuan, Sonoran, and Mojave deserts. To each of these ecosystems they impart not only a distinctive visual character but a distinctive scent as well, since all contain volatile oils that perfume the air.

Some desert shrubs are incredibly tough. Take creosote bush, which lives in some of the driest portions of the Chihuahuan, Sonoran, and Mojave deserts, and which boasts both deep taproots and wide, shallow roots. The age of one clump of creosote bushes in the Mojave (nicknamed King Clone, since these bushes are self-cloning) has been estimated at 11,700 years, making it the oldest known living thing on the planet. Not only can creosote bushes endure for a long time, they can also survive intense trauma. At the Nevada Test Site, where the United States military detonated nuclear explosions aboveground during the 1950s and 1960s, creosote bush is the predominant vegetation. In 1962, a nuclear test was conducted near a stand of twenty-one creosote bushes. All the aboveground portions of the plants in the area that were dusted by radioactive fallout died. Ten years later, however, ecologist Janice Beatley determined that all but one of these shrubs had resprouted. Plants don't get much hardier than that.

Shrubs are so prevalent in deserts because the combination of aridity, low soil fertility, and high winds makes it difficult for plants to grow to a large size. The same conditions mean that many desert tree species are typically low-growing, including mesquite in the deserts of northern Mexico and the American Southwest, eucalypts such as mallee in Australia, and acacias, whose thirteen hundred species are the predominant shrubs and trees in many of the world's deserts.

Other plants, seeking the advantage of growing low but responding to the fact that being tall is a good way to gain the notice of flying pollinators, spend most of their life huddled close to the ground, but then, before they flower, sprout tall stalks. The deserts of northern Mexico and the American Southwest are home to three widespread examples of this phenomenon: yuccas, agaves, and sotols. Each of these plants consists of a clump of basal leaves (and in some species of yucca a short trunk), and each reproduces by shooting up a tall stalk that bears one or more panicles of flowers, held high to attract the notice of passing pollinators. After their flowers have gone to seed, the withered stalks stand stark against the desert sky.

Agaves are an especially diverse genus, with more than four hundred species and many more subspecies. The word *agave* derives from the Greek *agaue,* meaning "noble" or "admirable," a name that honors the plant's many uses. For most of its life, the agave is a handsome cluster (sometimes roughly the size of a dinner plate, sometimes much larger) of gray-green, serrated leaves, each ending in a long sharp point. Thanks to these terminal spines, a common Sonoran and Chihuahuan desert agave, lechuguilla, is sometimes called "shin dagger," a phrase impaled hikers often utter preceded by an expletive.

Succulents with fleshy leaves that store moisture, agaves are protected by a waxy coating. They are long-lived but bloom only once, putting all of their life's energy into shooting up a rapidly growing stalk as high as fifteen feet and panicled with numerous upraised flowers that bloom at night and are pollinated by bats. After blooming the entire plant dies, although root runners often sprout clones, called pups, around the parent plant. Some large agave species are called century plants, giving the erroneous impression that they bloom only once in a hundred years, though typically they live for five to thirty-five years. Before the agave sends up its tall stalk, it concentrates carbohydrates and sugars in its body. If the plant is then harvested and the leaves chopped off, the rich juices stored in this body can be distilled into the alcoholic drinks mescal, tequila, or pulque. Large crops of the blue agave are grown commercially in Mexico for the tequila industry. A syrup can be made from this plant, and an industry has recently developed marketing agave syrup as a healthy sugar substitute for diabetics.

Like agaves, the roughly fifty yucca species are native to the arid parts of the Western Hemisphere. But being more cold-tolerant, their range is wider, extending into the American Midwest, Canada, South America, and the West Indies. Indeed, *yucca* is derived from a Taino Indian word (not to be confused with the similarly named but unrelated *yuca*, the widely cultivated cassava plant). One of the most widespread is the soaptree yucca, whose common name derives from the fact that its roots can be pounded and mixed with water to produce a soap sometimes used by Native Americans as a shampoo; it grows in many desert scrub communities and is an important component of semidesert grasslands, where it rises above the waving grasses. The largest yucca, the Joshua tree, can have a trunk thirty-five feet tall and flower stalks rising another fifteen feet. An indicator species for the Mojave desert, the Joshua tree was named by Mormon pioneers, who thought that its tall and contorted shape resembled the Prophet Joshua reaching his arms toward heaven in supplication. Individual plants live up to a thousand years.

Unlike agaves, most of which bloom only once, yuccas can bloom annually, and all of them reproduce in a famously complex symbiosis with yucca moths. In a relationship called coevolved obligate mutualism, female moths pollinate the yucca blossoms by gathering the pollen they find in one flower into a ball, which they then carry to a different blossom and deposit in the new blossom's stigma, fertilizing it. The moth then lays a few eggs in the flower's ovary. When her eggs hatch, they eat some, but not all, of the yucca seeds that have ripened, a beneficial arrangement for both parties. Without their specific moths, the yuccas cannot reproduce; without their specific yuccas, the moths have no place to lay their eggs. (In this instance, the word *specific* means particular to a species.)

Acacias are among the most common trees in North Africa and the Middle East and play an important role in the mythology of that region, having been venerated by Semitic cultures. Acacias are mentioned in the Book of Exodus and the Book of Isaiah, and some speculate that the burning bush of Moses was an acacia. Moses also had the Ark of the Covenant built of the wood of the shittah-tree, *shittah* being Hebrew for "acacia," and legends suggest that Jesus' crown of thorns was acacia, as was the wood of the cross. Egyptian mythology features acacia as the Tree of Life and the dwelling place for Osiris. Because of its importance in both Judeo-Christian and Egyptian mythology, acacia is also sacred among the Order of Freemasons. Along with whirlwinds, these thorn trees can serve as symbols for the combination of deserts and moments of intense spiritual or religious significance. In Johnny Cash's apocalyptic late song "The Man Comes Around," which is based mostly on the Book of Revelation, the refrain includes the words "The whirlwind is in the thorn tree." In contrast, when American desert writer Edward Abbey evokes these images, it is partly to reject traditional theology and the habit of reading nature as symbol: "Under the desert sun, in that dogmatic clarity, the fables of theology and the myths of classical philosophy dissolve like mist. . . . Whirlwinds dance across the salt flats, a pillar of dust by day; the thornbush breaks into flame at night. What does it mean? It means nothing. It is as it is and has no need for meaning. The desert lies beneath and soars beyond any possible human qualifications."

Because it is so common on the Serengeti plains, one species of acacia, the umbrella thorn, has become an icon of Africa with its easily recognizable spreading flat top. Flourishing in very arid conditions and extremely high temperatures, it grows well in sandy, stony, and highly alkaline soils found throughout northern and eastern Africa as well as on the Arabian Peninsula and in Israel, but it is most prevalent in savanna country and the Sahel. It protects itself with two different types of thorns, a long straight spine and a shorter hooked one. Both its leaves and its twisted seedpods provide forage for many animals, especially giraffes, whose long necks give them superior access. Elephants will shake the tree in order to knock seedpods to the ground, which they and other opportunistic animals then consume. As one of the few timber-size trees in much of its range, it is highly valued for its wood. Its shade, too, is valuable, particularly as individual trees tend to stand alone; in the heat of midday, one might find a pride of lions or a party of humans resting in an umbrella thorn's shelter. In aerial photographs, a solitary acacia may be revealed as the hub of a web of paths beaten by creatures in search of its gifts.

When acacia tree bark is ruptured, the plant oozes a sticky sap to protect the wound. Two African species, *Acacia senegal* (the hashab) and *Acacia seyal,* both of which grow in the Sahel, produce the valuable gum arabic, an important trade item for at least four thousand years. Because of its solubility and lack of flavor, this material is used in many food products, cosmetics, medicines, and industrial processes. The production of gum arabic, which can be tapped somewhat like maple syrup without killing the tree, is a mainstay of the economy in sub-Saharan Africa.

One of the largest and most striking desert trees is the baobab. This is the common name, derived from Swahili, for trees in the genus *Adansonia,* which consists of eight species, six native to Madagascar, one to arid mainland Africa, and one to northwestern Australia. Baobabs can be as tall as a hundred feet and have trunks some thirty-five feet in diameter—trunks made of spongy tissues that get saturated with water during the rainy season, making them the largest succulent plant in the world. These tissues can store more than thirty thousand gallons of water, enough to see the trees through the dry season, when they also shed their leaves. With their massive trunks and relatively short branches, baobabs look as if they are standing upside down with their roots dangling in the air, especially when their leaves have fallen. Many traditional folktales among diverse cultures in Africa explain this unusual appearance in a similar way, proposing that the tree offended or annoyed a powerful god, who turned the tree upside down as punishment.

Because of their great mass and welcome shade, baobab trees serve as meeting places in many villages, and many community marketplaces are located beneath them. Large baobab trees sometimes grow in quite desolate country where they can become the center of their own self-created ecosystem, supporting many other creatures, including many birds, mammals, and insects. Because of this effect, baobabs are worshipped as the Tree of Life by cultures in the African savannas, a fact that plays a part in the Disney film *The Lion King*. In Senegal, large baobabs are believed to harbor wise spirits that can be consulted for guidance, and that country has chosen the baobab as its national symbol. One of its most popular and influential musical groups is named Orchestra Baobab, a title taken from a Dakar nightclub, the Baobab Club, whose name in turn reflects the tree's role as a communal gathering place.

In the high mountains of the Sahara one finds the wild ancestors of many trees that have been domesticated for their fruits and nuts, such as pistachio and olive; in the Horn of Africa and the southern edge of the Arabian Peninsula grow the trees from which for thousands of years frankincense and myrrh have been harvested. Not surprisingly, however, the greatest number of large desert trees, such as palms, cottonwoods, and river red gums, grow around permanent water sources such as oases, rivers, and similar riparian habitats.

No doubt the most iconic of desert trees are the palms, Arecaceae. Within this family are more than two hundred genera and twenty-six hundred species growing in many diverse ecosystems. Most live in the wet tropics, and those that grow in arid places tend to be relict survivors of wetter climates that have managed to persist by clustering around the few permanent sources of water (as they do in the Mojave, for instance). The most important is the date palm, a plant that has been extensively cultivated for at least eight millennia, so long that its place of origin is unknown, though it was likely around oases in the Sahara. No doubt its dependable, nutritious, portable fruits greatly aided early human settlement and migration in the challenging desert regions of North Africa and the Middle East. Carried into Spain by the Arabs and then brought to the Americas by Spanish missionaries, date palms became widely distributed in California, where they are an important cash crop. And they are still widely cultivated as a food staple and export commodity across North Africa and the Middle East, where hundreds of varieties are still available.

Because of their importance to the cultures of the Middle East, references to palms are common in the sacred scriptures of Judaic, Christian, and Islamic cultures. The Christian celebration of Palm Sunday uses palm leaves to commemorate Jesus' entry into Jerusalem, when date palm leaves were strewn on the road before him. However, the date palm is most closely associated with Islamic culture. It is mentioned more times than any other fruit-bearing plant in the Qur'an; daily fasts during Ramadan are traditionally ended by eating dates; dates are placed in the mouths of newborn babies and eaten ceremonially at weddings. Indeed, so desirable are these fruits that paradise itself is pictured as containing date palms. More generally, in the Qur'an, references to heaven or paradise tend to portray a lush oasis: "God has provision for them of gardens with streams of running water, where they will abide forever." The word *paradise* derives from a Persian term meaning "walled garden," a sort of oasis. In a much more mundane illustration of the hold these desert-oasis plants have on the human imagination, they have also been widely introduced as plantings in resort communities, where they suggest luxury, ease, and an escape from the troubles of the world.

◇ ON THE SPOT: IN THE RED CENTER

Deborah Bird Rose

When we stepped off the plane in Alice Springs we were struck: by sun in our eyes, clear air on our cheeks, and sky that was bigger and brighter than anything. Alice is the main town in the center of the great expanse of inland desert known as Australia's red center. It is big sky country, and hugely red. It captures you. There is a funny story that expresses something of the power of the center's vividly unique self: the wealthy Texan stepped off the plane, looked around him, and said, "This country is so g-d darn spiritual, I've just got to buy me some!"

It was midwinter, and I was traveling with a philosopher, Jim Hatley. We decided to spend our time in the MacDonnell Ranges, a low band of hills stretching east and west of Alice Springs for a few hundred miles. The hills are cut by numerous gorges where ephemeral dryland rivers have sliced the sandstone and where springs sustain small bodies of permanent water. The ranges differ from the surrounding deserts by their water and by their topography, which have produced numerous endemic species, many of them endangered. Much of the land is therefore under conservation protection.

When European explorers encountered inland Australia in the 1800s, they expected to find a large inland sea or lake, set within a fertile wilderness that needed only the agricultural hand of civilized man to be brought into full potential as a land of plenty. With dismay they learned that the lakes were ephemeral at best, the soils were not fertile, rain was unpredictable, rivers flowed inland when they flowed at all, and life followed a boom-and-bust pattern. They called this "empty" land the "dead center" and started filling it with cattle. Social and cultural changes in Australian society over the past half-century have given us more complex geographies in which Aboriginal freehold land, cattle properties, national parks, nature reserves, historic sites, and sacred sites are located side by side, and the center has gone from dead to red.

Indigenous people's terms for this country expressed their sense of home. The land was owned and inhabited, and it was crisscrossed with the tracks, often called songlines, of creation beings known as Dreamings. The main creation story for the MacDonnell Ranges tells of the caterpillar ancestors called Yeperenye, and the hills do look exactly the way caterpillars would look if they had become the size of mountains and gone walking across the country nose to tail, just as the story tells us. Within this large story each gorge has its own groups of traditional owners and its own creation stories, only some of which are shared with the public.

Each gorge is a microcosm of life in the desert. One such place of water, life, and red beauty is Trephina Gorge Nature Park. From the parking lot Jim and I walked through an area thick with the scraggly acacias that Australians call scrub before coming out into a more open area and starting up a path that would take us to the top of the cliff on the western side of the gorge. We scrambled up a steep slope with loose stones, and it wasn't until the track opened out to ledges that formed the top of the cliff that we could stop and look. Suddenly we were face-to-face with the big exquisite red. Gazing across to the cliff face on the other side of the gorge we were struck again: that dark red stone, those purple shadows, those bands of black, orange, umber. These rocks were ancient, and their colors suggested that instead of fading with age, geological eons had intensified them. Deep colors, bands of strata, planes and folds, glittery brightness, dark and moody shadows: this place was shimmering with its own vivid presence. *Spirituality,* I thought, *is a default word—what we say when we don't know how to talk about soils and stones that are so alive.* Below us the riverbed was pale, pebbly sand, above us the sky was bluer than blue, and here and there we met eucalypts—most spectacularly, the white-barked ghost gums that articulate themselves so definitively against the red earth and blue sky. They can grow very tall and thick, but here on top they were only a few meters high. Their roots seek out water in cracks between stones, and their leaves hang delicately, each one positioned to minimize its direct contact with sunlight.

We followed the stony path through a patch of spinifex-hummock grassland as it wound around boulders. The country was dotted with bushes whose brilliant, ephemeral flowers shone in the sun. The desert fuchsias had mauve flowers and dusky sage-colored foliage; the yellow flowers of the cassias looked cheerful against their own dark green leaves. Nothing was crowded: we could see the red soil, the stones and boulders, the flowering bushes, the trees, and the spinifex hummocks. Everything articulated its own shape and place in this brilliant light.

From the top country Jim and I went down to the riverbed. We were walking on pale sand amid river gums, the extremely tall white eucalypt whose single trunk rises elegantly from the ground to branch out above the human level. River gums have a single great taproot that shoots down into the underground water; where they grow, there will be subsurface water. When I saw a willy wagtail, a small black-and-white insect-eating bird, I knew there must also be surface water nearby. We walked around a bend, and there it was: a permanent water hole, small, no more than a few yards across in any direction. A cliff face fronted one side of the water, and the sand surrounded it. The Trephina water hole was small relative to others in the MacDonnells, but like all water in the desert it was an attractor. Permanent water offers breeding grounds and refuges for water creatures that include fish, shrimp, and insects. Water holes sustain frogs and waterbirds; kangaroos and wallabies come in to drink around sunrise and sunset. By night the bats are active, and during the day pigeons, doves, and finches come and go, as do the willy wagtails.

In the desert one sees far more evidence of life than life itself. This is logical enough: many of the mammals are nocturnal to avoid the heat. Many of the reptiles hibernate in winter. There is also the logic of knowledge. Indigenous people have lived here for millennia, they know how to thrive. The scrubby acacias known as mulga, for example, produce a nutritious edible seed. Another scrubby acacia is called witchetty bush and is home to witchetty grubs, one of the high-protein staples of desert life. Another logic is that of the pulse: boom-and-bust flows of water produce flows of life. The sandy riverbed told us that water had flowed here many times, and would, we imagined, flow here again. We knew that water brings life, and we knew that in the long spells when the rivers do not flow, life waits. What we saw was only a fraction of the life that was there, and there was life to come that we would never witness. At the same time, the country was alive, and we knew we were in a realm of intersubjectivity. The philosopher responded by pausing under the shade of a river gum to sing. His voice, moving up and down the gorge, was life-affirming. As he sang, finches gathered and settled in the tree above him.

I sat quietly by the Trephina water hole listening to Jim sing and thinking about other visits to nearby places. I had been in the area in the 1980s and 1990s when I worked on many Aboriginal claims to land. The land-claim era was a great moment in decolonizing the relationships between Anglo-Celtic settlers' society and indigenous peoples. Legislation enacted in 1976 enabled Aboriginal people to make claims to parcels of land in the Northern Territory over the course of a twenty-five-year period. Each claim was brought before the Aboriginal Land Commissioner (a federal court judge). I worked for the judge, and every hearing required that we travel into the claim area to learn about the relationships between people and country.

Not far from Trephina Gorge we had visited a sacred site, a very small gorge whose permanent water was scarcely more than a trickle in the driest times. The Aboriginal women sang and danced. It was their home and their water; they danced life and fertility. They danced rain and the future: cane grass that grows with rain, butterflies that emerge, and new generations of humans, too. The women were dancing desert life and they were dancing water. It was all one dance. ◇

CLEVER CREATURES

In the heat of midday, a desert landscape can seem eerily still, silent, and devoid of animal life. In the cool of the morning, however, especially in sandy deserts, one will frequently notice a multitude of tracks left during the night by various types of animals: mammals, lizards, insects, arthropods, and even birds. Clearly, though they may be hard to spot during the day, these many tracks suggest that a lot of animals must be living in the vicinity. And they are.

Indeed, the numbers and variety of desert creatures can be difficult to imagine, especially given our usual assumption that these are deserted places. Just a few examples: between 1,000 and 1,500 native bee species live in the Sonoran Desert; the arid regions of Australia have 150 kinds of lizard, where they are the dominant predators; the hyperarid Atacama can lay claim to a mouse, a gray fox, a chinchilla species known as the viscacha, and two camel relatives, vicuñas and guanacos; the Sahara, similarly, hosts the jerboa and related gerbil, Cape hare, desert hedgehog, Dorcas gazelle, dama deer, Anubis baboon, spotted hyena, common jackass, sand fox, Libyan striped weasel, slender mongoose, and of course camel. Even desert birds are surprisingly varied, with more than 300 species in the Sahara and 350 in the Sonoran Desert. The mountains of southeastern Arizona, where the Sonoran and Chihuahuan deserts meet, is often considered the best birdwatching location in the United States. Uluru–Kata Tjuta National Park in central Australia records 178 bird species. Hearing the beautiful bell-like tones of the crested bellbird echoing off the rock walls is one of the more memorable experiences at the park—a sound suggested by the onomatopoeic name given this bird by the local Ananagu, *panpanparlala*. Elsewhere in Australia's deserts, many parrots live, as well as numerous cockatoos, the familiar cockatiel, and the budgerigar, which is also known as the parakeet.

Desert animals face many of the same survival issues that plants do. Because of the great variability of weather in many deserts, a diversity of strategies typically works best, and many animals combine more than one approach.

One of the most effective strategies is to simply avoid the extremes of temperature by burrowing or seeking shady shelter during the day and feeding only at night. So most desert animals live underground and are nocturnal or crepuscular. That's why a landscape seemingly lifeless by day can be teeming with scurrying critters in the dark. Other animals take this avoidance behavior to an extreme by estivating, the technical term for hibernating to avoid heat and aridity. As the hot, dry season comes on, these animals typically bury themselves deep in the soil and slow their metabolism, waiting patiently in the cool darkness for the weather to change, living off moisture and fat stored in their bodies. This strategy is most frequently practiced by amphibians and by reptiles such as the desert tortoise of the Mojave and Sonoran deserts, which spends most of its life estivating; but even some mammals estivate, including the various hedgehog and jerboa species of the Saharan and Asian deserts.

It may seem especially odd to find amphibians, whose eggs must be laid in water and whose early life stages are aquatic, thriving in very arid locations, but numerous species successfully do so by practicing estivation. In the Sonoran and Chihuahuan deserts, the spadefoot toad (so called for the handy shovel-like structures on its rear legs employed for digging) spends much of its life estivating peacefully three feet deep in the desert soil, far from the moisture-sucking power of the hot desert sun, rather like a tulip bulb. These toads take advantage of periodic heavy downpours to emerge, feed, drink, and mate, and the tadpoles that quickly emerge from their eggs must survive long enough in the pools to reach maturity before burying themselves as adults. It's remarkable how a quiet desert landscape can be transformed by a single rainstorm into a cacophony of tens of thousands of chirping, hopping, and madly mating spadefoot toads. Even city neighborhoods sometimes pulse with the wild vibrations of these scurrying creatures, who have been living secretly below unsuspecting city dwellers.

These North American spadefoot toads cannot survive more than two years of estivation. However, in many of the deserts of Australia, where seasonal rainfall is much less predictable and reliable, the appropriately named water-holding frog has evolved a method to outdo them. Before burying themselves, these frogs fill their bladder and special skin pouches with water. Then, once they're buried, they shed layers of skin that harden into a water-impermeable shell. Before emerging, they eat through this shell, regaining nutrients. Studies have found that up to a tenth of these frogs survived five years in estivation between periods of rainfall. Given their reproductive powers, this percentage is more than adequate to ensure the species' survival, and their population numbers remain high throughout their range.

Aboriginal people who live where water-holding frogs are found know how to locate them buried beneath the surface and will dig them up to take advantage of this hidden stash of water. A gentle squeeze releases the water—still fine to drink—into awaiting dry mouths, and the frogs are then returned to their burrows, apparently unharmed. In some Aboriginal Dreamtime narratives this frog appears under the name Tiddalik. In one version of the myth, Tiddalik hoarded all of the world's water supply and could not be made to give it up. Finally, one animal made him laugh, releasing a vast flood of water to quench the parched landscape.

So how do these spadefoot toads and water-holding frogs, buried underground, know when it's raining? Studies suggest they hear low-frequency vibrations generated either by rainfall or by thunder. Thus awakened, they begin their climb to the surface. Unfortunately, motor vehicles driving across the desert can create similar vibrations, triggering the toads or frogs to emerge into a dry landscape where, having expended their water and energy reserves in the upward climb, they die.

Other creatures also make clever use of the shelter the earth itself can provide, such as those most successful, diverse, and plentiful of desert creatures, ants and termites. Because the survival strategy of many desert plants includes the production of vast quantities of seeds, seeds are among the most dependable food sources in these places. So seed-consuming insects thrive in all of the world's deserts, constituting dominant elements in the arid-zone ecology of Australia, North America, and South Africa. Harvester ants, for instance, which generally forage at temperatures between 86°F and 104°F and retreat to their burrows in the hottest times, collect seeds when abundant and store them for later use. They are among the most common animals in Australia's arid zone, with an amazing density of one colony every one to two square yards, a fact not lost on travelers looking to set up a picnic luncheon on the ground. These ants contribute to the fertility of desert soils by discarding seed chaff and other debris around their nest mounds—in large enough amounts over a colony's fifteen- to twenty-year lifespan to make fertile islands that are even higher in soil nutrients than are the soils beneath shrubs. In the American Southwest, harvester ants play an important role in the stories of the Navajo Indians. In their creation story, ants are among the first creatures to appear on Earth, hold sacred powers, and should be treated with respect; neither the ants nor their mounds should be disturbed, and dire consequences can follow for those who violate this injunction. An important ceremony of the Navajo, the Red Antway, is based on this sacred mythology.

Like ants, termites typically build elaborate underground structures that keep them safe from both predators and dangerous heat. However, in semiarid regions that are subject to seasonally heavy rainfall, underground dwellings would get flooded, so some termites instead build towers as tall as thirty feet. Scientists estimate that some towers in Africa have been continuously inhabited for four thousand years. Such towers are oriented to minimize the heat they absorb from the sun and contain sophisticated channels to funnel air through the colony. At the same time, workers gather moisture from the damp ground at the water table below the tower and carry it to the aboveground tunnels, where it is applied to the walls over which rushing air passes, cooling the interior of the tower through a form of evaporative air-conditioning. By combining solar collection and evaporative cooling, these tactics keep the temperature deep inside these towers almost constant. Architects have lately been studying termite towers for ideas about designing energy-efficient buildings for human habitation. Perhaps termites will have a new role to play in human architecture, besides consuming it.

Larger mammals, of course, can't easily bury themselves, so they have developed other adaptations to desert heat. Oryx, for example—a strikingly handsome genus of antelope that can grow to five hundred pounds—possess a physiological adaptation known as selective brain cooling: they can keep the blood supply in their brain cooler than the blood supply in the rest of their body. They do this using a structure within the sinus cavity known as a carotid rete, a system of arteries through which hot blood from their body is air-cooled by their breathing before entering the brain in a process similar to that used in a car radiator. (Some other mammals also have this structure, including camels.) However, this cooling is not done in order to keep the brain safely cool (as was once widely believed), but rather to trick the brain's thermoregulation system into believing that the animal's body is cooler than it actually is. This trick inhibits the triggering of the sweat reflex, so the body retains moisture longer before beginning to perspire. Of course, the oryx must then be able to cope with a heightened body temperature.

Another key adaptation of desert animals is being able to survive with little water. Oryx, again, when it's necessary, can supply their moisture needs entirely from the plants they consume, though they will drink water if it's available. In the Australian deserts, only 5 percent of mammals, 10 percent of birds, and none of the invertebrates or reptiles need to actually drink water; instead, they too get what they require from the plants or animals they consume. Other mammals that don't drink water include jerboas, from North Africa and Asia, and kangaroo rats, found in the deserts of North America. Both animals possess a variety of physiological adaptations to conserve water, including extremely efficient kidneys. Kangaroo rat kidneys are five times more efficient than human kidneys, and this creature can also create its own water internally by a metabolic process that occurs during the oxidation of its food.

Other desert dwellers with less limited mobility also adapt quickly to changing circumstances—locusts, for example. Most people identify locusts with biblical times, but locust plagues still occur in the deserts of North Africa, India, and Australia. Locusts are ideally adapted to the variable amounts of food available in deserts, possessing a complicated life cycle with flexible timing of egg hatching and nymph maturity. When food is scarce, their numbers are small and they remain sedentary and solitary. However, when herbage becomes plentiful after adequate rainfall, locust populations explode, and under the influence of hormonal changes they adopt a gregarious and migratory behavior. A single swarm of locusts can be small, covering only a few hundred square yards, or it can be massive, composed of billions of locusts, with some 80 billion covering nearly four hundred square miles. And with the help of wind, such massive swarms can travel more than 125 miles in a single night, inflicting heavy crop damage and devastating subsistence farmers. Studies suggest they can endanger the livelihood of a tenth of the people on Earth.

Mobility is a powerful desert adaptation for other animals as well. Nomadic desert birds, for instance, can move quickly to areas where rain has recently fallen. Carnivorous birds can get the moisture they need from their prey, but seed eaters need water, usually daily. One common desert bird in Australia, the zebra finch (Anangu: *nyii-nyii*), typifies this nomadic life strategy. Subsisting on grass seeds and requiring water every day, large flocks linger around water supplies as long as possible. When the liquid runs out, they travel to find another source, although many may perish on the journey. Because of this bird's need to drink every day, desert travelers and Aboriginal people sometimes follow them to find water sources. The Anangu have an important ceremony, the Inma Nyii-nyii, related to the travels of the zebra finch.

Various desert mammal species have separately evolved a particularly effective form of arid land locomotion: hopping. Hoppers include such widely dispersed and unrelated animals as the numerous jerboa and gerbil species of the Sahara and Mideast, the kangaroo rats of the American West, the springhares of southern Africa, and, of course, the many hopping marsupials of Australia, such as the spinifex hopping mouse and various wallaby, wallaroo, and kangaroo species. This method of movement is well suited to desert environments: it works best in open country, is effective in sandy terrain, and is energy efficient; the upright posture also aids in cooling.

Those hopping animals par excellence, kangaroos, especially big reds, are well adapted to heat and aridity. Like many desert mammals, they have highly reflective fur. They can survive long periods on the moisture in the plants they consume, though the placing of stock tanks throughout arid Australia for cattle and sheep has also allowed them to expand into ever more arid country. As marsupials, they give birth to tiny embryos, called joeys, which suckle in a pouch. In times of extended aridity the mother can stop producing milk, killing the joey but ensuring her own survival. While nursing a joey in her pouch, a female may also have a fetus in utero whose development can be halted while she is nursing. Then, if the joey in her pouch dies and she stops lactating, the embryo in hiatus can resume growing.

The red kangaroo is the largest of arid Australia's marsupials, and its populations remain healthy. However, nearly all of the small and medium-size marsupials of this region have been seriously affected by European settlement, especially by the introduction of foxes and cats, which prey on them, as well as by cattle, sheep, goats, and rabbits, which compete with them for forage. Many members of this remarkable suite of desert-adapted animals, including the northern hair-nosed wombat and species of bilbies, bettongs, bandicoots, quolls, numbats, and wallabies, either have become extinct or are seriously threatened.

The most iconic desert mammal is undoubtedly the camel, another good example of an animal with multiple adaptations—adaptations that have very clearly helped us humans live in this habitat ourselves. Curiously, camels originally evolved in North America and ventured into South America and Asia about three million years ago. They went extinct in North America during the Pleistocene but have persisted in South America as llamas, guanacos, alpacas, and vicuñas and in Asia and Africa as the two-humped Bactrian camel and the one-humped dromedary.

Bactrian camels were once common in the colder deserts of central Asia, where they can withstand temperatures from as low as $-40°F$ to as high as $107°F$, in part by growing and shedding shaggy coats. Although domesticated Bactrians number roughly one million, they are now absent from most of their original wild habitat. A small population of less than a thousand survives in the Gobi Desert, on a former Chinese nuclear test site, where the Arjin Shan Lop Nur Nature Reserve has been established for their preservation. Recently discovered genetic differences have led to wild Bactrians being given a separate scientific name, even as their numbers are diminishing and they are deemed critically endangered.

The dromedary camel seems to be native to the Arabian Peninsula but is widely domesticated in deserts stretching from northern India through the Middle East and across the Sahara. Technically, wild dromedaries have been extinct for the past two thousand years, since all herds since that time have been either under herder control or feral. Indeed, these animals have been domesticated by humans for at least four thousand years and have played an essential role in most Middle Eastern and Saharan cultures.

Dromedaries thrive in hot, arid habitats, especially those with a long dry season and a short rainy season. They dislike cold and humidity and don't cope well in either temperate or tropical climates. They can go for two weeks without drinking, and when they do drink they can consume more than twenty-five gallons of water in ten minutes, a feat that would so dilute the blood chemistry of most mammals as to prove lethal. Contrary to the popular notion, their humps store fat, not water. Their lips are thick, enabling them to consume thorny plants like acacias. Their wide, pad-shaped feet are adapted for traveling on sand, but these feet can be easily injured on sharp stones and provide little traction on wet and muddy ground, making travel during rainstorms treacherous. All camels stand tall enough that their heads rise above the thickest level of blowing dust. They can close their nostrils to keep from inhaling sand, and their eyes are protected with a double row of eyelashes. Like oryx, they conserve water by allowing their body temperature to fluctuate throughout the day (from $93°F$ to $107°F$), minimizing water loss through sweating. And they can tolerate a water loss of more than 30 percent of their body weight, whereas most mammals die after losing just 15 percent; a loss of even 12 percent is fatal to humans.

Because of central Australia's exceptional aridity and the scarcity of surface water, European explorers and settlers found that horses, bullocks, and mules could not endure the harsh conditions. So between 1840 and 1907 several thousand dromedary camels were introduced, along with handlers often called "Afghans," though most were from Pakistan. These animals were so widely used by explorers and to deliver supplies and mail to remote settlements that it is fair to say that the European settlement of central Australia (such as it is) would have been nearly impossible without them. With the arrival of motor transport and the railroad, however, the camels were supplanted. Rather than kill their beloved but expensive and now-unneeded animals, their handlers released them into the wilds. From a population of roughly four thousand in the 1940s, camels in Australia now number upwards of a million. A similar story in the United States ends differently. In the 1850s, imported camels helped survey a wagon road across Arizona (along what became Route 66), but the Civil War and expanding railroads led to their abandonment. During the 1880s and 1890s, small feral herds roamed the Mojave, shocking travelers and spawning tales that sober citizens dismissed as the products of mirage, heatstroke, or bad whiskey. Their numbers did not increase, and the last reported sighting was in 1905.

The camel's odd mixture of extinction, rarity, abundance, and transplantation is certainly not typical, but it is also not unique. The oryx is in a similar situation. The Arabian oryx had gone extinct in its wild homeland by 1982, and reintroduction efforts in several countries have met with mixed success; the scimitar oryx of the Sahara also went extinct in the wild, a few survived in captivity, and a reintroduction effort has begun in Tunisia. But the East African oryx and the closely related gemsbok are widely distributed in eastern and southern Africa, where their numbers are stable at an estimated 275,000, in good part because of the protection of national parks. And, in a smaller echo of Australia's feral camels, one of the largest populations of oryx is in the Chihuahuan Desert of southern New Mexico, where they were imported for the benefit of sport hunters. Taking to this dry, isolated landscape, where coyotes and mountain lions do not prey on them as packs of African lions and hyenas would, they have increased their numbers to perhaps as many as four thousand. Although efforts have been made to fence them out of the gypsum dunes at White Sands National Monument, where they overgraze the fragile ecosystem, visitors are occasionally startled by the appearance of these impressive animals along the roads to the monument.

◇ **ON THE SPOT: IN THE NEGEV DESERT**

Ellen Wohl

The Sonoran Desert is lush with dozens of varieties of cacti and succulents. Elemental blocks of color—vertical cliffs of red sandstone
beneath azure sky—characterize the high desert of the Utah canyon-
lands. The crystalline rocks of the Sinai Desert form orange-colored
mountains as rugged as a clenched fist. By comparison, the Negev
Desert appears subdued. Weathering and erosion dismantle sedimentary rocks that come in hues of cream, tan, and brown, leaving
great piles of broken blocks everywhere and a landscape that seems
to be a massive crumbling. The desert is nearly completely devoid of
vegetation, leaving the skeletal structure of the ground clearly visible; faults disconnect the clean lines of the rocks, and folds contort
them.

It seems unlikely that living organisms could survive in this landscape of rock where only a few inches of rain fall each year. Yet
life abounds. Slender, delicate gazelles and ibex move easily across
steep slopes in search of sparsely growing grasses and the leaves
of scattered shrubs. Hyraxes—rodents the size of corgis—live in
colonies sheltered among the big boulders lining dry streambeds.
Migratory birds sweep through on their way to Europe or Africa.
Springs burst forth from the apparently unyielding rock, creating
oases such as En-Gedi, where pink flowers of Persian cyclamen
bloom among sheets of tufa covered with emerald-green moss and
pale green maidenhair fern. I taste the fresh, cool, clear water, and I
believe in miracles in the desert.

Perhaps most impressive are the endolithic algae that grow just
beneath the surface of cobbles lying on the hillslopes; one quartz
grain deep and the algae can still photosynthesize while receiving
protection from desiccating desert winds. Snails less than half an
inch long gouge tiny grooves into the rock to get at the algae, but
the snails are not very efficient processors. They digest only about
5 percent of what they eat, excreting the remainder in small coils of
powdered rock and undigested lichen. Approximately twenty snails
graze each square yard of rock. Not so many, given the snail's dimin-
utive stature. Yet together these snails graze up to 7 percent of the
total rock surface area in the Negev each year, turning about eight
hundred pounds of rock into sediment and adding three pounds of

nitrogen to each acre of the desert. On such invisible processes all life rests.

As in most deserts, the silence and immobility of the Negev dominate first impressions. Closer inspection reveals evidence of change. Boulders lie strewn along dry streams. Pockets of silt carried suspended in the waters of past floods are tucked up into little alcoves high along the canyon walls. All of these signs make more sense when we watch a heavy winter rain rush over the stepped cliffs in ephemeral waterfalls and tumble rocks down the hillslopes in debris flows. Cobbles rolling hidden in the brown water rumble with a sound like bowling balls. The dry channels drain eastward to the Dead Sea, which sits blue as a sapphire in a setting rimmed by white salt and golden rock. The brown plumes of floodwaters surge far out into the sea. On another occasion, a day of blue sky and little wind, we turn to look upstream at a slight rushing sound like a breeze coming down the canyon. A little flood wave, born of rain clouds far to the south in Egypt, rushes down the channel in a hissing yellow froth. At first the water is shallow and flows in orderly streamlines, but it quickly deepens into a roiling mass of thickened brown in which we can hear the cobbles bouncing along.

Human history is layered into the rocks of the Negev. Nabatean steps worn smooth by generations of feet descend steep bedrock canyon walls to shadowed pools where travelers from the capital of Petra stopped to refresh themselves. Oddly geometrical lines of rocks on the desert floor resolve into the foundations of a Roman fort when seen from an adjacent hillside. The spire of rock reputed to be Lot's wife turned to salt rises up from a broad valley near where Sedom once stood. And among hills darkened by a layer of cobbles, an oddly assorted line of partially wrecked cars and tanks goes from nowhere to nowhere, targets for the Israeli fighter jets that erupt over the horizon and leave jagged-edged silence in their wake.

The fighter jets symbolize the unrest that continues to strain this land. We spend our days studying ancient floods amid the spare, uncluttered beauty of the desert. Following a dry stream, we descend through golden walls of coarse sediment, drop suddenly into a sinuously walled gorge cut into maroon sandstone, and then emerge into a wide badlands eroded into soft, cream-white rocks. Other days we climb up through steep, narrow canyons cut into polished white limestone pitted with bathtub-size potholes and deep plunge pools. In morning light the rocks glow with warm hues of gold and orange. Green acacia trees spread horizontal parasol-like layers along the stream channels. At dusk the acacias stand darkly among the pale rocks. The sun sets in a rose wash over a land of black hollows and blue hills, and a full moon glows brilliantly over cliffs of milk and dark chocolate brown. At sunrise the mountains of Moab form black paper cutouts against the orange sky while the water of the Dead Sea reflects a pale gold.

One of the great joys of being in the desert is sleeping without shelter under a sky overflowing with stars, but this we cannot do in the Negev. My Israeli colleagues and I retreat each night to our fortifications. Sometimes it is a desert field school rimmed by double layers of tightly coiled barbed wire and guarded gateposts. Other nights we make the long drive back up into the hills of Judaea, where the white stone walls of Jerusalem sit like a fortress in the sky.

With deeper knowledge, there is nothing subdued about this desert. ◇

THE HUMAN DESERT

One species has managed to thrive in nearly all of the world's deserts: us. Like plants and animals, we *Homo sapiens* need to find and store water (and food), accommodate ourselves to changing conditions, and shelter ourselves from heat, cold, and drying winds. Because part of our evolution occurred in semiarid savanna country, we have some adaptations suited to desert living, particularly our bipedal locomotion, which gets us above the hot ground and facilitates nomadism, and our lack of fur, which allows for evaporative cooling. However, we lack many of the adaptations of our fellow animals: our kidneys are inefficient, our eyes and nostrils fill with blowing sand, though we may siesta we can't estivate, we need to drink water frequently, and no other animal perspires as profusely as we do. However, while our large brains suffer heatstroke easily, they also enable us to overcome our many natural deficiencies with a variety of clever cultural adaptations.

When reliable water can be found, humans have thrived in nearly all deserts, sometimes at very low densities and sometimes in quite large numbers. Excluding the Nile Delta region around Cairo (with more than ten million people) and the heavily populated Mediterranean coast (Algiers alone has more than three million residents), the Sahara has roughly two and a half million inhabitants—a sizable number, but still less than one person per square mile, one of the lowest population densities on Earth for the simple reason that large areas lack surface water and so have no permanent settlements. Likewise, in Australia's western deserts, human population density over vast areas can be close to zero, especially as formerly nomadic Aboriginal people have clustered into permanent settlements along the desert margins. In fact, since most people living in deserts today are concentrated in growing urban areas (around sometimes declining water supplies), the vast majority of desert dwellers are now urbanites. In the United States, roughly twenty million people live in these demanding landscapes, mostly in cities.

Humans in deserts have developed a variety of lifestyles. As a general rule, where water is in short or irregular supply, nomadism is the traditional norm. Where water is more abundant and reliable, more sedentary lifestyles involving agriculture and mercantilism have evolved into modern urban life.

Nomads can subsist as hunter-gatherers, herders, traders, raiders, or some combination of these. Although we sometimes use the word *nomad* to mean one who wanders at random, traditional nomads follow regular, usually annual migration patterns, seeking to make the most effective use of water and food resources as those become abundant in various locations thanks to seasonal changes. For hunter-gatherers these resources would be mainly water, game, and edible plants. For pastoral nomads, the migration involves water and forage for their animals, typically goats, sheep, camels, or cattle. Pastoral nomadism has been a common lifestyle in the Sahara for more than seven millennia, through wetter and drier periods, with populations moving north and south along with the monsoons and the desert-Sahel boundary. For raiders, these seasonal resources might be the ripening of crops in the fields of agricultural societies or the wealth of other nomads carrying merchandise along trade routes. For many centuries, traders crossed the central Asian deserts along the Silk Road and the Arabian Peninsula along the Frankincense Road. Nomads such as the Tuareg, Mande, and Fulani ran several extensive and lucrative trans-Sahara caravan trade routes for the empires of Ghana, Mali, and Songhay, transferring gold, ivory, slaves, and salt between the Senegal and Niger rivers and the North African coast, with caravans of up to ten thousand heavily laden camels.

Nomadism has been in decline for centuries, if for no other reason than that it conflicts with modern culture. Nomads typically cross international borders and don't always ask permission or acquire visas. National governments don't like this, and they also typically dislike wanderers of any sort, who are hard to govern, are often independent sorts, and may raid settlements and travelers. For these reasons, modern governments in much of the world, including China, North Africa, the American West, and the Australian Outback, have suppressed desert nomadic cultures. In many locations nomads have been displaced by sedentary pastoralists such as ranchers, a shift enabled by such water schemes as well drilling and dam building, which bring water to the pastoralist—in contrast to the way nomadic pastoralists take themselves to the water.

Humans have also invented smaller-scale cultural adaptations to the challenges of deserts, including two contrasting styles of clothing. In some desert cultures, such as that of the Ju/'hoansi (Bushmen) of the Kalahari and Namib and the numerous Aboriginal cultures of the Australian deserts, people wear little or no clothing. However, an unclothed person can absorb twice as much heat as a person dressed in lightweight clothing, so in most desert cultures people take the opposite route and wear many layers of long, loose clothing. Although it may seem counterintuitive, a person thus garbed will stay cooler, and retain more water, than will a person wearing no clothes. These clothes greatly reduce the loss of water to sweating, and perspiration soaked into the inner layers provides evaporative cooling as air flows between the loose layers. Head coverings also protect against the sun, and many desert dwellers wear long scarves and fabric headdresses that can be wrapped around their faces to keep out sun and blown sand. Sahara-dwelling Arabs and bedouins like the Tuareg dress this way, and the Tuareg are notable for their distinctive indigo head scarves, which flow behind them as they ride their camels across the Sahara.

For nomadic desert cultures, housing is usually a temporary affair composed of sticks and hides; nomadic cultures with pack animals can build sturdier but still transportable structures, like yurts and large tents. Among sedentary desert cultures, the most widespread buildings are composed of mud bricks widely referred to as adobe, an Egyptian word that traveled through Arabic culture into Spanish and finally English. Adobe is made with a combination of sand, clay, water, and sticks or straw to aid binding; the wet clay is poured into molds and then sun-dried to create bricks. A structure can often be built out of the very ground it is being built upon, greatly reducing transportation needs; and depending on the design, little timber needs to be used, an additional asset in environments that typically lack large trees.

Traditional adobe structures have thick walls with small doors and windows, keeping interiors cool in the day and retaining absorbed heat into the night. This can provide heating during cold months, but during the hot season many adobe dwellers sleep outdoors, often on the flat roofs characteristic of these dwellings. Although the phrase "mud brick" conjures images of shanties and hovels, in fact adobe buildings can be quite elaborate. And some truly magnificent buildings, notable in the history of world architecture, have been constructed this way. In arid West Africa, where adobe is called *butabu,* such architecture includes the UNESCO World Heritage Site mosques of Sankore, Djinguereber, and Sidi Yahya, which together comprise the University of Timbuktu. Other notable *butabu* structures in West Africa include the mosque in Komio, Mali, and the decoratively painted village of Sirigu on the savanna of northern Ghana. Iran's Arg-é Bam, or Citadel of Bam, a fortress on the Silk Road built before 500 B.C.E., was the largest adobe building in the world until it was almost completely destroyed by an earthquake in 2003; it too has been designated a UNESCO World Heritage Site and is being restored. Because of a growing appreciation for adobe architecture's vernacular aesthetics and its high degree of energy efficiency, this kind of construction is undergoing a revival.

In the Americas, New Mexico is home to some remarkable adobe structures, including the famous church of San Francisco de Asís at Ranchos de Taos. The Pueblo Indians of the Rio Grande constructed their multistory apartment dwellings out of adobe, a process they had independently developed long before the arrival of the Spanish. Taos Pueblo, also a UNESCO World Heritage Site, is a notable example of this architectural style. Adobe structures must be regularly maintained by being replastered with mud; when abandoned, they quickly return to the soil from which they were made. This natural process fits well into the cosmologies of Puebloan people. As Santa Clara writer Rina Swentzell explains, "The belief that the *Po-wa-ha* [water-wind-breath] flows through inanimate as well as animate beings allows buildings, ruins, and places to have life spans and to come and go, as do other forms of life. . . . Traditional Santa Clara Pueblo with its soluble mud structures is an organic unit expanding, contracting, and changing with other life-forms and forces."

One modern invention has been especially influential in promoting increased human residence in deserts: the air conditioner. The invention of evaporative coolers and air-conditioning has certainly made life in hot deserts more comfortable for many people. These tools of human ingenuity have greatly increased the population in some deserts, especially in the United States. Without the cooling breezes generated by air-conditioning, the population of the Phoenix metro area would likely be only a fraction of its current total of more than four million. The same can be said about air-conditioning in cars, an innovation that also allows visitors to enter and cross deserts in comfort, safe from the hazards that threatened and sometimes killed earlier travelers.

Given the difficulties desert environments create for human bodies, joined with their visual spareness, it is not surprising that our attitudes have typically been similarly extreme, often polarized between loathing and love. Cultural historians have suggested that the development of urban life has led city dwellers, disconnected from the rich biodiversity of arid lands, to see the desert as a vast, empty, and threatening void; and that on the other hand, nomads and other indigenous desert dwellers typically have great affection for their desert homeland. We might also think about the range of attitudes in terms of the common perception that deserts are deserted.

Often we have regarded deserts as empty, hellish (as suggested by the many "hells" and "devils" found in desert place-names), and therefore expendable wastelands. Perhaps no human activity more typifies this attitude and its practical results than the use of deserts as sites for the detonation of nuclear weapons. Of the more than two thousand nuclear test explosions that have ever occurred, most have taken place over or beneath arid landscapes. (Most others have been conducted at sea or in the Arctic.) The very first nuclear explosion was detonated in the Chihuahuan Desert, and the eventual permanent site for American nuclear weapons testing was on the boundary between the Mojave and Great Basin deserts. The Soviets used the steppe deserts of Kazakhstan's Semipalatinsk (Semey), while the Chinese used the Lop Nur site in the Taklamakan Desert. India has detonated its nuclear bombs beneath the Thar Desert, and Pakistan has done its testing in holes drilled into the Chagai Hills in a desert region of Baluchistan. Countries that lacked their own deserts sometimes used the ones in their colonies: the French detonated seventeen in Algeria before departing under duress in 1962, and the British detonated nine in Australia's Great Victoria desert. In a quintessentially modern irony, because such places are off-limits to other human activity, and because they are of necessity quite large, some have served, despite lingering radioactivity, as de facto wildlife refuges, and they retain thriving levels of biodiversity. The persistence of Bactrian camels at Lop Nur, or the resprouting of creosote bush shrubs at a nuclear ground zero in Nevada, should give us some hope.

Sometimes, more benignly, we have seen deserts as empty and therefore needing to be improved, redeemed, changed into something else. Hence our many dams and irrigation projects—and the familiarity of the phrase "to make the desert bloom."

But we have also often seen their emptiness as a positive characteristic. An empty space can seem to offer us an uncluttered, direct route to some kind of fundamental truth: to the true nature of the world, to true wildness or wilderness, to our true selves, to God. According to a traditional proverb of the nomadic Berbers, "The desert is the Garden of Allah, from which the Lord of the faithful removed all superfluous human and animal life, so that there might be one place where He can walk in peace." Many religious traditions share this image of the desert as a place devoid of distractions and superficialities, and therefore as a place where one might more readily commune with the divine. Moses, as we have noted, sees the Lord in a burning desert bush and receives the Ten Commandments atop a desert mountain. And the Israelites he led wandered for forty years in the Sinai desert before reaching their promised land, as though undergoing a cleansing trial. According to tradition, Muhammad was orphaned as a youth and raised among nomadic bedouins, serving as both a herder and a trader. As a middle-aged man, he grew discontented with the decadent life in Mecca, so he retreated into a cave on a desert mountain for meditation and reflection. As the narrative states, it was here that the Angel Gabriel appeared to him in 610 and began to dictate the words that would become the Qur'an.

John the Baptist, a prophet in both Christianity and Islam, is reported to have preached in the desert wilderness near the Dead Sea, where he famously subsisted on a diet of locusts and wild honey. Many contemporary Christians cringe at the idea of a revered prophet consuming insects and so have interpreted "locust" as referring to the bean pods of the locust tree rather than to the insect. But in fact many desert cultures consume locusts and other insects, and most scholars accept that John the Baptist did so as well. Indeed, in both Jewish and Islamic dietary law, locusts are the only insects considered kosher or halal.

John the Baptist's example generated a tradition of desert hermit monks who retreated to the edges or interiors of deserts across a wide swath of Egypt, Syria, Asia Minor, and Judea. (They came from several religious traditions, including Christian, Gnostic, Jewish, Neoplatonic, and Egyptian.) St. Anthony of Egypt (251–356) became the most influential of these "desert fathers," retreating gradually from the edge of a Nile River village first to a place near Wadi el-Natrun (west of Alexandria), then later to a spring-fed oasis (complete with date palms and wild animals who spoke to him) southeast of Cairo, not far from the Gulf of Suez. (Both of these sites are still home to living monasteries, as are other very early desert monasteries around the larger region.) The notion of emptiness was important to many of these desert monks: the process of becoming emptied out, a sense of divinity as spacious and beyond language, the power of silence and solitude to transform the soul. Some of the same characteristics of deserts led to the notion that these landscapes offer refuge from the complexities of ordinary life for the purpose of relaxation or recreation—which we might in this context read not just as the chance to go mountain-biking or rafting but also, more seriously, as spaces for re-creation.

Curiously, the three major monotheistic religions, Judaism, Islam, and Christianity, are all desert-born faiths that evolved from a small group of desert pastoralists under the leadership of a herder named Abraham (Ibrāhīm in Arabic), and as we have seen, all three faiths are permeated with desert references. Scholars speculate about the role of the desert environment on the origins of monotheism. Some suggest that the vast openness of the landscape served to produce a unifying vision of theology. Others speculate that the unobstructed view of clear night sky vivid with stars called forth the idea of a single deity. Still other scholars look to the sociology of desert pastoral nomads, whose tendency toward powerful patriarchal clan elders may have suggested a similar divine pattern.

On the other hand, although monotheism began in the desert, the fact is that most desert-dwelling cultures have been and remain polytheistic; even those that have been ostensibly converted to Islam or Christianity retain polytheistic elements. The many Dreamings of Aboriginal Australians illustrate the richness such landscapes can have to those who live in them for long periods and come to know them intimately, whose cultures emphasize the ways humans are intertwined with the desert's rocks, waters, plants, and other animals. Dreamings, which tell of the journeys of ancestral spirits across the landscape, constitute a narrative form of sacred geography that aligns closely with the physical geography; they explain how the physical geography came into being, and they tie human stories to the stories of other living things. Because it is forbidden to harm the plants and animals that figure in these stories, sacred Dreaming sites function as wildlife sanctuaries at the same time as they preserve information valuable to human survival, such as the location of water.

And as this chapter should have made clear, we whose acquaintance with deserts is younger or shallower need not see deserts as empty either.

We can see them instead as places that are full in their own way. These themes arise over and over in literature written about these landscapes since the late nineteenth century, but for just one example, we will turn again to American writer Edward Abbey, who both captured and helped to establish and spread many of our contemporary attitudes in his classic book, *Desert Solitaire* (1968). For Abbey, deserts are places that offer "the shock of the real" where one can "see, as the child sees, a world of marvels." There we might hope to confront "the bare bones of existence, the elemental and fundamental, the bedrock which sustains us." Yet these are also landscapes full of "much to see and marvel at, the world very much alive in the bright light and wind, exultant." In his desert wilderness, Abbey—like many a visitor after him—feels himself to be "at the center of things, where all that is most significant takes place." As he and many others have recognized, the dry places we call deserts possess, and can teach us to see, their own kinds of beauty, richness, and wonder, and they offer lessons for us all of adaptation, flexibility, toughness, and resilience.

The Complexities of the Real

Mostly, so far, we've been looking at some of the largest forces that shape the earth. Plate tectonics, climate, water, sunshine—powers like these are the Himalaya among mountains, the blue whales and red-woods among living things, the bulldozers among tools. We've also been looking at change, fluidity, adaptation, curiosity, imagination, wonder: massive forces all, though of a different sort. Along the way, you will have noticed, we've frequently had to simplify and generalize, to qualify our statements with such words as *typically, sometimes,* or *roughly*—frequently, though not nearly as much so as we could have done. For if there is one quality that permeates our world, it is surely complexity. Everywhere we might look about us, there will be more going on than first meets our eye, more even than we might imagine.

For this final chapter, let's take our imaginations on a journey from the center of the planet out toward space, moving quickly through places and topics we've already explored, slowing down for some we haven't, particularly grasslands and forests, and glancing as we go at some of the kinds of complexities that saturate everything we'll pass through.

UNDERFOOT

The bulk of the earth we can cover in no time, having already lingered there: core, mantle, magma, imaginary interior caverns, floating crust, all moving all the time, consuming the old and creating the new. As we rise toward the surface, we start to encounter other intersections between inside and out. About seven miles beneath northwestern Russia, we might meet the Kola Superdeep Borehole, the deepest hole yet drilled for geological exploration. Liquid water occurs about six miles deep there, probably not from old rainfall but freshly made from the hydrogen and oxygen in magma. About two and a half miles under South Africa, we could find ourselves in the deepest mines, dug in this case for gold, and just a little higher, about two miles short of the surface, we would be at the level of the deepest life forms yet found: bacteria and other minuscule creatures that can make their frugal living in amazingly hot, cramped, airless conditions. Some of them live in water that has permeated far beneath the crust; others live squeezed into dry pores among grains of rock. Elsewhere we'll encounter mines for coal and lead, silver and uranium; wells for oil and gas; the long, thin holes where sediment cores have been removed, marking sites of scientific curiosity; tunnels for trains, cars, water, power, communication; and, finally, skyscraper foundations and basements in ordinary houses.

We might find caves, too, especially if we're rising through limestone, with its sometimes-vast networks of absolutely lightless rooms and tunnels. Creatures live there who have lost the use of their eyes. Cavers visit in search of adventure, rock art, fabulous stalactites or stalagmites, drawn by mystery, beauty, and scientific curiosity. If we were to come up beneath Mexico's Chihuahuan Desert, we might be lucky enough to find ourselves in the Cave of Crystals. There, a little less than a thousand feet beneath the surface, in a mine, we could marvel at a maze of white selenite crystals, each one giant enough to dwarf a team of human explorers.

Some of what we pass through will be wet, such as the large aquifers below the western Sahara, Australia, the North American High Plains, South America—even, perhaps, Darfur, where remote sensing has detected a large ancient lakebed beneath which lifesaving water may still lie. Or we might encounter underground rivers of various kinds, including those that run through conduits we've built for them, shadow rivers spreading beneath and beside visible rivers, rivers vanishing and reappearing in the karst landscapes of the Yucatán (where they fill the cenotes that helped the Mayans survive the dry seasons), Ireland's Burren, the sinks and springs of Florida.

We'll pass through recognizable signs of the geological past: not just volcanism (or metamorphic rocks, transformed by heat and pressure) but also layers of sediment, mountains and hills taken apart and brought low by cold and heat, water and wind, mixed with the remains of plants and animals, neatly layered—at least relatively neatly and at least for a while—waiting to be exposed to our vision in sediment cores, on the walls of mines, road cuts, and canyons, and sometimes right out in the open air. Rising through a deep cut in Earth's surface like Arizona's Grand Canyon, we could begin surrounded by black rock some two billion years old, climb through rock layers of green, purple, red, tan, cream, and gray, and finish with our feet on fresh soil and fallen ponderosa pine needles. Such a climb will also take us past fossil trilobites, brachiopods, armored fish, sea lilies, reptile tracks, ferns, conifers, fish teeth. Most of the world's fossils will always remain buried, and so we must call on our powers of observation, reason, and imagination to read the stories hinted at by the few we see.

In some places, we'll rise straight from rock into air: through California's Half Dome and Australia's Uluru, Utah's slickrock and Dartmoor's tors, Wyoming's Devil's Tower and the stony tops of high mountains. It was such a spot that led Thoreau to one of his major insights about reality, in an experience many other climbers have had of what we might call, paradoxically, transcendent immediacy. Nearing the top of Maine's Mount Katahdin, Thoreau was stunned by the abrupt meeting of raw granite with atmospheric drama. "What is it to be admitted to a museum, to see a myriad of particular things, compared with being shown some star's surface, some hard matter in its home!" he wrote. "Talk of mysteries!—Think of our life in nature,—daily to be shown matter, to come in contact with it,—rocks, trees, wind on our cheeks! the *solid* earth! the *actual* world! the *common sense! Contact! Contact! Who* are we? *where* are we?"

More often, the transition from stone to sky is much less abrupt. Sometimes our final layer of rock will be loose glacial debris or sand, and it may be hard to tell where soft rock ends and soil begins, as where volcanic ash or windblown silt lie hundreds of feet deep. We may be able to see clear lines dividing the final layers below the surface, or, as soil scientists say, horizons: solid then broken rock, older then newer subsoils, what is currently topsoil. Or we may instead encounter dirt that has been thoroughly mixed by such tunneling agents as farmers, roots, termites, and prairie dogs—and by earthquakes, floods, landslides, windstorms, and other major disturbances. Where more dirt has been blown or washed out than in, where erosion has outpaced the creation of new soils from the magical interactions of animal, vegetable, and mineral, we'll reach the surface more quickly. Where the opposite has occurred, we may pass through buried human settlements and artifacts thousands of years old.

Earth's soils are far from monotonous, and their variety has earned them several major categorizing systems. Nor are they static. What is where depends on a host of factors, including, among other things, the "parent rock" (lava, granite, sandstone, and so on) and its chemistry, the soil's age, the effects it has felt from past and present climate and weather, and its experiences with plants, animals, and humans. According to writer William Bryant Logan, there are fourteen thousand soils with their own proper names. And, as he says in his lovely book *Dirt: The Ecstatic Skin of the Earth,* while the soils of Eden might be forever fertile, "East of Eden, a soil would always start and end infertile, but in the maybe 100,000 years in between, it might go through many permutations of fertility, depending on the combinations of climate, mineral matters, organic matter, and slope." Perhaps most surprisingly, soils are also full of life: a single handful of topsoil may hold billions— that really is *billions*—of living individuals: microbes (including those that have given us penicillin, streptomycin, tetracycline, and neomycin), invertebrates, and vertebrates.

It is in this active, permeable transition zone where we first begin to encounter such direct products of sunlight as roots. Here, where the lithosphere merges into the biosphere, the globe of rock into the globe of life, our journey will be shaped most strongly by the kind of ecosystem we find about us. We've looked carefully at wetlands and deserts, so here we'll slow down for two other kinds of places, grasslands and forests.

◇ ON THE SPOT: IN ANTARCTICA'S DRY VALLEYS

Diana Wall

Looking down from the helicopter window, I follow our tiny shadow on the glacier ice. We spiral down to land in an immense valley of brown soil, surrounded by mountains with huge glaciers that have fed the ice-covered lakes on the valley floor. After nine months of preparation and nearly four days of international travel, I'm flying from the U.S. science town of McMurdo Station to my study plots in Antarctica's Dry Valleys. The copter leaves, and I stand for a minute taking in the overwhelming silence and stillness.

My companions and I are the only humans in a valley few people will ever see. We hear no wind, no tiny whisper, no buzz, no trickle of water; we see no waving green leaves, no flying bugs or birds. Nothing much has changed here since Robert F. Scott found this valley about a century ago. High on the mountain walls around me is a bathtub ring left ten or twenty thousand years ago by ancient Lake Washburn. At my feet, dark pebbles polished by wind-driven sand cover the ground, their smoothness a contrast to the sharp peaks and ragged glacier surfaces. The world is the colors of rock and ice—brown, beige, gray, black, white, blue, and green.

Suddenly the wind rises, a glacier cracks like a rifle shot, and it turns colder. Time to move quickly to stay warm and start work. Weather changes here are abrupt and dangerous. Thus, our first task is to assess what we're wearing. Aside from survival bags, we each carry personal gear in case we get stranded in the field by bad weather. Today we think it will be "sunny and warm" for a few hours, so we exchange the very warm boots we wear while flying for the field boots that will let us move more quickly. Then we organize backpacks, large coolers that hold our sterile sampling bags, scoops, and smaller backpacks. We gather a snack (we eat while we work for energy), bottles of water, and empty pee bottles.

We head down the path we staked out many years before to our experimental plots on the edge of frozen Lake Hoare. As we walk, we see the footsteps we left behind last season or earlier.

This is the coldest, windiest, and driest place on the globe, and until the 1990s, NASA scientists and others thought of these valleys as analogues to Mars. Certainly it looks like another planet here: lots of barren soil, no plants, lichens, mosses, or cryptogamic crusts, none of the plants I've seen in hot deserts. Just lots of rocks and soil. But over the past twenty years, we have learned that this extreme desert is complex and perhaps more exciting because many kinds of life are hidden to the eye.

Ephemeral melt-streams, for instance, hold algae, mosses, and bacteria, plus invertebrates like nematodes, rotifers, and tardigrades (also called water bears). Where a liquid moat surrounds a permanently frozen lake, an amazing food web of viruses, bacteria, Archaea, huge cyanobacterial mats, and small invertebrates can live because the ice lenses warm up the water.

I'm here with my team to examine soil life. Our research continues to identify new species that contribute to soil food webs. The invertebrates, nematodes, live in tiny water films around soil particles; the less abundant microarthropods are more patchily distributed and occur beneath pebbles. The soil also holds other life—soil yeasts, Archaea, bacteria, and protozoa. Within porous sandstone rocks at higher elevations, cryptoendolithic microorganisms form a thin green line. These tiny creatures have developed multiple capacities to endure life without water, eight months of darkness, and cruelly cold temperatures. Nematodes, rotifers, and tardigrades stop metabolizing when they get too dry, are blown to new spots by the wind, and then revive when conditions improve. Desiccated algae are blown from aquatic habitats to soils where they serve as a carbon source for the soil food web.

Today we can predict the distributions of each invertebrate species in the many soil habitats in the valleys, data that will be important as climate change alters habitat characteristics and the range of species. We know that the nematode species here are sensitive to temperature change and that they contribute significantly to the turnover of organic matter in the soil. And we are beginning to apply climate-change scenarios to this complex ecosystem.

Less than ten minutes of walking brings us to our experimental area, where we have a spectacular view of a large frozen lake and the Canada Glacier rippling down the mountainside. Our experiment, which looks like a small graveyard, is about a hundred yards from the lake edge in a relatively flat area. A white PVC pipe stands in each corner, and rows of large nails distinguish the corners of the plots. Some plots hold translucent cones that act as greenhouses to warm the ground. Climate-change models are predicting warmer temperatures in this region of Antarctica, and we want to know how the animals that live in the soil will be affected when the temperature and moisture of that soil increase.

We have planned how we will work before we come to the field, and so we split up right away to work in pairs. One person, the director and map reader, stands on the outside of the plot holding a field book and directing the second person, the sampler, to the correct plot number. The samplers become awkward dancers in their heavy clothes and boots, stepping from rock to rock, bending to the soil, trying to be light-footed so they don't trample the experimental soils. Sampling soil itself is fairly easy: the dancer simply takes a fist-size scoop of undisturbed soil and puts it in a labeled plastic bag, closes the bag, mixes the soil gently, and returns it to the director, who carries it to a cooler and then directs the sampler to the next plot. We repeat this practice for each plot year after year, taking the samples back to the laboratory at McMurdo Station to ana-lyze and compare what happens to soil life in the warmer chamber-covered plots.

All around me, voices call out numbers—"next plot is TW4," or "sample ready!" We work steadily, sometimes adding or removing a jacket. Occasionally, we stop to use our pee bottles: what we bring here, we carry out. We don't forget to drink water, a necessity in one of the driest deserts on Earth.

For every hour of sampling work, we have six hours before us in the laboratory at McMurdo, where we'll wash the species from the soil and examine each individual under a microscope to reveal the complex biodiversity in the soils from these frozen ecosystems. The answers to compelling questions still await us: what is there; how many are males, females, or juveniles; have there been changes since last year; are the warmed soils different from the untreated soils?

A few hours later, having hauled the sample-heavy coolers to the landing area, we hear the distant chop of the helicopter. We double-check that our gear won't blow away in the rotor draft, load it up, and lift off, leaving the valley and its hidden life to an immense silence. ◇

OCEANS OF GRASS

If we're emerging from underground onto land, not water, there's a good chance that like a family of prairie dogs popping up from their subterranean towns, we'll find ourselves in some kind of grassland. Landscapes that can be collected under this name cover roughly a third of the land's surface, with the precise percentage depending on how we set our boundaries. In every case, we'll find the "true" grasses (members of the family Poaceae), along with other grasslike plants such as reeds (Juncaceae) and sedges (Cyperaceae). We'll generally find a good mix-ture of what biologists call forbs or herbaceous plants (and the rest of us often call wildflowers), sometimes in very high concentrations, such as in South Africa's fynbos, the alpine and subalpine meadows of the Rocky Mountains, or the flowery meadows where cows, sheep, and goats have long grazed in England, Iceland, or the Alps. And we'll find some shrubs and trees, widely or thickly scattered, clustered along waterways or in scattered groves—but not too many trees, as their absence or scarcity is one defining characteristic of a grassland.

The spectrum of "grassland" possibilities stretches from deserts to marshes, from the uplands of New Mexico and Arizona and the spinifex grasslands of Australia to the "river of grass" of the Florida Everglades. It may include both Arctic and alpine tundra and the tundralike *páramo* of the Andes, places where tussock grasses and cushion plants grow in soils that are cold, perhaps permanently frozen, thin and rocky, or peaty. It might include a grassy heath or moor, where soils are acidic, or the sagebrush stretches of the American West, where they are alkaline. The chalk downs of southern England, the hilly Palouse of eastern Washington state, the seaside machairs of some Scottish islands, and the tiny grassy balds of the Great Smoky Mountains are grasslands, each with its distinctive natural and human history. So, of course, are the vast stretches of the South American pampas and llanos, the Asian steppe and Tibetan Plateau, the acacia-dotted savanna and veldt of eastern and southern Africa, and the great prairies and plains of North America— the vast grassy landscapes we'll mostly be considering here.

As always, our words for these places are too simple, yet they suggest their own stories of cultural history and attitudes. In his 1832 poem "The Prairies," William Cullen Bryant rhapsodizes over the tallgrass prairies of Illinois but notes that for them "the speech of England has no name." The French *prairie* originally referred to something like a "meadow," implying a limited scale and a decidedly human value for the production of hay or the grazing of stock. (Today, in England, *prairie* often means "monocrop.") *Plain,* like *desert,* notes what is not there rather than what is—a plain is flat, not mountainous, hilly, or obscured by the thick green tangles of forests. (Indeed, during the nineteenth century, the Great Plains were sometimes called the Great American Desert.) The Dutch colonists of South Africa called their grasslands *veldt,* meaning "field" but often connoting something more like "outback" or "bush." Some names are indigenous rather than colonial: *steppe* comes from a Russian term for a flat, dry landscape, while *pampa* derives from Quechua. Environmental historian Dan Flores adopted an elegant translation from Navajo to name the southern plains of North America Horizontal Yellow.

These places suggest, overwhelmingly, simplicity and vastness: the clean lines of grass and sky; often relatively flat or gently rolling ground that allows long views in all directions; the greens, golds, and browns of the grasses, which trace with their waving arcs and shifting tints the slightest movements of the air. The sky and the light change all day long. At sunrise and sunset, or when a storm is building, the vast dome of space may fill with intricate and dramatic color and texture, towering explosions of clouds, jags and sheets of lightning. In the darkness of night, lightning bugs may flicker a mirror of the starry sky, and the stars themselves may seem strikingly bright and close. In her classic and largely autobiographical children's story, *Little House on the Prairie*, Laura Ingalls Wilder describes her young heroine falling asleep in the open air, imagining that her father could pluck one of the stars "from the thread on which it hung from the sky, and give it to her." On another night, listening to her father's fiddle music, she thinks the stars are singing with him, those "great, bright stars swinging so low above the prairie." Like most who write about these landscapes, Wilder returns again and again to the prairie's "grasses blowing in waves of light and shadow," the "great blue sky above it," its clean, vast emptiness.

As Wilder's word *waves* suggests, the most frequent motif in writing about these places, at least in North America, is probably the comparison of prairie to ocean. Here is Bryant again:

> As if the ocean, in his gentlest swell,
> Stood still, with all his rounded billows fixed,
> And motionless forever.—Motionless?—
> No—they are all unchained again.

Similarly, in *The Oregon Trail* (1848), Francis Parkman writes, "we saw the green, ocean-like expanse of prairie, stretching swell beyond swell to the horizon." Parkman's narrative influenced Herman Melville, who in *Moby-Dick* wrote of the whaling grounds as "these sea-pastures, wide-rolling watery prairies." In *My Ántonia* (1918), Willa Cather's character Jim Burden remarks, "The grass was the country, as water is the sea." Those who are new to the light, space, and openness of grasslands may experience feelings of being overwhelmed, exposed, and disoriented—as we sometimes say, they may feel "at sea." But at least as often, perhaps more so, such descriptions attend positive responses to the particular kinds of beauty in these landscapes—and sometimes a recognition of their ecological richness.

The surface of the ocean hides far more than it reveals, and the same is true of grasslands. The apparent simplicity of a savanna or prairie is entirely an illusion, but one that takes real effort to dispel. Much of what happens in these places does so underground or at scales of time and space that we cannot perceive without concentrating our attention on details, sometimes with the help of scientists' technical knowledge and equipment. And they are intricately complex: the specific character of each patch of grassland, at each moment, results from one of many possible combinations of dozens of variables—large and small, fast-changing and slow, wholly natural and partly or largely cultural.

THE SHAPES OF COMPLEXITY

The largest variables, the forces that shape and alter grasslands, are those we've already considered: tectonic forces and topography, the sun's energy and our current and changing climate. The largest of these ecosystems occur in places with enough precipitation for grass but too little for trees, often in the rain shadows of mountain ranges, along the edges of the subtropical desert zone, deep inside continents, and on high plateaus. They typically receive between ten and forty inches of moisture a year. Some are just dry enough that grasses stay comfortably in charge. Some are so dry that a few years of drought will turn them into a sort of desert, at least temporarily; some are moist enough (especially in wet years) that they may be overtaken by upstart forbs, shrubs, and trees—unless, that is, some other factor such as fire returns them to their treeless state.

Biologists distinguish two major types of grassland. The first, tropical grasslands, are often called savannas, a term that came to English through Spanish from Carib. Originally the word indicated a treeless expanse, as in Savannah, Georgia, where salt marshes dominate the views. Now, though, it generally implies a relatively high number of trees, as in East Africa's Masai Mara and Serengeti (from the Masai term for "endless plain"). These grasslands, on the wetter half of the precipitation spectrum, tend to be geologically stable and are relatively old. The second, temperate grasslands, occur in the midlatitudes, are often at higher elevations, have cooler weather, and are drier. In North America, the shortgrass prairie or steppe immediately downwind of the Rockies sees as little as twelve inches of annual precipitation on average, and its grasses reach heights of only about sixteen inches, quite often less. (Grassland scientists now tend to prefer the term *steppe* for these shortgrass ecosystems.) Farther east, tallgrass prairie (what little remains of it) is likely to receive more than twenty inches of moisture a year (sometimes a good bit more than this) and produce grasses over five feet tall.

Smaller related factors also affect the character of grasslands. Some grass species are warm-season, some cold-season. The timing as well as amounts of moisture, heat, and light matter: growth is different depending on whether doses of rain are smaller and more frequent or larger and rarer—a matter researchers are actively investigating, since such details are quite likely to change as the climate does. Basalt, tuff, limestone, glacial till, loess, the silty bottoms of vanished ice-age lakes: these substrates offer different minerals and drainage patterns and thus support different suites of plants and animals. The Serengeti Volcanic Grasslands differ from their surroundings. In England, on the thin limestone soils of the chalk downs, a grassland today because of human logging and farming some four thousand years ago, nutrient values are poor and sheep continue to graze, but a unique and diverse plant community has developed, one that is rich but precariously balanced. In some places, grasslands exist mostly because the soils don't allow the right kinds of drainage and root penetration for trees, such as in the Hayden Valley of Yellowstone National Park, where the fine-grained silt left from the bottom of a vanished ice-age lake supports not trees but grasses and bison.

Subtle aspects of topography are important, too. Slope and aspect—the steepness and direction of hill faces—dramatically affect plant variety. Equator-facing slopes tend to be hotter and drier, while pole-facing slopes are cooler and wetter; slope dictates the angle at which the sun hits, how much plants shadow themselves and each other, and the speed with which rainwater runs off instead of seeping in. In eastern New Mexico and the panhandle of Texas, on the famously flat and featureless Llano Estacado (Spanish for "staked plain," perhaps for the yucca stalks that rise from it), playa lakes need be only inches lower than their surroundings to fill seasonally and create critical habitat for migrating birds and other wildlife. Other grasslands in the American West are known for breaks, gullies, draws, buttes, bluffs, coulees—all manner of wrinkles and ruptures that give form and definition to the grassland. In the Serengeti, rock outcrops called kopjes, Afrikaans for "little hills," are small islands in the sea of grass and island niche ecosystems as well, harboring plants and animals that do not exist on the surrounding plain. Even topographic features the size of termite mounds can create significantly different microclimates and thus microhabitats. All these variations significantly influence grassland biodiversity.

Many smaller variables help shape grasslands, too, in ways too multiple and complex to detail here. Ecologists are likely to be most excited about what goes on underground in these places, by things that surprise because they are buried, or too small for our eyes, or unexpectedly intricate.

They are interested, for instance, in the myriad of ways that specifics about soils, root systems, aboveground plant growth, animal life, cli-

mate, and geology are linked in complicated chicken-and-egg relationships. Nitrogen, phosphorus, potassium, calcium, magnesium, sulfur,
silicon, salts, bacteria, algae, mycorrhizae, nematodes, protozoa, earthworms, beetles, termites, ants, shrews, mice, prairie dogs, fungi, decaying root systems and other plant tissues, the depth, amounts, and timing of soil moisture—all these ingredients (and others) can affect which
plants grow where, and how well. A deep, vigorous root system makes
room in the soil for more roots to grow and water to penetrate; com-
pacted soils are less hospitable to water, to roots, and therefore to plants.
In the dark, fertile mollisols of relatively young temperate grasslands, the
decay of roots creates chemical conditions that bind nutrients instead of
letting them be leached out by rainwater, and then evaporation draws
those nutrients upward to become available to new plants; by contrast,
the iron-red oxisols of older tropical grasslands have had their nutrients
leached out for millions of years.

Such underground matters are important because some 60 percent
to 90 percent of the biomass in grasslands is in root systems. Generally
speaking, drier grasslands have higher percentages of roots to shoots.
But even in the relatively moist tallgrass patches that remain in North
America, one square yard of bluestem prairie may contain some twenty-
five miles of roots. Frequent fire can increase this root mass; grazing
may increase or decrease it; deeper soil moisture increases it but more
moisture in general decreases it. Rain that falls in fewer, heavier episodes appears to be good for shortgrass prairie and bad for tallgrass
prairie. This sort of information emerges after painstaking study that
may include high-tech laser optics, radioactive tracers, underground tunnels of glass that hold cameras, hand-controlled microclimates, and the
slow, careful, mind-numbing separation—by hand—of soil and root in
a clump of shoveled earth.

Variations in how individual plant species operate and compete with
one another also contribute to grassland complexity. What is in bloom
in a single patch of prairie may change radically from one week to the
next, as the niche structure of these ecosystems is connected to timing.
Sod-forming grasses and bunchgrasses operate differently, as do annuals
and perennials. Many grasses and forbs need to be primed by low temperatures before they flower. Plant parts grow at different rates during
the season, sometimes according to day length, sometimes temperature.
Some plants reproduce by dispersing seeds (wind, animal digestion and
fur, ants and other insects, gravity), some by spreading rhizomes (underground) and stolons (aboveground). Many plants produce chemicals that

do things like protect against being eaten and either stimulate or slow growth in their neighbors.

The difference between C3 and C4 grasses is also important to grasslands. Some plants have developed a complicated but effective method of absorbing the carbon dioxide they need for photosynthesis that allows them to operate in warmer temperatures. Cool-season C3 grasses, which include many human food sources such as wheat, barley, and rice, usually grow and go to seed earlier in the summer; warm-season C4 grasses, which comprise about half of all Poaceae, including corn (maize), grow later, stay active in hotter and drier weather, and can better resist long periods of low moisture. C4 grasses dominate tropical grasslands and allow grasslands where otherwise desert would occur; they are also common in warmer parts of temperate grasslands. In most cases, the two exist in mixed communities, and such mixtures make richer habitats and more nutritional forage. These two kinds of plants react differently to increases in atmospheric carbon dioxide and to heat, though not in straightforward ways. More carbon dioxide will make C3 plants (including unwanted weeds) grow more quickly, but not necessarily C4 plants; and this faster growth occurs only up to a point, since after a while plants reach some other limiting factor like the availability of nitrogen, phosphorus, or water. Scientists are working hard to figure out what changes rising carbon dioxide will bring to grassland composition and fertility, and to rangelands and croplands, the humanized versions of these landscapes.

◇ ON THE SPOT: ON THE CHALK DOWNS

Richard Kerridge

Rounding the shoulder of the down, I see the sea. I am walking a narrow chalk path, white on the green slope. On either side the downs rise and curve away. Ahead of me the sea shimmers.

What are the things I love about the chalk downs of southern England? Their rounded and long oval shapes are gentle, their green surfaces smooth. White cliffs and gashes stand out against the green. Chalk gleams through the thin soil and grass. The flanks and long backs of these ancient hills show all their scars, old and fresh. It's a naked landscape, familiar and open, well known and well trodden, patient.

Complexities of light and shadow pass over the downs. This

landscape seems to acknowledge the procession of mixed emotions. The downs understand disappointment; their weathered, marked, enduring surfaces soothe anger and fear. They open up joy. With the weather, they shift from drab solidity to deep green to radiant pale gold. Always in view above the towns, villages, and roads, they draw your gaze up, but not to the intimidating height of mountains. The landscape does not overwhelm. You can easily walk up here, it tells you. Look, it's not far.

But once you do, you are in a powerfully different place. There is a stillness, accentuated by the murmur from the road—the normal world feels nearby, but definitely somewhere else—and the faint, incessant calls of skylarks, pinging like computer-game gunfire. Here where I am now, in Dorset, I can hear both of these, and the sea's murmur too. The sense of airiness and spaciousness is wonderful; the bareness and smoothness of the hills give you long, unimpeded, rolling views. On windy days there is nothing to hold on to; you could be whirled away. In rain, the porosity of the rock makes the whole hillside weep: runnels everywhere, crumbly chalk turned to a slippery milky gray.

Flints are embedded in the chalk: coated with a dull gray encrustation, but shiny, bluish, and knife-sharp when chipped open. As the soft chalk crumbles and rubs off, many are left loose; they lie on the white paths and are thrown up by ploughing. Often they have strange jointed shapes, like twisted human torsos. Because of its hardness, flint was the rock most favored for axes, knives, and arrowheads. I look out for worked flints, chipped at carefully to form a cutting edge, but find only naturally broken ones that look possible at first.

Hot days in the downs: heat haze crinkling the air. Butterflies dance from plant to plant, tumbling about each other in the air— tiny fawn skippers, earthy meadow browns, marbled whites with charcoal markings on cream, and blues like glittering shards of sea or sky: the common blue, the chalkhill blue, and the rare, iridescent Adonis blue. They flutter away from the walker's feet, across the springy turf, the thyme and marjoram, the yellow vetches, oxeye daisies, orchids, red clovers, delicate blue harebells, lilac scabious flowers and deep purple rampions, the woolly thistles too, and the grassy disused anthills that pimple the turf. On the paths in summer the earth dries to a flat, cracked surface, dusty and littered with leathery dry cow pats, sheep droppings, rabbit droppings. Dung flies, fawn and velvety, gather on the cowpats, rising in clouds as you pass. Grasshoppers sing in the turf. Sheep on the high slopes look like white maggots.

Twilight makes the chalk paths and patches luminous against the shadowy green. I am waiting for that time now, lingering up here on the downs at White Nose cliffs, sitting on a stile, watching the July sunset fade, waiting for the moths and glowworms. It isn't time yet. A concrete fortification from World War II faces the sea. Did they fear that the Germans would land in one of the hidden coves, sheltered by the chalk cliffs, and advance up these soft valleys? Not far off are other fortifications—Iron Age hillforts, whole hills dug into concentric rings. Albion, the name of the mythic Celtic kingdom of Britain, means "white cliffs."

The air is still warm. A thin, high bleat is answered by a hoarse one. The sheep are indistinct shapes now. I stand up. The green is deepening, and soon the downs will have their ghostly, silvery look, especially mysterious on midsummer nights when the air is full of flower scents, herb scents, and drifting moths. Fat black slugs have come out onto the path. A fox on the slope pounces on something; another runs up to it: youngsters. I find them in my binoculars. They are snapping up slugs, I think, shaking their heads and chewing hard, as if on something glutinous. When I lower my binoculars, I can hardly see. For a moment it is quite dark. Then the shapes of stile and bushes, and the outlines of the slopes, emerge again—just. I begin to follow the shimmering white path, down the hill, into the new-cut fields, toward the village and the pub. ⬦

EVOLVING TOGETHER...

Other key factors and variables that shape grasslands lie in the zone where we sometimes try to distinguish sharply between ourselves and the rest of the natural world, especially other animals, and where we always run up against the impossibility of doing so, at least in biological or ecological terms. These factors are grazing and fire, both of which may or may not involve humans. Both are crucial to grasslands because they hold woody plants at bay; without them, upstart shrubs and trees are likely to move in and take over. (A related but wholly human equivalent is mowing, which has some of the same effects.)

The explanation is evolutionary. Grasses evolved along with grazers (grass-ers) and in the presence of frequent low-intensity fires, and several aspects of their structure are adapted to these disturbances: as anyone who has ever had to maintain a lawn by mowing knows, cutting grass doesn't kill it. Nor do grazing or fire. Indeed, both, if they're not too severe, may well stimulate growth and increase diversity. Burned grasses grow back with remarkable speed: a visitor to, say, the northern half of North Dakota's Theodore Roosevelt National Park, just a week after a prescribed burn, will see a robust wash of green overtaking the dark sweep of ash. In African savannas, fire consumes coarser grasses, makes space for more succulent grasses, and fertilizes fast-recovering plants with the ash of dead ones. It keeps woody plants in check, too: while shrubs and trees may also have evolved ways of resisting (thick bark, for instance), fire nevertheless keeps them from growing too closely together by killing seedlings. After fire season in the savanna, the new growth of native bunchgrasses provides an intense period of high nutrition that coincides with birthing periods for antelope species. Fire also helps control invasive species, which are less likely to be adapted to that disturbance than are natives.

Coevolution, naturally, affects both sides of a pair. Indeed, grasses and large mammals in general came to dominance over the same periods, and while not all mammals are associated with grasslands, many . are. This may be because they eat grass itself: think of cows, sheep, goats, and bison, as well as mammoths, mastodons, and many other extinct creatures. It may be because, like us, they eat grass seeds such as corn, wheat, barley, rye, or rice. Or it may be because, like lions, wolves, saber-toothed tigers, and again us, they eat grass-eaters.

Some grazing adaptations are quite specific. In the case of ruminants like bison and cattle, ever-growing teeth and multiple stomachs allow them to chew through the silica and digest the cellulose in grass. (Kangaroos, which are not ruminants, have similar adaptations.) Other grassland animals may not be true grazers but instead may browse on leaves of trees and other plants, including the easy-to-digest tender young grasses: deer and pronghorn, giraffes (who prune acacia trees) and elephants. Grassland mammals are also adapted to cover great distances so they can find enough nutrition in what can be sparse browse and follow seasonal changes in precipitation and growth. Some are speedy, like gazelles, antelopes, and pronghorns, an adaptation that also protects from predators and fire; indeed, the North American pronghorn, which is often, though incorrectly, called an antelope, is thought to have evolved when there was a North American cheetah to chase it. Some plod or trot (with brief bursts of speed) and endure, like the wildebeest and its many hoofed companions in the great Serengeti migration.

It is factors like these that make it so hard to separate wholly "natural" from seminatural, semihuman, "managed" grasslands. Human-set fires and human-herded grazers have for millennia helped maintain the North American prairies and the African savanna, and the proper management of grazing and fire remain crucial to grassland conservation. The grassy balds of the Great Smoky Mountain National Park are an interesting case. Although nobody is certain, one leading theory is that they originated as tundra during the last ice age (they do contain some relict Arctic/alpine plants whose nearest kin are far to the north). It is thought that as the forests returned, they were kept open by wildfire and whatever grass-eaters were around—first by such Pleistocene grazers as mastodons and giant ground sloths, later by bison and elk, and then by the domestic livestock herded by early European settlers. Now they rely entirely on human management—fire, mowing—for their survival. An abandoned grassy bald soon becomes a tangle of shrubs and then joins the surrounding forest.

In general, patterns of grazing and fire need to match a grassland's long-term evolutionary history to be truly productive. In contrast to grasslands in the northern Great Plains, those at the arid end of the spectrum, such as in Arizona, did not develop in the company of enormous herds of bison, and too much grazing pressure from cattle or sheep will damage them much more easily. Many studies indicate that native grazers fed and interacted with the landscape in very specific ways, and that domesticated grazing animals like cattle, with their quite different feeding patterns, may diminish habitat and biodiversity in the grassland. For instance, bison shape grasslands by creating wallows where they roll, depressions that can become water potholes or small topographical features that offer ecosystem niches. Cattle do not wallow. Bison do not congregate near water, while cattle do. More significant than these differences, however, are the economic pressures that often lead ranchers to stock their range to or beyond its perceived capacity. Overgrazing creates openings for exotic species and allows wind and water erosion of soil—unhealthy forms of disturbance.

Put all these variables together (with many others we could add) and the result is landscapes that may look like simple seas of grass but are instead intricate, subtle, dynamic mosaics.

◇ **ON THE SPOT: IN THE TALLGRASS PRAIRIE**

Bruce Campbell

If you sit down in a tallgrass prairie in midsummer you won't be able to see more than five feet. You walk *on* a shortgrass prairie and can scout for birds or coyotes in the distance, but you walk *in* a tallgrass prairie. You have to pay close attention to what is directly in front of you and at the same time look twenty or fifty yards ahead, or, more important, to the top of the next hill or that bad patch to your left. You see the hawks only when you stop for a breath.

The best way to walk through a tallgrass prairie is to follow the deer paths. Let's say you want to get to that patch of different grass near what might be a pond. If you go straight at it, you will get off the deer trail and have to plow through the big bluestem with your arms out in front of you and your knees lifted up high. Wear long pants or you will bleed. But if you try to think like a deer (an art that for a predator takes learning), you just relax and follow the trail. Sure enough, if you go with the flow of the land and think about hiding quickly if the need should arise, you end up at the pond. I did that once and found a large buck and doe in a heavy thicket five feet in front of me. But if you want to see anything or get anywhere in a modern sense, you need to ride a horse.

My daughter Elaine says this about being a child in a prairie: "Running through a prairie is one of the most beautiful experiences you can have. You can't see more than two feet in front of you, and what you do see is just grass. I pull my hood over my head so that my hair doesn't get caught. The grass whizzes past. It is gold and you feel magical. You can get completely lost with no trail. Just wandering is lovely, though, and you almost will yourself to get lost. You can stand next to the impossibly tall stalks, spread your arms out wide, and fall backward. Don't worry about hurting yourself, the grass will catch you. When you look up, you see all the golden lines reaching up and meeting in the blue of the sky, and you hear the sound of the grasses brushing against each other."

Of course, the prairie is more than grass. I get a catalog with sixty-five pages of prairie plants for sale—hundreds of flowers, sedges, grasses, and such for gravelly soil, wet soil, clay soil. A real tallgrass prairie, which is hard to find these days, is a complicated set of organisms, relationships, soil, rain, and sun that have been doing their thing for about ten thousand years. When you find a real one, you can tell right away. Like awkward teenagers at a dance, the plants clump together for reasons only they understand. You are walking along in little bluestem or switchgrass and bang, you are in a fifty-foot-diameter patch of some different kind of grass or flowering thing. "Well, why did you guys decide to live here?" you ask.

A prairie is amazingly diverse, not only over space but also over time. I haven't spent enough time in the prairies near my home to tell you that I know what I will see when I go there, but I do know that every week it is different. In late winter it can be a bit grim, all those dead plants, but when you look closely through the dirty snow, you can see the green race beginning. Tracks of rabbits, voles, squirrels, and deer are everywhere. Hawks watch from above. Last spring I saw some flowers that shot up about three feet higher than the new grass, bloomed, and then probably turned into one of those things that make your legs bleed. When I came back two weeks later, other plants had taken over. They're all racing for the sun.

They're also racing to reproduce. As in most long-term relationships, a lot of what happens in a prairie is underground. Plants here can go ten or twenty feet into the soil, so when the tops get burned or eaten or dried out or frozen, they just wait for another day or year and come back. In midsummer when you're walking in the prairie, you need to understand that you are on a mirror. What is above you is also below you. Since you don't understand most of what is going on anyway, the fact that you can't see below you shouldn't bother you much.

It certainly doesn't bother the more mobile organisms flitting or hiding around you, and there are a lot. You've got insects, arachnids, amphibians, mammals, birds, all this just aboveground. I once saw a coyote running across a prairie with a large, brown, furry thing in its mouth, something bigger than a squirrel, maybe a mink or a muskrat. This was a patch of prairie smack in the middle of a very large suburban area.

I should tell you about the sound. In the winter, or really any season when the grass is high, I love to hear the wind through the grass, the millions of impacts of grass blade and stem on other grass blades and stems, parts sliding across each other. It sounds like the hiss of rain. There are birds if you are quiet—sparrows, warblers. There is the "move, don't move, just shut up!" sound that small prey mammals make, and they are all around you. In the spring, summer, and fall you have more sounds around you than you can hear.

Then there is the light. I am a physicist, not a biologist, so I know what light does to things and how it reflects and goes through things, but I don't know why grass plays with it, why grass has such fun with it. Look through the grass toward the sun: depending on the season, you'll see a gorgeous luminescent green or an eighteen-carat yellow gold. (I put my wedding ring up to a grass stem, and it matched.) Now look the other way at the same grass: you see dull yellow or green-blue. That's why the major grasses for this prairie are named big bluestem and little bluestem.

To be in a prairie and appreciate it, you have to stop, be quiet, and look around. That's what I'm doing now, writing these notes in a reconstructed prairie at the Volo Bog in northeastern Illinois in early winter. ◇

. . . AND MOVING APART

Just as life originally emerged from the sea, a grassland may have been the scene of a defining moment of human evolution, the landscape in which our species developed upright behavior so that we could better see long distances and give chase. About fifty thousand years ago, a hotter and drier climate led to the expansion of deserts and grasslands. This shift favored *Homo sapiens,* who took to the savanna, running down hoofed mammals in our earliest marathons. Some believe that a kind of genetic memory of the African savanna explains our love for grassy areas interspersed with occasional shade trees and bodies of water, the pattern we usually imitate when we landscape our public parks and our houses. Certainly grass built human civilization as we know it, as the biblical story of Cain the farmer and Abel the herder indicates. From the first domesticators of cattle and rice to today's producers of hamburgers, soy milk, and breakfast cereal, from the empire of Genghis Khan to the global markets in futures for grains and meat, our lives are based to a significant degree on the products of grasslands. Fittingly, the Chicago Board of Trade, home to one of the world's main commodities markets, is topped by a tall statue of Ceres, Roman goddess of agriculture, whose name and our word *cereal* may share a root with the words *create* and *increase.* It is also the case, on the other hand, that in the process of developing civilization, we have destroyed natural grasslands.

As with other natural landscapes like wetlands and deserts, some human cultures have developed and survived in grasslands in ways that we would now call sustainable. Traditional grassland cultures tend to be pastoral and nomadic, with small groups of people following herds of wild or domesticated grazers, along with changing seasons and the growth cycles of grasses, taking with them their housing in the form of yurts, tepees, and other such portable structures. The Mongols of the central Eurasian steppe; the yak herders of the Tibetan Plateau; the Masai of the Serengeti; the bison-based Plains Indian tribes like the Lakota and Crow, made much more mobile with the arrival of horses from Europe (traded northward from Mexico well ahead of European humans): these are classic grassland peoples.

Perhaps because of the visual sweep of their home landscapes, these cultures have often lent themselves to epic stories and films, some of which have strong links to colonial and national identities. The original grassland epic may be the story of Genghis Khan and his thirteenth-century horde of horsemen, creators of what may be the largest empire ever and inspiration for a veritable library of movies, each likely to be described as "a sweeping epic." One from Japan employed its own horde of more than twenty-five thousand extras. Another, *The Conqueror* (1956), stars John Wayne as Genghis and is often considered one of the worst movies ever made, largely because of this casting. In a layer of rich historical and mythic irony, its filming was dusted by the fallout from atom bomb tests at the nearby Nevada Test Site, and it is said that some 40 percent of the cast and crew members developed some kind of cancer by 1981, more than would be demographically likely, even considering the fact that several of them were also heavy smokers—a high human cost for a failure. Still, the casting of Wayne as Genghis has its own absurd appropriateness, which may be why he wanted the role, given the grassland pastoral nature of cattle ranching in the American West, the usual setting for this actor's epic movies and certainly central to America's mythic sense of itself. The more recent *Dances with Wolves* (1990) tells an updated version of this nation's grassland epic, one whose contemporary mixture of romanticizing nostalgia, scenic accuracy, and attempt to shift the historical balance of perspectives away from those of the European American settlers seemed to many viewers a moving and valuable addition to the national mythology.

Australia's grassland epics include the book by Mary Durack (1959) and television film (1997–1998) *Kings in Grass Castles,* the tale of Irish immigrants who create a sheep-grazing dynasty in the Kimberley region in the latter part of the nineteenth century. Danish writer Isak Dinesen's memoir *Out of Africa* (1937) and the 1985 film of that story are a colonialist version of a grassland epic especially notable for their powerful verbal and visual depictions of the grasslands of Kenya, called British East Africa when Dinesen was there. A more recent epic about East Africa's savannas is the 2006 film *The Rain Warriors,* which—though it is a French production—features Masai people speaking their own language. This film retells a traditional (precolonial) story, in which a band of young men goes in search of a legendary black-maned lion in order to lift the vengeance of the Red God and end a serious drought. Significantly, the youth who actually slays the lion, and gives his own life in the process, is not the most promising warrior or son of the most eminent chief, but the son of a poor family, a lowly herder. In a recent adaptation to ecological imperatives, while it was once considered a valuable rite of passage for a young Masai man to kill a cattle-eating lion, now, because of the scarcity of these top predators, this practice has largely been replaced with agreements with environmental organizations to accept reparations for lost livestock.

Because its cold, arid climate is not amenable to agriculture, Mongolia is home to one of the few remaining truly pastoral economies. The country is still 80 percent grassy rangeland, herders make up 40 percent of the population, and herds have been increasing even in recent years, as economic woes have sent some urban immigrants back into the steppe. Mongolian pastoralists have for many centuries tended sheep, cows, goats, horses, yaks, and camels, and they continue to do so now with the help of some modern technologies. Unfortunately, the most damaging species of grazing animal, the Inner Mongolian goat, is the source of the most profitable product, cashmere. Still, though it is now threatened by mining and other kinds of exploitation, the long-lived pastoral economy of the region demonstrates that a reasonably sustainable human-land interaction is possible on the grasslands.

Some of the tensions between traditional and modern economies and lifeways are the subject of the 2006 docudrama film *The Cave of the Yellow Dog,* in which Mongolian director Byambasuren Davaa shows us a modern family working to maintain a nomadic, pastoral lifestyle, herding sheep, goats, and yaks. (Her other films include one about camel herders, *The Story of the Weeping Camel* [2003], and one [2009] involving a song called "The Two Horses of Genghis Khan.") Economic pressures have sent many such families to the cities, leaving the steppe more desolate—and more dangerous, we learn, as former herd dogs go feral and join packs of wolves, increasing the threat to the flocks. Nansal, the elder daughter of the family, who attends school in the city, travels back home for the summer and, despite her father's protests, befriends a dog she finds in a cave, an act that resonates with a story about a yellow dog she hears from an old woman. When her father tries to abandon the dog when the family moves camp, it saves the family's toddler son, who is approaching a large and ominous-looking group of vultures. The steppe landscape in this film is luminously beautiful, apparently shaped by the massive vanished forces of glaciation, one dominated by the sky and the weather (the wind blows nearly all the time, and the family uses a portable windmill to generate some electricity), by great soft sweeps of green grass and colorful flowers. It seems, at the same time, to be an intimate place, perhaps because of the round, warm spaces of the yurts; perhaps because of the tenderness among the family members and between Nansal and her dog, Zochor; perhaps because these people are so thoroughly at home in this landscape.

Although it would be impossible to include them all in a single telling, all real-life grassland stories include layers of complexity involving human behavior. The story of North America's Great Plains offers some representative examples. We might start such a narrative with centuries of nomadic native people, first without and then with horses, often with economies that included both movement-based hunting (most notably of bison) and part-time crop-tending near rivers. During the nineteenth century, European American settlers looked at all the grass and grazers, thought they might as well grow the grasses and grazers of their own choice—wheat, corn, and cows—built herds, put up fences, and plowed up (and built houses from) the prairie sod. Then they established towns and cities with suburbs with grasslandlike lawns and parks, all the while developing national and world markets for their products—and displacing the indigenous people and cultures. These changes were mostly gradual and cumulative, of degree rather than of kind. So was the damage they did to the grasslands: only about 4 percent of the native tallgrass prairie of a couple of centuries ago remained in this region at the beginning of the twenty-first century.

One ingredient in this process was wholly cultural, a western European attitude about the superiority of settled to nomadic people and a fear of too mobile a population. In a 1775 speech to the House of Commons on conciliation with the American colonies, British philosopher Edmund Burke worried that the expanse of the American continent would cost England its colonies, using a comparison to the Mongol—or Tartar—hordes:

Many of the people in the back settlements are already little attached to particular situations. Already they have topped the Appalachian Mountains. From thence they behold before them an immense plain, one vast, rich, level meadow; a square of five hundred miles. Over this they would wander without a possibility of restraint; they would change their manners with the habits of their life; would soon forget a government by which they were disowned; would become hordes of English Tartars; and, pouring down upon your unfortified frontiers a fierce and irresistible cavalry, become masters of your governors and your counsellors, your collectors and comptrollers, and of all the slaves that adhered to them.

Empire-builders in the United States had similar attitudes about Native Americans. William Emory, whose role as a topographical engineer was to oversee the United States–Mexico boundary survey during the 1850s, distinguished between "semi-civilized" settled tribes and nomadic "wild" peoples. The semicivilized, he thought, could be tolerated, but the nomadic tribes were beyond redemption and had to be "exterminated," an opinion he published in his 1857 report to Congress.

But European Americans also cherished the chance to make their own new lives in this prairie landscape. Perhaps the leading writer about this subject is Willa Cather. Particularly in *O Pioneers!* (1913) and *My Ántonia* (1918), Cather captures the immigrant homesteader experience of the prairies of Nebraska. While in general her work celebrates the new agricultural empire on the plains, it also exhibits a grasslands aesthetic and celebrates aesthetic and spiritual connections between settlers and their new landscape. On the one hand, in *My Ántonia,* we see her admiring that symbol of industry and conquest, the plow:

> There were no clouds, the sun was going down in a limpid, gold-washed sky. Just as the lower edge of the red disk rested on the high fields against the horizon, a great black figure suddenly appeared on the face of the sun. . . . On some upland farm, a plough had been left standing in the field. The sun was sinking just behind it. Magnified across the distance by the horizontal light, it stood out against the sun, was exactly contained within the circle of the disk; the handles, the tongue, the share—black against the molten red. There it was, heroic in size, a picture writing on the sun.

On the other hand, Jim Burden, who narrates the novel and finds much to admire in the newly arrived Ántonia, daughter of Bohemian immigrants, associates his feelings for her with the magic of the plains landscape as he walks along with her at sunset: "In that singular light every little tree and shock of wheat, every sunflower stalk and clump of snow-on-the-mountain, drew itself up high and pointed; the very clods and furrows in the fields seemed to stand up sharply. I felt the old pull of the earth, the solemn magic that comes out of those fields at nightfall. I wished I could be a little boy again, and that my way could end there." For Cather, farming is not a mechanized death sentence for the prairie, but rather an intimate activity, a juncture where humanity and earth interact.

A more ambivalent narrative is Terrence Malick's masterful 1978 film *Days of Heaven*, set in the Texas panhandle around 1916. Bill (Richard Gere), his lover, Abby (Brooke Adams), and Bill's sister, Linda (Linda Manz) escape nightmarish labor in Chicago for brutal work in the wheat harvest. Posing as her brother, Bill encourages Abby to accept the romantic interest of the wealthy but apparently ill farmer for whom they work (Sam Shepard), with an eye to inheriting his land. The deception may be seen as mirroring the folly of turning grassland into wheatland, as a swarm of grasshoppers followed by wildfire destroys the crop just as the love triangle explodes into violence that leads to the death of both men. The human tragedy plays out in an enormous landscape, and many of the most vivid shots are of people involved in passionate conversation seen at a distance, dwarfed and silenced by the sky and wheat fields. The narrator, the young girl Linda, frames the tale in terms that feature fiery apocalypse and an aimless wandering that is an impoverished imitation of deeply rooted nomadism.

In a larger-scale tragedy, plowed, planted, and cattle-grazed prairies have not turned out to be sustainable, especially in the drier western half of the Great Plains. There, notoriously, decade-scale climate shifts and the drought they brought combined with disturbed ecological systems to create the massive erosion and human disruption of the Dust Bowl in the 1930s, events from which the region has not yet fully recovered, and may never do so. Indeed, grassland writer Richard Manning has suggested that the damage done by industrial agriculture to the North American Great Plains constitutes a kind of ecocide. The combination of natural and human causes for this disaster, and its essential underlying violence, were understood very early, as is evident from Pare Lorentz's 1936 documentary film *The Plow that Broke the Plains*, whose narrator notes the drought but also offers a refrain that reminds us that the natural limits of the land had been exceeded: "without rivers, without streams . . . a country of high winds, and sun . . . and of little rain." Flashing back and forth from World War I tanks to Great Plains tractors and harvesters, the film shows the mechanized war waged against the landscape and its destructive result. (*Days of Heaven*, too, ends with a scene of soldiers boarding a train, on their way to join the war in Europe.)

John Steinbeck's more familiar novel (1939) and film (1940) *The Grapes of Wrath* begins with a lyrical but painfully vivid depiction of the drought in Oklahoma, the dying plants, searing sun, growing dust clouds, banks foreclosing on loans and land. His characters feel they have little choice but to turn themselves into nomads, at least for a while, and move to the wetter fields of California. Timothy Egan's *The Worst Hard Time* (2006) tells the story of those who did not leave and offers some bone-chilling descriptions of the dust storms, the sometimes fatal "dust pneumonia" they caused, the way static electricity built up so much that a handshake or grasped door knob could knock a grown man to the ground. People who experienced one of these storms—or even saw it from a distance—never forgot it. Those who didn't can find photographs (of farms for sale, dust dunes in fields and along small-town streets, and black billows covering the sky) at the American Library of Congress or its American Memory Web site, home to the massive collection of Depression photos taken for the Farm Security Administration. Visitors to the region can still see the effects of this combination of overuse and drought in dying towns, abandoned farmhouses, and impoverished soils, and, more positively, in the ubiquitous lines of trees planted along roads and around houses as windbreaks. The scattered national grasslands are also artifacts of the Dust Bowl, land bought back from failed farms in an effort to foster better land conservation.

When artist George Catlin traveled the prairies in the 1830s, he painted a gallery of portraits of Native Americans. Full of admiration for these plains dwellers, he advocated that the prairies be kept as a nationally protected wilderness park where Indians might be allowed to hunt in their traditional manner. Today Catlin's vision seems dated, treating indigenous people as antiques to be kept in an unchanging display. But notions of large prairie preserves are still alive, now focusing more on land and wildlife preservation. There are many small preserves and sections of restored prairie scattered across the center of North America, and several larger areas are protected (and managed with the help of bison and fire), as for instance in South Dakota's Custer State Park and Badlands National Park, North Dakota's Roosevelt National Park, the Tallgrass Prairie National Preserve in the Flint Hills of Kansas, the Nature Conservancy's Tallgrass Prairie Preserve in Oklahoma, and Saskatchewan's Grasslands National Park.

And then there is the ambitious Buffalo Commons proposal from

Rutgers University geographers Frank and Deborah Popper. Pointing out that the agricultural experiment on the Great Plains has failed and that the region has continued to lose population, they proposed in 1987 that the lands—a mixture of public domain, privately held farms and ranches, and Indian reservations—be put back into grass and that buffalo herds be reintroduced. With its keystone species returned and its ecological health improving, the resulting landscape would provide a tourist and pastoral economy to sustain communities of the region. While they have been perceived as elitist outsiders, enemies of private ownership and progress, the Poppers have maintained that their work is not a prescription but a description of a process already under way. Indeed, the proliferation of bison ranching, led by the efforts of Ted Turner, suggests that such a future really might be possible.

Certainly humans and grasslands have a long history of changing and developing together. We need not conclude that all transformation of the grasslands is ecologically ruinous—it is no simple thing, after all, to draw the line between changes that happen as a course of nature and changes that are caused by human alteration. As in all ecosystems, the crucial factors are likely to be the speed and scope of change. When a landscape is too quickly transformed, its established species have no time to evolve to fit the new conditions; exotic species accompany human development, seizing the opportunities offered by disturbed soils. The effect on biodiversity—the typical barometer of ecological health—can amount to an impoverishment of the landscape. It may be that grasslands and their traditional peoples will play a significant role in teaching us how to prepare for and adapt to global climate change, for the ability to live lightly on relatively arid ground may soon be at a premium.

◇ ON THE SPOT: ON THE TIBETAN PLATEAU
Julia Klein

The Tibetan Plateau is our planet's third pole, its highest body of land, home to its largest persistent ice mass outside the Arctic and Antarctica, source of the water that sustains more than half the world's people, and one of the great shapers of both regional and global climate. The legendary Himalaya dominate its southern edge, but other equally spectacular but more obscure mountain ranges cut its expansive grasslands into a patchwork of high valleys, places that have for millennia been home to human populations. I have come to the northeastern region of the plateau to study how climate change and changing land-use practices may affect the pastoral resources on which local livelihoods depend.

By the calendar it is early spring, but winter lingers tenaciously. Temperatures flirt around freezing, and snow still blankets the surrounding mountains. This morning I peered from my sleeping bag and saw a flattened and amorphous landscape through the window of my barrack-style accommodations. Now, outside, I inhale pungent smoke from the herders' dwellings as they too start this snowy day slowly.

I leave the Chinese compound, pass through a tall iron gate that exudes power and authority, and enter the herders' winter rangelands. I search for familiar, distant peaks, but today snow covers the valley floor and low clouds narrow my often boundless field of vision to nearby minutiae and a heightened awareness of my own musings. Soon I am joined by my field assistant, Tsering Thondup, who tends a small flock of sheep nearby. He is neatly dressed in a white hand-knit wool sweater, roomy and worn pants carefully mended with patches, and a knockoff Chicago Bulls baseball cap that he purchased at the nearest market, several hours away. The Bulls logo bears a striking resemblance to a yak and is popular in these parts.

We walk along a dirt footpath worn by the comings and goings of yaks and sheep. The few inches of fresh snow are crisp beneath my boots. Tsering walks softly with his hands clasped behind his back, in characteristic herder position, poised to hum a tune for the day's walk or throw a rock at a wandering animal. Along the path are slumbering yaks, their curly brows peppered with snowflakes.

When the temperature rises over the next few weeks, the vegetation waiting just below the hard soil will become green and nutritious. The yaks and sheep will eagerly nibble at the tiny, green shoots, even as they have barely pushed up through the soil. Monsoonal rains will then thunder in, turning this marginal and subdued landscape into luxuriant grasslands, decorated with alpine flora in midsummer glory. But even during the height of summer, it may snow.

The herders' winter homes nestle in the lower-elevation valleys where Tsering and I walk. Some are constructed of reddish-brown clay, typically one or two rooms with a simple stove. Dung fashioned like pancakes sticks to the outside walls to dry so it can be used for heating and cooking fuel. The herders are busily preparing for their migration to their summer camps higher up, their way of finessing life in this marginal and changing landscape. Closely tracking the green and nutritious wave of summer vegetation sweeping up the mountains, their animals also will give the lower-elevation grasslands some rest.

Tsering and I are passing two middle-aged women who are squatting, chatting, and sifting barley, the main ingredient in *tsampa,* the staple of the Tibetan diet, when one of the women waves us over. As we consider this invitation, I notice that in the shadow of their courtyard, an older, silent, deeply weathered man diligently repairs a panel from a yak-wool tent. The rich, chocolate-colored tents decorate the landscape, as people air them out for summer. Yesterday, I watched in awe as some herders wrested the yaks onto their backs, tied their legs, wrapped the yak fur around a wooden dowel, and pulled out the shaggy winter coat by hand. The yaks grunted in protest against this forced molting. Later, the herders will spin the wool by hand and then weave the woolen panels for tents and other necessities.

We duck away from bright glare into a dimly lit room smelling of burning yak dung and pungent yak butter. I notice a small altar adorned with a feather, bowls of rice offerings, and pictures of a local lama. I greet a stooped-over grandmother, her long hair tied in a braid of black laced with gray. As I sit down on a corner of a firm bed, she hands me a small porcelain bowl, which she quickly fills with yak-butter tea. I sip respectfully and hastily to prevent a viscous barrier of cream and butter from forming on top.

Then I notice something or someone else on the bed—a baby, tightly bundled in layers of woven blankets, happily dozing. The grandmother peels away the layers, and the baby coos as she pees onto the dirt floor through the open slit sown into the crotch of her bright red knit clothing. We all laugh.

After we politely excuse ourselves to get to work, Tsering and I cross an icy stream and begin to climb a slope that leaves me breathing hard. Elegant and regal purple irises, which dare to bloom so early in this uncertain environment, poke through the snow's fleece. There is a predictable rhythm to the flowering of plants here. After the irises comes a wave of yellow generated by flowering potentilla shrubs. By midsummer, these will be superseded by the cottonball heads of bistort. Then indigo gentians, whose roots are valued for their medicinal use, will bloom.

Tsering collects dung fuel from the ground to take home with him. The dung is an ecological service provided by this land. Livestock and other rangeland plants and animals provide life's necessities: food, fuel, clothing, shelter, medicines, and transportation. The relative warmth of the summer and the monsoon rains, combined with the sun's energy, support the vegetation and animals that form the basis of pastoral existence. Climate here is a factor that defines and enables life.

Soon we squeeze between two fence wires and enter my experimental plots. I am examining how climate warming affects these grasslands and the services they provide. I simulate warming by using clear, round, knee-high fiberglass chambers the width of my arm span. The chambers mimic the greenhouse effect, warming the air and soils within them. Today we'll readjust the chambers and batten them down with wire against the plateau gusts. We'll check and download the data loggers, machines that record air and soil temperatures every few minutes, year-round, on computers in waterproofed cases. Meanwhile, the clouds have lifted, revealing the stunning peaks to the north. I take a deep breath to acknowledge the powerful allure of this magnificent land.

As Tsering and I head back to camp, the sun fades behind emerging peaks and the moon rises in the boundless sky. Tibetan mastiffs are roaming the grasslands now, so I gather a fistful of rocks in case one of them ventures too close. They begin their howling concert, and will not stop until morning.

Almost ten years later, I will return to the same place with my husband and my ten-month-old son. While my graduate student and I resample the vegetation after a decade of experimental warming, my son will conduct his own experiment. He will let go of my husband's hands and take his first shaky steps on this uneven land. One of his first words will be "yak, yak, yak." And the little baby with whom I laughed in the grandmother's home will be a girl who helps milk the yak and serves tea on these resilient grasslands and the mountains that nourish them. ◇

AMONG TREES

If we sometimes think, mistakenly, that grasslands are monotonous, we're less likely to make this error in considering forests and woodlands. Although any one patch of trees might seem uniform, it is easy to call to mind a large number of forest types—or, in terms of the story we're following in this chapter, a wide variety of journeys we might find ourselves taking as we rise upward from beneath the ground—for these landscapes are varied, complex, and dynamic. What they look and feel like, what they contain, how they sustain themselves and change, how much organic mass they create through growth, their carbon budgets (how much carbon dioxide they absorb or release): all these things, again, depend on a host of factors, and these factors combine to create another realm of great complexity.

As with grasslands, many variables are related to climate, such as available water, solar energy, temperature, and wind; in each case, as we have seen before, both amounts and timing matter. Latitude and elevation are involved, as is topography at every scale. So are soil types and the richness and specific components of soil bacteria, fungi, and microfauna. High-latitude trees get their light in seasonal binges, while equatorial trees see steady light all year long; because they are sunnier, equator-facing slopes of hills and mountains dry out more quickly than do pole-facing slopes; tiny microclimates and their partner microhabitats follow creeks and hollows and rises; extreme air and soil temperatures restrict growth to more benign seasons; forests near oceans generally experience smaller climate swings throughout the year than do those farther inland. And—a factor that is particularly important in forests, especially the tallest of them—these conditions change with height above the ground.

The histories of particular locations contribute to forest variability, too. The very slow movement of tectonic plates accounts for the close resemblance of Chile's and Tasmania's trees; oceanic islands like Hawaii develop distinctive forests with their own species. Long-term climate changes can leave forest relicts like the bristlecone pines in the mountain ranges of the Great Basin; where species live today may depend on where their ancestors survived during the last ice age and how quickly they were able to migrate when the ice melted. And while ecologists don't know how forests will alter as the climate changes again, they do know they will change, that new associations of species will form as one kind of tree dies off, a second hangs on where it is, a third moves in.

Shorter-term disruptions come in many forms: hurricanes and microgusts, wildfires and floods, insect or disease outbreaks, volcanic explosions and mudslides, slash-and-burn agriculture and housing developments. The eruption of Mount Saint Helens toppled giant trees more than a dozen miles out, all in the same direction. In an old-growth forest, every now and then a single tree will fall, taking down whatever lies in its path and opening a small sunny glen. Such disturbances open gaps in the forest cover that can change the composition of the community and relationships among its members, for short or long periods, in predictable or unpredictable ways. It matters what sorts of insects, birds, and animals are present to pollinate some species, spread the seeds of others, eat or ignore particular plants. It matters whether seeds move by wind or squirrel stomachs or a deer's fur; whether they germinate readily or rarely; whether they need special conditions to flourish or can make themselves at home under many different circumstances. The addition or subtraction of a single species can have major repercussions. If a so-called keystone species vanishes from a region, say a crucial pollinator or a top predator that keeps the plant-eaters in check, or if a particularly aggressive new species moves in, many additional changes are likely to ensue, and the character of the forest may change entirely.

Patterns of succession also contribute to variety. Some trees grow quickly; some need shade. Some root in mineral soil; some in organic duff; some, it seems, in solid rock. Some seeds can lie dormant for many decades before springing to life when conditions are just right; some are released only by the heat of forest fires. Some trees love disturbed places; some need uninterrupted centuries in which to flourish. Many plants (and not just in forests) thrive only in the presence of particular soil microbes; every tree that survives does so with the help of underground fungi whose filaments (called mycorrhizae) wrap and penetrate its roots and help provide food for their host. Because these soil communities are themselves subject to change, what happens to them affects the trees they support. And evolution always has a hand as well, fashioning new species and ecological relationships to cope with changing conditions and to fill whatever niches are available.

The result of all these factors put together is a world of forests that are multiple and complex in every dimension, and not only because they can contain many, many species.

Everything will change around us as we travel upward, along a vertical plane. From roots to crowns, through changes in the amount and quality of light, we'll pass through many distinctive layers: The zone of muscular roots, filigreed rootlets and mycorrhizae, the burrows of voles and mice. The protected boundary-layer world of mosses and ground lichens, duff and mast, tangles of dead and decaying wood, where we find chanterelles, boletes, truffles, morels, porcini, and matsutakes, the fruiting bodies of those mycorrhizae. Low-growing grasses and flowers.

Tiny seedlings. Shrubs and small trees, some of them perhaps growing in straight lines out of the fallen trunks that serve as nurse logs in wetter places. Middle-aged and midsize trees that may or may not match their neighboring seedlings and elders. In all these trees, vines and lichens and mosses and bird nests and insects and small mammals. And, finally, the tops of the tallest trees, which in the largest, oldest forests can have canopies of incredible ecological richness. Some small animals spend their entire lives there, sometimes in the top of a single tree. Leaves high above the ground send out roots into tiny pockets of soil they collect from the wind. Crumpled green sheets of lichen collect nitrogen from the air, then later fall to the ground to fertilize the whole forest.

And everything will change around us as we travel sideways, along a horizontal plane. As in many other ecosystems, thanks in large part to the continuous processes of disturbance, forests are patchy at all scales. Their ingredients are unevenly distributed across the landscape: sparse and dense, tall and short, young and old, sun- and shade-loving, dead and living. If we study one square yard of ground, we may find small clumps of moss, each clump itself a patchwork of several species; one or two different wildflowers and a couple of grasses; a little pile of bark; a small anthill; a mushroom; a scatter of needles and leaves, some more decayed than others. If we study a square mile, we'll find the same thing: an uneven distribution, one patch of this kind of tree, another patch of that, a bit of open ground where an old tree recently fell, taking some of its neighbors down with it. A grove of aspen will be broken up by sprinkles of young lodgepole pines, a small meadow quilted in multicolored wildflowers, bits of boggy ground, higher bits of dry ground, and maybe a creek lined with water-loving willows, alder, blue spruce, or watercress. And if we look at a hundred square miles, we'll see the same pattern again. Occasionally we'll encounter a relatively uniform swath, of single-aged lodgepole pines for instance, all dating from the same forest fire, but such swaths will be patches in the larger landscape.

Along with this patchiness comes a proliferation of boundaries, where one patch encounters or becomes another. These boundary zones are complex places in their own ways. Sometimes, as when the edges begin to outweigh the interiors, or where too much forest gives way to cornfields or parking lots, habitat fragmentation results and edges become the source of forces that reduce native species or overall biodiversity. Some birds and other animals thrive only in the middle of large forest extents, far from grassy edges or clearings; near forest-grassland edges in the American Midwest, cowbirds lay their eggs in the nests of forest songbirds, and the baby songbirds can't compete with the larger and more aggressive cowbird hatchlings.

At other times, though, these boundaries are highly productive ecotones and ecoclines, where the proximity to multiple habitats and food sources nourishes especially rich mixes of species and some particular creatures that require such habitats. Where a stream cuts through a grove, say along the line between a south- and a north-facing slope, with a small meadow opening nearby, the total biodiversity may well be more than what we'd expect from the sum of the parts. The end result of all this complexity is a proliferation of niches. The presence of many niches fosters biological diversity and sometimes the development of new species. And diversity underlies flexibility, resilience, and the balance of stability with change that characterizes healthy ecosystems.

⬥ ON THE SPOT: IN A EUCALYPT FOREST
Kate Rigby

For a full minute or two, we find ourselves held in her gaze as she peers inquiringly at us from the arched trunk of an old tree fern no more than a few feet away. Surprised by our sudden appearance in her sheltered domain, she had left off scratching around in the thick leaf litter and flapped up there in her ungainly way to check us out as we approached. The lyrebirds of Sherbroke Forest in the Dandenong Ranges east of Melbourne have grown accustomed to human presence, but they usually skip off into the bush if you encounter them on one of the main trails. We have ventured into a less frequented spot, though, and this plucky creature seemed initially more curious than alarmed by our visit. She is soon back fossicking for minuscule crustaceans around a bend in the creek at our feet. It seems we are not so interesting after all.

Released from her attention, I can now turn mine to the space that surrounds us. Filtered through the fronds of the giant tree ferns that embower us, the thin, wintry light has a faintly greenish hue, and although the air is cold, it is oddly humid in this moist green gully, and tangy with the smell of rotting plant matter. In its steamy density, the tangled vegetation feels more like a jungle than the dry forest and woodland that most associate with the Australian bush. This sliver of cool temperate rain forest is a rare find on a continent that has become over the millennia ever more arid and fire-adapted, recalling as it does the lushness of more ancient times when Australia was still wedded to Antarctica.

Take these crazy tree ferns, for example: they are Gondwanan survivors, and according to local Aboriginal people, their soft hearts make good eating at this time of year. A friend of mine used to liken them to limbo dancers in appearance. They start out straight and tall, topping forty feet in the case of the rough tree fern that predominates in Sherbroke. As they age, though, their big, heavy crowns gradually drag them earthward. Instead of falling, they bend from the waist, assuming all sorts of eye-catching poses as they do so, availing themselves of a nearby tree as a crutch if they can, and sometimes looping right over to bury their heads in the ground and begin again from there. This slow dance is particularly character-istic of the smaller soft tree fern that abounds in the gullies. These engaging characters appear to have dressed for the occasion as well, their supple trunks all ruffled and furred with decorative epiphytes: feathery filmy fern, mainly, and the misleadingly named veined bris-tle fern, which is lusciously lacy and soft to the touch.

Here and there are handfuls of pointy finger fern, too, along with larger tufts of mother spleenwort and, higher up, the occasional kan-garoo fern (so-called for its resemblance to macropod paws, though the fronds look more like fish bones to me). Something rather dif-ferent has sprouted on this one, though: it looks like a banyalla, a small, bushy tree of the Pittosporum genus that has recently hybrid-ized with an introduced variety, one of the many garden escapees to have "gone bush" here. Like some other tree species of Australia's wet sclerophyll forests, the banyalla has learned to evade the hungry maw of the swamp wallaby (whose ancestry is also Gondwanan) by germinating up above the forest floor in the accommodating trunk of the tree fern. The mighty plumwood of southeastern New South Wales does this too, as does the southern sassafras, such as the one I noted on the way down here: it had nearly overgrown its host entirely, securing for it a kind of afterlife in the course of its ascent into the light of the middlestory.

When the sassafras is in bloom, its small yellow-centered white flowers lend a hint of nutmeg to the mingled aromas of the forest. Looking more closely at the densely interwoven throng of small trees and shrubs sloping up from the gully, I am beginning to dis-cern many that will soon begin to flower: that banyallà, for instance, will unfurl its yellow bells brushed with maroon, and there will be creamy plumes on the hazel pomaderris and orange- and purple-dotted sprays of wedding-gown white on the Victorian Christmas bush, along with flashy yellow clusters on the blanket leaf, whose big, droopy leaves hang off the ends of its branchlets like ponderous bunches of unripe bananas. Right now, though, the show has been stolen by the wattles, the largest trees of the middlestory. Not long after the winter solstice, first the silver wattles, then later the blackwoods, throw off their accustomed incognito and set the forest alight with a million trillion tiny suns, as if beckoning the spring to return. For once the towering mountain ashes don't have a monopoly on charisma. Even so, now that I gaze upward, I am once more entranced by those smooth white trunks, surprisingly slender still for all their great height, gleaming where golden shafts of sunlight catch them aslant, rising up and up to the clouds of glittering grayish-green leaves shimmying in the faint breeze against an azure sky.

Alluring though they might be, it is well to think twice before you enter a forest of ash. "Widow makers"—that's what the pale-skinned settlers learned to call Australia's sky-scraping, shallow-rooted, loose-limbed gum trees, and the mountain ash is the tallest of them all. Potentially reaching up to 330 feet when fully grown, most of the trees here are youngsters, regrowth from a big blaze in the 1920s, though a few are considerably older. It's not only falling timber you've got to watch out for, either. In summer there are the snakes. Above all, though, you have to remember that these trees are time bombs. In dry times their deep leaf litter turns to tinder, and those long ribbons of bark dangling from the branches where they have snagged function like fuses, ferrying sparks up to their open and oil-laden crowns.

Unlike the dry sclerophyll on the other side of the Dandenongs—a name derived from an Aboriginal phrase for "hills on fire"—the mountain ash forest does not burn easily. When it does, though, the conflagration is more explosive than that of any other forest fire. Perversely, it is only by means of a really violent crown fire, preferably once every three hundred years or so, that a new generation of mountain ash can germinate. More frequent cooler fires will destroy this forest, but so too will the absence of a major blaze for more than four hundred years. It is a delicate balance, attuned to catastrophe, and one that has been disrupted since British settlement. How this transient and unstable wonderland will fare as the planet warms remains to be seen. But if, as seems likely, we are in for more frequent summers like those of 2008 and 2009, when large swaths of Victoria, including several townships, were consumed by fire, places like this, and the species they support, might not survive too much longer.

For the moment, though, all is calm and cool and deliciously damp in this corner of Sherbroke Forest, which was graciously spared in the Black Saturday blaze, and I am reminded that the nearby township of Monbulk takes its name from another Aboriginal phrase, meaning "healing place," perhaps on account of the crystalline springs that well up here and there in places like this throughout this part of the forest. Reluctantly clambering back out of the gully, leaving lady lyrebird to feed in peace, we find our steps arrested by the clarion call of one of her flashy mates, cutting the air with his strident renditions of crimson rosella, yellow-breasted robin, magpie, whip bird, cockatoo, and kookaburra, echoing so many voices of this remnant rain forest that pulses still with life. ✧

ZOOMING IN

All these kinds of complexity explain why both scientific and cultural responses to forests always zoom in on smaller parts of the bigger picture. Scientific texts written for specialists, texts that typically attempt to order their subject into categories and often taxonomies, display a startling variety of divisions, all inevitably imperfect, split according to the writers' professional interests, to the scale of divisions, and to the characteristics that govern the dividing. Twenty tomes about forests will offer twenty kinds of categories—if not more. The massive series *Ecosystems of the World* includes six volumes on five major types of forests (coniferous, temperate deciduous, temperate broad-leaved, forested wetlands, and tropical rain forests), as well as volumes on other ecosystems that include forest components. Tropical rain forests take up two entire volumes; the second contains thirty-five chapters; the chapter on American tropical rain forests recognizes twenty major subset forest types and variations within those twenty. In the volume on coniferous forests, one chapter focuses on Eurasia's boreal forests and offers six forest types in Siberia alone. Even so, such texts can never reach complete coverage of their topics. They can only gesture toward such a goal.

Writers and artists who wish to say something about forests to less-specialized audiences have the same problem, and they too need to zoom in on small pieces of the larger picture. Because they so clearly and sometimes literally display their methods of focusing, such books, photographs, and films show us a few things about how we humans cope with the world's complexity when we encounter it in the form of overwhelming amounts of information. Artists and writers with literary leanings are typically very aware of the problems of selection involved in creating representations—that is, using a part to represent a whole, as happens in the rhetorical figure synecdoche (one gives one's hand in marriage) or in representative forms of government (my senator or member of parliament should stand in for me and my neighbors). They are also likely to want to represent or evoke some of the effects of a forest on human senses—the feel of mist or wind on skin, soft moss underfoot or rough bark beneath the palm, the blended scents of what are literally thousands of chemicals in the air. They're always limited by their medium: photos can represent only sight, a tape recorder only sound. But they have other tools that let them activate the less-concrete but potentially very powerful senses that lodge in our memories and imaginations.

Standing behind a camera or an easel, you are at every moment standing in a single place, looking in a single direction. In any one instant, you can capture wide-angle, telephoto, or close-up views, but not all three; you can zoom in or out, pan or stand still, but not both; you must focus on some things and not others. Even the most globally oriented documentary films, such as the *Planet Earth* series narrated by David Attenborough (2007), are subject to these restrictions. Considering tropical rain forests, for instance, *Planet Earth* offers regular visual and narrative overviews of large forest expanses, from the point of view of someone in a plane or helicopter, but except when they're abstracted to a turning globe, even its largest views can show us only one stretch of forest at a time, African or Amazonian or New Guinean. Most of the time this film focuses on smaller topics and accompanying examples, such as the amazing forms taken by species in these highly diverse ecosystems (half the world's species on just 3 percent of its land) and the mating displays of three (out of nearly forty) New Guinean birds of paradise—film footage that took the cameraman 120 hours hidden in three blinds to capture.

Similarly, a soundscape artist must set up microphones in a single place, as when musician David Dunn recorded the sounds of one tree for his CD *The Sound of Light in Trees: Bark Beetles and the Acoustic Ecology of Pinyon Pines* (2006)—the movement of its branches in the wind, its circulatory system, the chewing of the mountain pine beetles in its interior. (This native beetle and its relatives, flourishing because of drought and warmer winters, have been killing enormous areas of forest across the western half of North America. The same areas are experiencing dramatic increases in wildfire damage, another apparent effect of the warming climate.) Or when Nalini Nadkarni, a pioneer in forest canopy ecology, asked a small troupe of modern dancers to visit the Costa Rican cloud forest with her and compose a dance in response, those dancers had to choose a particular location in which to perform and a small number of characteristics of rain forest trees to trace with their bodies. They solved these challenges in part by dancing up the center of a giant hollow tree, mimicking the climbing gestures of jungle plants that are always twisting and reaching toward the light.

Sometimes writers and artists will focus their attention on one or more individual trees. In such cases, the featured tree, like a character in fiction, is likely to be, simultaneously, an idiosyncratic individual, a type or representative of others of its kind or neighborhood, and a single focus point around which a larger inquiry or meditation can be developed. In *A Sand County Almanac,* Aldo Leopold traces the environmental history of his Wisconsin farm by considering the rings of an oak killed by lightning, as he cuts it down and then splits it for firewood. In *Nature Cure* (2005), Richard Mabey organizes part of his meditation on the creative and healing powers of place, nature, friendships, and art around the wind-toppled body of an ancient yew in the village of Selborne, where early naturalist Gilbert White lived. And a single juniper tree provides one of the more resonant images of solitude in Edward Abbey's *Desert Solitaire.*

Photographers and other visual artists can focus on individual trees as well. James Balog's *Tree: A New Vision of the American Forest* (2004) offers one compelling collection of such portraits, including such things as a composite digital image of a giant sequoia so tall that nobody could ever see it whole except through such an artifact. Balog generated this image by rappelling for four hours down a neighboring tree, shooting 451 separate still photographs. Then he spent six weeks at the computer stitching the parts together to make a whole that is at once actual and virtual, documentary and artistic. He notes, too, that this process of reconstructing individual trees from fragmentary views made him think about the real history of fragmentation and loss in American habitats. It allowed him to feel that he was in some small way helping "to put the lost forests of America back together again."

Another option is to concentrate on a particular type or group of forest inhabitants. One lovely instance is botanist Robin Hall Kimmerer's

Gathering Moss: A Natural and Cultural History of Mosses (2003). Like Leopold, Mabey, and Abbey, Kimmerer (who lives in upstate New York) writes in her own first-person voice and uses her own experiences and thoughts to unify her essays, but she also adds information she draws from her Potawatomi cultural heritage and her experience as a bryologist, a scientist specializing in mosses. This mixture allows her to plunge into considerable detail when she wishes. She tells us, for example, how she figured out the way that chipmunks both create and help fill tiny gaps in moss expanses, and she lists the inhabitants of a typical muffin-size clump of forest-floor moss: 286,900 protozoa, water bears, springtails, rotifers, nematodes, mites, and fly larvae. It also allows her to make much larger points, as when she compares the complex structure of a rain forest with its miniature replica on the level of a moss forest or meditates on the tensions among ownership, control, and ecological caring.

For the Sierra Nevada, where forests stretch more than ten thousand vertical feet from the lower to the upper tree lines and where elevation determines which trees dominate the woods, John Muir's extended descriptions are classic. Noting that "to gain anything like a loving conception of their grandeur and significance . . . one must dwell with the trees and grow with them," he writes, for instance, that light in ponderosa pines (*Pinus ponderosa;* he calls them silver pines) "seems beaten to the finest dust, and is shed off in myriads of minute sparkles that seem to come from the very heart of the tree, as if, like rain falling upon fertile soil, it had been absorbed, to reappear in flowers of light. This species also gives forth the finest music to the wind . . . the vibrations that give rise to the peculiar shimmering of the light are made at the rate of about two hundred and fifty per minute." This is Muir's signature mixture of careful observation (and often measurement) and ecstasy, the same mixture that appears in his writings about glaciers, meadows, squirrels, mountain ascents, painterly aesthetics, and the many other subjects to which he gives his attention.

Famously, too, the adventurer Muir describes climbing a midsize Douglas fir (*Pseudotsuga*) during a windstorm in 1874. For hours he "clung with muscles firm braced, like a bobolink on a reed" and gloried in the swirl and waves of vast fields of treetops, the "wild exuberance of light and motion," the music and the fragrances carried by the wind. By studying trees, writers are apt to draw conclusions about what is real, about how we are related to the rest of creation, about what matters. Muir's conclusion is this: "We all travel the milky way together, trees and men; but it never occurred to me until this storm-day, while swinging in the wind, that trees are travelers, in the ordinary sense. They make many journeys, not extensive ones, it is true; but our own little journeys, away and back again, are only little more than tree-wavings—many of them not so much."

Well over a hundred years later, Nalini Nadkarni tied small paintbrushes to some branch tips and let them paint a couple of minutes of their travels on sheets of paper. The results, she says, looked like Chinese calligraphy, and she was able to calculate the typical travel patterns of several species by extrapolating from the travel rate of a single twig to a whole treeful of twigs: "A hundred-foot tall Douglas-fir, for example, moves the astonishing equivalent of 186,400 miles in a year, or over seven times the circumference of the earth"—an odd tidbit of information that pertains to soils, water, and tree physiology as well as to our imaginative, aesthetic, and kinetic responses to a tree's dance. Coincidentally, but in a wonderful metaphor for the dynamic nature of trees, forests, and our world, this is just over the speed of light. In a single year, in the company of millions of companions, the outer tips of a single Douglas fir travel about as far as a photon travels in a second—while staying firmly rooted in its native soil.

Some of the best literary portraits of tree species are in the work of Donald Culross Peattie, who in two thick books surveyed many of the native trees of eastern, central, and western North America. (He planned but never wrote a volume on southern trees.) Each essay offers a sensory and emotional sketch, botanical information, and a survey of a wide range of human interactions with and uses of the tree. Although in general these essays are so full of specifics that they are hard to characterize in brief quotations, we can glance at a few of the more lyrical notes in his *Natural History of Western Trees* (1950) to see the kinds of characterizations he offers.

In a grove of coast redwoods *(Sequoia sempervirens),* Peattie writes, "Time, the common tick-tock of it, ceases here, and you become aware of time in another measure—out of an awesome past. . . . But this solemnity is not like that of church or tomb; it is enlivened by the soft dispute of a stream with its bed, or the swirling, blurred whistle of the black-throated gray warbler so high in the clerestory of the woods that he cannot be seen." High-altitude limber pines *(Pinus flexilis)* "are lifting their voices in a tale of their life that is all fierce endurance." Western trembling (or quaking) aspens *(Populus tremuloides),* larger and more brilliant than their eastern kin, food for beavers, moose, rabbits, hares, porcupines, deer, grouse, elk, and bears, light up the landscape in the fall, when "the dry crystalline atmosphere lets the sunlight through them like a clear, sustained blast on angelic trumpets. The foliage shines for miles; thus as you cross the deserts in October, still hot and sere, you can see the gold band of Aspens in the sky, where they top the mountain peaks." Desert-dwelling mesquite *(Prosopis juliflora),* which sends its roots fifty or sixty feet down to groundwater and also spreads them widely just below the surface to catch every possible drop, grows "where you would not think that any sort of tree could exist, in such arid soil, such flaming heat" and "when it comes to leaf, is a blessing on the earth that bears it." Like many others who take the time to study them, Peattie recognizes that each kind of tree has its own character, its shapes and gestures, textures and light, motions, sounds, and fragrances, its own history of interactions with human and other animal explorers, visitors, and inhabitants.

It's when we start to think about whole forests that the picture gets really complicated and we have to resort to big generalizations, scatter-shot sampling, and impressionistic lists and litanies. For just as every tree, every species, has its own character, so does every forest. On our journey outward from the center of the earth, we might emerge into forests that are swampy or bone-dry, equatorial or subarctic, monotonous or varied, short or tall, dark or sunny, small or large, a thin strip along a river through a desert or grassland or an expanse it takes days to cross by train. We might find ourselves among leaves or needles or bare branches, tangles of green or clouds of pale gray, in groves or forests of palm, tree fern, saguaro (a cactus), Joshua tree (a yucca), bamboo (a grass). We might be able to walk up to the closest tree and hug it, one pair of arms reaching all around one trunk. We might need a village to encircle just one tree. Or we might not be able to walk up to the trunk at all but instead encounter dense barriers of branches, sharp needles and thorns, buttresses taller than we are.

Our eye-level view—never mind everything lower or higher—might be dominated by glowing emerald mosses or gray-green lichens; by dark green scales studded with small silvery-blue spheres; by disks of bright yellow or fancy cutouts in red; by tangles of vines, thick pyramids of aerial roots, clumps of air plants, delicate, waxy orchids; by seed cones as small as a fingernail or as large as a logger's forearm; by acorns, luscious, colorful fruits, pairs of linked seed-wings, dangling pods. We might hear hornbills or hoopoes, chickadees or owls, frogs or cicadas. We might encounter a wolverine or a jaguar, squirrels or monkeys, elk or elephants, snakes and salamanders.

Some of the planet's most impressive forests fall in its wetter areas. One of these is the equatorial zone where (as we saw in chapter 2) the sun's intensity creates giant convection cells that suck water into the air from the oceans and from the vapor created by the forest itself, and then release it again in great torrents of rain, and where the trade winds collide with their humid loads. Here we find the forests of Central and South America, central Africa, Southeast Asia, and Oceania, forests with their own range of variation, from parts of the Amazon that are deeply flooded for months each year to the jungles of Cambodia and Thailand, where elephants carve out tunnels in the undergrowth, trees bury vast complexes of temples, and sacred groves survive in the midst of giant clear-cuts, or to the tangled mountain forests of Costa Rica and Kauai, the haunts of beautiful rare birds.

Another productive wet zone is the oceanic coastal edge (mostly west coasts) between 40 and 60 degrees of latitude, where air currents have had many miles in which to collect the ocean water they release as they encounter land, especially on the windward sides of the high mountains marking the convergences of tectonic plates. Here we find the forests of giant trees, the largest surviving stretches in North America and smaller remnants in Chile, New Zealand, and elsewhere: trees hundreds of feet high and hundreds (even in some cases thousands) of years old, everything saturated with moisture, the emerald fur of mosses, decaying branches and trunks, the cool air, every layer packed with life both vegetable and animal.

The characteristics of such wet forests merge imperceptibly into those of the more temperate forests scattered around our globe. Here, even if we limit ourselves to those whose histories are relatively free of large-scale human disturbances—a fast way to cut our options over most of the planet, not just its middle zones—it is difficult to envision so many varieties. In the remarkable uncut forests of Bhutan, squeezed between India and China along the monsoon-watered southern slopes of the Himalaya, nightingales sing before dawn and temples, prayer flags, and flowering rhododendrons emerge from the green. The forests of Europe are vast in myth and history, but now much reduced in reality. The inland forests of North America differ strikingly in character from region to region. The bamboo forests of Asia, with their pandas, their periodic massive blooms and die-offs that bring rat plagues and famine, have inspired centuries of paintings and poetry. At the middle elevations of Hawaii's Big Island, on the slopes of volcanoes, the forested kipukas—areas spared by recent lava flows—are oases of endemic birds and plants, of the red flowers and curly lines of ohia trees and the yellow flowers and tall, straight trunks of koa trees.

It's a challenge even to envision just the single most varied spot in the long (though by now much fragmented) chain of forests from Maine to Georgia, the Smoky Mountains of North Carolina and Tennessee. There the glaciers never reached, enough pockets have escaped logging to preserve an enormous number of species, moisture is abundant, the climate is mild, the elevation spans some 4,500 feet, and—perhaps now most important—the land is protected by the National Park Service. These forests face their share of hazards: a blight that killed all the chestnuts some decades ago, insects that are killing the hemlocks, acid rain and air pollution, wild pigs and human plant poachers, climate change. Still, the ongoing Great Smoky Mountains National Park's All Taxa Biodiversity Inventory lists dozens of tree species that grow in fifty-four different kinds of ecological communities in the subcategory "forest" alone. (The other categories are woodland, shrubland, dwarf shrubland, sparse vegetation, and herbaceous.) Visiting in the spring or early summer, we might be particularly struck by the tiny white fringed phacelias, which carpet the cove hardwood forests before the leaves are out. It would be easy to turn up salamanders staying cool and damp under logs and rocks (thirty species have been inventoried). Water would be falling everywhere in creeks and cascades. Misty blue air would soften the rounded old mountains and steep slopes cut by ancient waters.

But forests also grow where conditions are much less benign. In Death Valley, as we have seen, small groves of palms tap into scattered pockets of water, while on the Arabian Peninsula, fog-drinking trees produce frankincense. Where water is available just with seasonal rains, vanishing to evaporation the rest of the year, forests merge into mixed and open woodlands, shrublands, and savannas: wiry, thorny acacias offer scant shade to giant iguanas in the Galápagos and support the dangling nests of weaver birds in Africa's Rift Valley. Along warm seacoasts, forests of mangroves plant their roots in salt water and salty mud and sand, creating a rich habitat for birds and sea creatures and buffering the power of waves and storm surges. In Siberia, Canada, and Alaska, where trees growing in bogs or not-quite-solid permafrost may tilt at crazy angles to make what sober scientists call drunken forests, forests contain just a few tough species: black and white spruce that never get very tall and may look rather like large bottlebrushes, paper birch, balsam poplar, quaking aspen, larch, jackpine, alder, and many kinds of willows that can't quite reach the size of trees. Here, mosquitoes torment all warm-

blooded visitors. Some of the oldest rock on the planet gleams with glacier polish and brightly colored moss and lichen. Ankle-deep tangles of
reindeer lichen look like fields of piled-up antlers or silvery crushed baskets, and white-crowned sparrows sing their clear, ebullient notes.

Often, high-latitude and high-altitude treeline occurs as a fairly quick
thinning—fewer clusters of trees, or trees that are gradually reduced to
the size and shape of bushes that may be only ankle high, and more of
the open spaces of meadow or tundra. But sometimes a more dramatic
edge occurs, as it does along Trail Ridge Road in Colorado's Rocky
Mountain National Park. Here white-crowned sparrows also sing, this
time among meadows where bighorn sheep and elk graze, marmots and
pikas look at the views from their sunny rockpile perches, and white-
tailed ptarmigans nibble on low willows and wildflower blossoms. At
11,400 feet above sea level, the Engelmann spruce, subalpine firs, and
occasional limber pines grow thinner and shorter, begin to take the
shape of flags with all their branches extending downwind, and then
retreat into low, dense islands of krummholz, German for "bent wood."
A single tree might be three feet tall, ten feet wide, and many hundreds
of years old, and the cushions they make are strong and dense enough
to walk on; indeed, they are often easier to walk on than to crawl under.

When conditions permit, such a tree will venture to send up a vertical shoot that the cold winds of winter will trim back with their abra-
sive load of sharp ice particles, and so the trees grow low rather than
toward the sky. As our climate warms, more of these experimental new
shoots will survive. Treeline in this area is predicted to rise some 350 feet
per degree Fahrenheit of warming (or 160 meters per degree Celsius), in
which case the forest will advance its realm upward, pushing the tundra
ahead of it right off the tops of all but the highest peaks. For those who
love the tundra, this prospect is a cause for grief, but for the trees and
their forest it is an opportunity to seize. And we too might want to pay
attention to what this high forest has to say; if some forests remind us
that life is fecund and diverse, here, as in other difficult environments,
we can be intensely aware that it is also, and equally, tenacious.

RETURN TO WONDER

This world invites endless wonder. Infinite fields of questions offer themselves to be asked.

Suppose you were to undertake to say everything possible about a
single place, one that you know well and are willing to study. Here are

some of the things you would have to consider. The evidence of your own senses and the fit of your body in this place. Your emotions, memories, and experiences. All the other people who have been here, ever: what they have sensed, felt, remembered, experienced, imagined, made, done. The records they might have left in footprints, paintings, photographs, stories written and told. All the other animals that have lived here or passed through—mastodons and mice; mosquitoes and dragonflies; pterodactyls and hummingbirds. The plants. Tiny creatures like microbes, bacteria, fungi, and algae. How the ground has drifted, risen, eroded, flooded, dried. The quality, chemistry, and physics of the air. The light. Seasons and climates past, present, and future. All that is hidden or invisible, what has vanished or is yet to come. All the motions, the intertwinings.

You could never finish such a task, not even for just one place, certainly not for the entire world. (Never mind the zillions of planets, stars, and galaxies beyond our own.) Our perceptions and our minds are limited, and the world is infinitely complicated, layered, various. The real exceeds what we can know, even what we can ask.

It's not just forests that are patchy at all scales: nearly everything is. All things are connected, too, in a myriad of ways, and every thing, every relationship changes all the time. We ourselves are woven into this intricate, mutable fabric. The result? A world whose richness lives also in our vocabulary: *variegated, pied, dappled, filigreed, pieced, quilted, brindled, webbed, multiple, chaotic, patterned, mottled, scattered, lopsided, irregular, asymmetrical, askew, spiraled, turbulent, bunched, gleam, glitter, sheen, shine, tumble, swirl, wave, eddy, cascade, vortex, mosaic.*

It can be easy to lose sight of all this. Ordinarily we focus on daily, human-centered matters like family, friends, pets, work, gardens, art, music, books. And we must give proper attention to darker matters, to illness, disaster, poverty, death, environmental destruction, the urgent problem of climate change. Many of our responsibilities lie in these directions, and to meet them, we have to filter out much of the world around us—even, ironically, if we are working to preserve its richness, its diversity. Sometimes, too, especially if our vision has been shaped in part by the traditions of Romanticism and Transcendentalism, we look to nature for respite from complications: we crave there what seems to us simple, clean, fundamental, and timeless.

Yet Earth's complexities can also nourish our spirits and give us strength. This is true, again, at all scales, and it is true in both ordinary and extraordinary places. An hour with a microscope and a drop of water—or a magnifying glass and a patch of moss or lichen—may amaze and energize us, as may learning a new "gee-whiz" fact from a magazine, television show, or Web page. So may just happening to see a female broad-tailed hummingbird zip into a backyard crabapple tree, land on a thin twig, hone her beak on the wood, and then stretch out her neck so she can probe her shimmering green feathers with that long, thin tool. Or watching a thunderstorm settle in, piled-up billows of indigo and gray, gusty chill wind and rain, sheets and zigzags of fire in the sky, the scent of wet pavement, sagebrush, grass, trees.

As Thoreau noted near the end of *Walden,* in the chapter "Spring,"

> At the same time that we are earnest to explore and learn all things, we require that all things be mysterious and unexplorable, that land and sea be infinitely wild, unsurveyed and unfathomed by us because unfathomable. We can never have enough of Nature. We must be refreshed by the sight of inexhaustible vigor, vast and Titanic features, the sea-coast with its wrecks, the wilderness with its living and its decaying trees, the thunder cloud, and the rain which lasts three weeks and produces freshets. We need to witness our own limits transgressed, and some life pasturing freely where we never wander.

This from a writer who each year spent more of his time considering in detail exactly how "Heaven is under our feet as well as over our heads," studying the mysterious details of the woods and meadows where he lived.

Or consider the 1877 poem "Pied Beauty" by Gerard Manley Hopkins, whose vision and language were both deeply religious and profoundly ecstatic:

> GLORY be to God for dappled things—
> For skies of couple-colour as a brinded cow;
> For rose-moles all in stipple upon trout that swim;
> Fresh-firecoal chestnut-falls; finches' wings;
> Landscape plotted and pieced—fold, fallow, and plough;
> And all trades, their gear and tackle and trim.
>
> All things counter, original, spare, strange;
> Whatever is fickle, freckled (who knows how?)
> With swift, slow; sweet, sour; adazzle, dim;
> He fathers-forth whose beauty is past change:
>   Praise him.

All this dazzle and dappling! Many other writers and artists have also worked to capture this quality of the real: sometimes from a religious or spiritual perspective (as with Hopkins, who felt an agonizing conflict between his poetry and his Catholic faith, or the contemporary American writers Pattiann Rogers and Annie Dillard, two other ecstatic chroniclers of natural detail); sometimes, as in the case of Thoreau, from a perspective that we might (borrowing from Edward Abbey) call "eartheist."

If we return to this chapter's journey from the inside out, we'll find that a pied beauty characterizes the earth even from high above the ground. There are many ways, now, that we can see from such elevated points of view: not just the ancient prospects from the tops of trees, hills, and mountains; not just in our endlessly creative imaginations; but also from skyscrapers and airplane windows, documentary films and Google Earth. So we'll close our journey with three collections of photographs of our planet seen from above. Each occupies in a slightly different way the space bounded by science, art, and action on behalf of the natural environment and all its inhabitants. Each alters and enlarges our perspectives, both literally and conceptually, to illuminate new aspects of the place where we live. And each shows us a planet that is anything but simple, one that says, with poet Walt Whitman, "I am large, I contain multitudes."

To see as a bird sees on the wing, we can study the images made by French photographer Yann Arthus-Bertrand for his traveling exhibition, Web site, and large-format book *Earth from Above* (2002). Working with the visual artist's key elements of light, color, texture, and design, Arthus-Bertrand photographed (from helicopters) all the kinds of landscapes we've considered—volcanoes and rift valleys; glaciers and icebergs; mangrove swamps and marshes; sand dunes and thorny deserts; savannas and forests. And as we have tried to do here, he keeps human relationships with the rest of nature firmly in sight. Most of his photos include people or signs of people, and always, in some way, these human elements are embedded in the landscape or visually interwoven with nonhuman lives, with crops and gardens and plant-based fabrics and dyes, with animals both domesticated and wild.

Sometimes he focuses our attention on art—England's Uffington White Horse, the Nazca hummingbird and the Paracas candelabrum in Peru, a decorated courtyard in India. But more often, he shows us the beauty that comes with such activities as drying carpets and fabrics, plowing fields and planting crops. He shows the damage we and nature can do to each other: volcanic eruptions and hurricanes, toxic wastes and clear-cuts. We see ourselves dwarfed by our surroundings: a tiny person on an ice floe, a slender line of camels and their herders crossing some dunes. And we see our own proliferation: tan houses covering tan ground in Morocco, a bright spill of shanties down a hillside in Brazil, a sandy graveyard for tanks in Kuwait.

Over and over, Arthus-Bertrand shows us that complex patterns are everywhere, endlessly varied mixtures of repetition and similarity with change and difference. The more time you spend studying his photographs, the more echoes you'll feel you're seeing, and yet the more you'll doubt the accuracy of your perception. Are the shapes really similar? What accounts for the echoes that aren't exactly echoes, the parallels that don't quite line up? Those shanties in Rio de Janeiro are followed by a karst forest in Madagascar, two images whose intricate patterns are startlingly similar, though not quite similar enough to make a comparison comfortable. A garden in France and a marshland in Turkey offer some of the same shapes—but again, not exactly the same. One photo from Kenya shows a web of animal tracks heading for the shade of a solitary acacia tree; the next image, from the same country, shows a market clustered around another lone tree, an apparent twin to the first. The same earth and plant colors appear in both (but not quite the same, for the greens in the first are more yellow), but in the second photo they are overlaid with the bright hues of used clothes donated by people in richer nations. All his images are beautiful, but as the accompanying notes and essays make explicit, many of them are also disturbing, evidence of human poverty and environmental degradation. Here is where his work acquires its ethical, political, activist weight. Not surprisingly, two of his subsequent projects have focused on reducing carbon dioxide emissions and on seeing and listening to the "6 billion Others"—in this case, other human beings.

When we rise even higher above the surface, our perspective changes again, and in ways we might not always expect. We can do this on our computers with Google Earth and other Internet sites that offer us views from higher and higher vantage points, letting us zoom in and pull back as we wish, sometimes in to very high resolutions, sometimes back out to views as broad as a hemisphere. We can learn a lot from such images. But to combine information with visual pleasure and stimulation, we might instead consult a wonderful collection of images published by Oxford University Press in 2004 as the *Satellite Atlas of the World*.

These images are drawn from such sources as the American Landsat program (now free to the public via the Web site of the United States Geological Survey), whose cameras (Multispectral Scanners) can record things human eyes can't see. This is so not just because they're so high, 437 miles up for most of this book, with resolutions of 15 to 60 meters per pixel, but because they can "see" longer waves than we can. Thus clear water is black while shallow or silty water, which reflects more light, is blue and green; vegetation types are distinguishable by the wavelengths they absorb and reflect, so that, for instance, the healthiest rain forests and mangrove swamps are rich, dark reds. The resulting data can also be manipulated to make certain features more easily visible, in "enhanced" or "false color" images. Like a microscope or telescope, such technology expands our human perceptions beyond the limitations of our own physical organs and shows us new things about our world.

We see here again the landscapes we've been considering: volcanic cones, lava flows, and ash plumes (stretching toward the west in the trade-wind bands, toward the east in the band of westerly winds); glaciers and ice sheets with their surface meltwater ponds, adjoining lakes, calved-off icebergs, and ice floes; the variety of dune types, salt flats, and rocky mountains in deserts—and the skirt of deserts around the subtropics; the vegetation of grasslands, wetlands, forests, and agricultural fields; cities that look like gray splotches clustered in valleys and along rivers; crop fields that look like quilts made of very tiny pieces of cloth. Earth, air, fire, water: no big surprises here, though plenty of pleasure. But this atlas also contains some really startling and beautiful images.

One such cluster is of major river deltas: the Irrawaddy in Burma/ Myanmar, the Lena in Siberia, the Yukon in Alaska, the Paraná in Argentina. These images look as though they were produced during a wild party attended by painters Gustav Klimt and Jackson Pollock, artisans of millefiori glass and marbled endpapers, an amorous but feathershedding peacock, a manic puzzle-maker, an artist of chaos physics and fractals, and several biologists fascinated by nerve pathways, arteries and capillaries, the convolutions of brains and the veining of leaves. The Lena, for instance, is an image of nearly hallucinatory beauty, a fanshaped carnival of black, purple, and blue water branching into smaller and smaller squiggles, surrounded by tiny splotches, dots, and curls of greens, black, rust, coffee brown, tans, and pinks. The Irrawaddy (taken before the cyclone of 2008, which altered this landscape) has a smaller color palette, just blues and greens, a few oranges, and the deep reds of mangroves, and its lobe shapes are patterned like erratic puzzle pieces. Similarly, an image of the northern coast of Alaska, along the Beaufort Sea, shows an area of wet tundra where there is no dominant drainage pattern but where the blacks of small ponds make a messy lace or cobweb with the lavenders, burnt oranges, and greens of vegetation and the pale blues of ice.

Another amazing cluster of images reveals the patterns made by different rock types in mountain ranges and in deserts, visible here in ways that they could never be to unaided human eyes or to ordinary cameras. Patterns of geological movement show clearly: the long crumpled and saw-toothed arcs of colliding tectonic plates, the split of faults, the piled or nested onion layers of compression, deposition, erosion. The western Zagros Mountains in Iran and Morocco's Anti-Atlas Mountains reveal their complex underlying structures in bright turquoise and aqua zigzags and ovals, clumps and blotches of maroon, purple, royal blue, orange, yellow. Drainages stand out, too, like intricate, multicolored trees: just west of Riyadh, for instance, in pastel shades of green, blue, rose, and lavender, crossed by streaks of yellow sand and smaller patches of russet, teal, and a very pale gray. Iran's Great Salt Desert looks like the swirls and eddies of a fast-dropping creek, or like colored oils spilled on uneasy water.

Perhaps the two richest, most complicated images are of a pair of the most forbidding landscapes on Earth, two nearly lifeless places we'd ordinarily think of as surpassingly bleak, revealed in this book as sites of wild, extraordinary complexity. The high, frozen, volcanic desert at the intersection of Chile, Bolivia, and Argentina is a splashy, chaotic kaleidoscope of white and blue ice, indigo water, gold and yellow sand, black and red lava, patterned in blobs, squiggles, splotches, and cracks. And the area around Lake Disappointment in the Gibson Desert of Australia makes an image that is even wilder, even more like an inspired abstract impressionist painting, ragged brushstrokes scattered over a dappled ground, colors and shapes too many and too amorphous to name, all of it scratched by a fine-toothed comb from upper left to bottom right.

For just over half a century, we have been able to rise even higher by means of photographs taken from space. By now, for many of us, at least a few of these photos have become familiar enough to be part of our internal image collections; they have helped shape our most basic understanding of our planet. In December 1969, the astronauts orbiting the moon on *Apollo VIII* caught on film the image called *Earthrise*: the surface of the moon curves across the bottom of the photo, lifeless, pale gray, lightly pocked, close to the camera's eye, while our planet floats in the sea of black above, just over half of it visible, bright in sunlight, a soft swirl of blue, green, and white. And four years later, eighteen thousand miles into their trip to the moon, *Apollo XVII* astronauts made the photo called *The Blue Marble:* Earth alone and complete, again a swirl of blue and green and white, the whole shining disk surrounded by the dark of space. If you look closely enough—zoom in on your computer screen—you can see the fine layer of atmosphere that wraps us around and allows us life. (Two newer NASA projects share the name *Blue Marble:* composites created from relatively cloud-free images taken over long periods of time.) Curiously, the original orientations of these images are not the ones we ordinarily see, for the astronauts watched Earth rise to the left of the moon's surface and saw Antarctica at the top of the blue marble, the deserts of North Africa and the Arabian Peninsula at the bottom. Perhaps we can take in only so much visual strangeness at one time. Combining science and art, these photos have had enormous force for environmental activism. They show us in a way that is both intellectual and emotional that our world is not limitless, that its pied beauty requires our care.

But photographs like these are static. Our planet is not. As we have said over and over, everything changes all the time. And so we'll end by setting the blue marble in motion. It streaks through space with the galaxy and with the solar system. It loops around the sun. It moves through sunlight, around and around, north to south to north. It spins, wobbles, and tilts. Its atmosphere roils and flows in great waves and whirlpools. Hurricanes and tornadoes spin across its surface; rainstorms, blizzards, and summer breezes form, move, hesitate, and disappear. Plants grow green, sometimes red and gold, brown, and green again; stalks and leaves materialize from nowhere, then vanish. Waters slosh back and forth across vast basins, circling in gyres, splashing against the edges of the land. Caterpillars wrap themselves in silk and emerge as butterflies, drink nectar, pollinate flowers, lay eggs, then die. Pieces of land slide around the surface, tear, collide, vanish; new rock rises from deep below, flows like burning water into the open air. Ice and snow expand and melt. Forests burn and grow again. Cities grow and crumble. Birds and airplanes fly between continents; mammals walk or drive hundreds of miles; sea creatures rise through fathoms, then sink. Hummingbirds hone their beaks in crabapple trees, both breathing, oxygen or carbon dioxide in, carbon dioxide or oxygen out.

This—this wonderful moving kaleidoscope—is the face of the earth.

Epilogue

In a High Flower Meadow

SueEllen Campbell

Meadows are everywhere in the Rocky Mountains, and they're at their best in the summertime. On sun-baked flats and sandy hillsides, silvery sagebrush shines over ivory buckwheat, pale pink evening primroses, and orange sunflowers. In forests of conifers or aspens, where the trees have been disturbed by fire, disease, or cutting, or where vanished ponds or lakes have left soils especially fine or impermeable, soft, green lawns offer air, light, and food for grazing wildlife. In valley bottoms alongside meandering creeks or chains of beaver ponds, or where water simply collects as the snow melts, palettes of saturated greens and purples mark small fens, marshes, and the wet meadows that ranchers may expand by irrigating to produce hay for their cattle. Mountain meadows are flat and sloping, wet and dry, tiny and miles across, among, below, and above the trees, low and high on the peaks. They may be called parks, holes, flats, valleys, prairies, even simply meadows. Or—like this one, the one I love most—they may be nameless.

It's high summer, mid-July, midday, and I'm halfway up a steep slope in a wilderness area in central Colorado. Some five hundred feet below me gleams a tiny ribbon of water. Above me, sky meets red ridges and peaks in the long arc of a glacial cirque. That sky is the intense blue that comes with dry air and high elevations—I'm about 11,500 feet above sea level. A few tiny cumulus clouds rest on the western rim, a reminder that afternoon thunderstorms are likely here. The sun feels hot against my skin, the shade is cool, and the air is crisp and sweet. I hear the susurrus of a breeze, the buzzing of bees, the high metallic whir of a male broad-tailed hummingbird swooping close to investigate my red T-shirt.

And I'm afloat in a bright sea of wildflowers—sweeps, splashes, exclamations of intricately shaped and varied blues, purples, reds, oranges, yellows, ivories, and greens. Creamy celadon spires of green gentian rise out of cobalt pools of lupine and fire-red Indian paintbrush; shoulder-high cow parsnip's lacy white umbrellas shade hip-high blue larkspur. If I tried—I've done it before—I could list more than sixty species in this one flower meadow, and I know some of the particulars of micro-climate—glacier lilies and marsh marigolds where the snow has just melted, ground-hugging alpine forget-me-not and phlox near the ridge-lines where tundra plants thrive, fringed gentian and elephant heads on flatter spots where water pools into small marshes, tall purple monks-hood along the many tiny rivulets, scarlet fairy trumpets in drier spots.

All around me, Colorado blue columbines hover and sway above the bright carpet. With their butter-yellow centers, five white petals, five longer sky-blue petal-shaped sepals, and five long thin spurs, they are irresistible to hummingbirds, whose long tongues perfectly fit the nectar pools in the spurs. I can't resist them, either—the way they float on invis-ible stems, the delicate elegance of their shapes and gestures.

This meadow is a feast. I drink in the colors, forms, and textures, the light and air. Hummingbirds move from columbine to penstemon, fairy trumpet to columbine, their wings an invisible whir, scarlet throats and green bodies darting forward, back, holding still. Bees hang from every lupine blossom, fat, striped, fuzzy, and dusted with pollen. I hear an occasional fly or mosquito and remember that nonhuman eyes see infrared and ultraviolet markings on the blossoms that are invisible to us. I've worn red on purpose, in fact, so that I'll be less visible to hungry female mosquitoes, who don't see the color well, and attract the hum-mingbirds, who do.

I've seen elk and deer here, enjoying the nourishing greenery, and herds of domestic sheep grazing the meadow above me. Yellow-bellied marmots eat these flowers, and in a couple of weeks, a bit higher up, pikas will be harvesting winter sheaths of grass. Pocket gophers eat bulbs and roots, moving tons of earth every year and keeping these gar-dens aerated and fertilized—a bit earlier in the summer, before new growth hides the ground, I've seen these meadows covered with their eskers, long ropes of soil they've displaced under the snow, named for their resemblance to the rock and gravel eskers made by rivers beneath glaciers. Together with the steepness of the slope, the high snowfall and likelihood of avalanches, the very short, cool summers (some two months from most of the snowmelt to the first frost), and the intense sunlight, all these hungry gardeners keep trees out of the meadow and nourish flowers so lush and luscious that except in a very dry year it's hard to see any bare soil at all.

The trail, though, has worn deep into the ground during the decades I've been coming here. Because this is a designated wilderness area, old mines, Jeep roads, off-road vehicles, and mountain bikers stop at the valley below me, and dozens of day hikers, backpackers, and sometimes horse-packers use its trails every summer day, helping spread the meadow's riches through the photography, outdoor gear, and tourism industries. This place is a bit of a laboratory, too: one valley over is a biological center where scientists study high-elevation ecologies to learn how they work when they're healthy and how they respond to pressures like acid rain and climate change. Among the ecosystems most vulnerable to global warming, they say, may well be the meadows of the Rocky Mountains.

If someone I loved became ill with a disease likely to be fatal, I'd want to spend time with her, less to mourn than to *live* with her—to absorb her particular vitality, the forms passion and beauty have taken in her body. And so here, too, I find a place to settle in for a few hours, a meadow nest where I might simply soak in the brilliant flourishing around me. Sinking down so that my head is at the level of most of the flowers, I open myself to their exuberance, their vigor and resilience, their ability to hold nothing back, their sheer joy in being here, now.

Sources

We have consulted many sources in writing this book, far too many to specify, including many books, journal articles, Web sites, maps, and brochures; we have also talked via e-mail and in person with numerous specialists in different topics. In some cases, we have included in the text enough information to enable an interested reader to find our sources. In the case of something like a quotation from John Milton's *Paradise Lost* or Henry David Thoreau's *Walden,* a simple Internet search with the author, title, and a few of the quoted words will find the passage. The Web site of the Gutenberg Project (gutenberg.org) is a good place to look for many of these. Below, for readers interested in finding out more about some of our topics, are some of our major sources and other books we recommend.

First, these are a few of the books we've used as general references: Andrew Goudie, *The Nature of the Environment,* 4th ed. (Oxford: Blackwell, 2001); Anthony R. Orme, ed., *The Physical Geography of North America* (Oxford and New York: Oxford University Press, 2002); *National Geographic Atlas of the World,* 7th ed. (Washington, DC: National Geographic Society, 1999); Armin K. Lobeck, *Things Maps Don't Tell Us: An Adventure into Map Interpretation* (1956; repr. Chicago: University of Chicago Press, 1993); the series of volumes *Ecosystems of the World* (Amsterdam: Elsevier); Eric G. Bolen, *Ecology of North America* (New York: John Wiley & Sons, 1988); the *Compact Edition of the Oxford English Dictionary* (Oxford and New York:

Oxford University Press, 1971); Julia A. Jackson, ed., *Glossary of Geology*, 4th ed. (Alexandria, VA: American Geological Institute, 1997); Robert K. Barnhart, ed., *The Barnhart Dictionary of Etymology* (New York: H. W. Wilson, 1988); Walter S. Avis, ed. in chief, *A Dictionary of Canadianisms on Historical Principles* (Toronto: W. J. Gage, 1967); and Frederic G. Cassidy, ed. in chief, *Dictionary of American Regional English* (Cambridge, MA: Belknap Press of Harvard University Press, 1985–).

For material on England and on considering cultural landscapes, these books are useful: Oliver Rackham, *The History of the Countryside* (London: J. M. Dent, 1986); Jennifer Jenkins, ed., *Remaking the Landscape: The Changing Face of Britain* (London: Profile Books, 2002); W. G. Hoskins, *The Making of the English Landscape* (1955; repr. Harmondsworth, Eng.: Penguin, 1970); Tim Ingold, *The Perception of the Environment* (London: Routledge, 2000); and Iain Robertson and Penny Richards, eds., *Studying Cultural Landscapes* (London: Hodder, 2003).

For material on Australia and New Zealand, see Geoff Park, *Nga Uruora/The Groves of Life: Ecology and History in a New Zealand Landscape* (Wellington: Victoria University Press, 1995); Eric Pawson and Tom Brooking, eds., *Environmental Histories of New Zealand* (Oxford and New York: Oxford University Press, 2002); *Australian Encyclopaedia*, 6th ed. (Terrey Hills, NSW: Australian Geographic Pty, 1996); Peter Atwill and Barbara Wilson, eds., *Ecology: An Australian Perspective* (Oxford and New York: Oxford University Press, 2003); Don Garden, *Australia, New Zealand, and the Pacific: An Environmental History* (Santa Barbara, CA: ABC-Clio, 2005); David Johnson, *The Geology of Australia* (Cambridge: Cambridge University Press, 2004); Mary White, *After the Greening: The Browning of Australia* (East Roseville, NSW: Kangaroo Press, 1995); Tim Flannery, *The Future Eaters* (Sydney: Reed New Holland, 1994); George Seddon, *Landprints: Reflections on Place and Landscape* (Cambridge: Cambridge University Press, 1997); Eric Rolls, *Australia: A Biography* (St. Lucia: University of Queensland Press, 2000); and Tom Griffiths, *Forests of Ash: An Environmental History* (Cambridge: Cambridge University Press, 2001).

CHAPTER 1, LANDSCAPES OF INTERNAL FIRE

On the history of ideas about the structure of the planet: Henry Faul and Carol Faul, *It Began with a Stone: A History of Geology from the Stone*

Age to the Age of Plate Tectonics (New York: John Wiley & Sons, 1983); Susan J. Thompson, *A Chronology of Geological Thinking from Antiquity to 1899* (Methuen, NJ: Scarecrow Press, 1988); Andrew Geikie, *The Founders of Geology* (orig. 1897; repr. of 2nd ed., New York: Dover Publications, 1962; Geikie's other books are also interesting); and David Standish, *Hollow Earth: The Long and Curious History of Imagining Strange Lands, Fantastical Creatures, Advanced Civilizations, and Marvelous Machines below the Earth's Surface* (Cambridge, MA: Da Capo Press, 2006).

On volcanoes themselves and our cultural and imaginative responses to them: John Calderazzo, *Rising Fire: Volcanoes and Our Inner Lives* (New York: Lyons Press, 2004); Haraldur Sigurdsson, *Melting the Earth: The History of Ideas on Volcanic Eruptions* (Oxford and New York: Oxford University Press, 1999); and Haraldur Sigurdsson, ed. in chief, *Encyclopedia of Volcanoes* (San Diego: Academic Press, 2000).

On Yellowstone: F. V. Hayden, text, and chromolithographs of watercolors by Thomas Moran, *The Yellowstone National Park, and the Mountain Regions of Portions of Idaho, Nevada, Colorado, and Utah* (orig. 1876; repr. Tulsa, OK: Thomas Gilcrease Museum Association, 1997); Windham Thomas Wyndham Quin, the 4th Earl of Dunraven, *The Great Divide: Travels in the Upper Yellowstone in the Summer of 1874* (orig. 1876; repr. Lincoln: University of Nebraska Press, 1967); Peter Nabokov and Lawrence Loendorf, *Restoring a Presence: American Indians and Yellowstone National Park* (Norman, OK: University of Oklahoma Press, 2004); Joseph O. Weixelman, "Fear or Reverence? Native Americans and the Geysers of Yellowstone," *Yellowstone Science* 9, no. 4 (Fall 2001): 2–11; Lee H. Whittlesey, "Visitors to Yellowstone Hot Springs Before 1870," *GOSA Transactions: The Journal of the Geyser Observation and Study Association* 4 (1993): 203–211 (this article contains the letter to Thomas Jefferson); and Scott Herring, *Lines on the Land: Writers, Art, and the National Parks* (Charlottesville: University of Virginia Press, 2004).

On art and literature: The quotation from Jorge Luis Borges is from *Other Inquisitions 1937–1952*, trans. Ruth L. C. Simms (Austin: University of Texas Press, 1964). For the connection between Keats's poem and the "year without a summer" and other English literary responses to this event, see Jonathan Bate, *The Song of the Earth* (Cambridge, MA: Harvard University Press, 2000), chap. 4. For reproductions of many of Dr. Atl's marvelous paintings, see *Dr. Atl: El paisaje como pasión*, text by Beatriz Espejo (Mexico City: Fondo Editorial de la Plástica Mexicana, 1994).

On wonder: Lorraine Daston and Katharine Park, *Wonders and the Order of Nature 1150–1750* (New York: Zone Books, 1998); Steven Greenblatt, *Marvelous Possessions* (Chicago: University of Chicago Press, 1991); Lawrence Weschler, *Mr. Wilson's Cabinet of Wonder* (New York: Pantheon, 1995); and the Museum of Jurassic Technology (http://mjt.org) in Culver City, California, which in January 2008 featured an exhibit about Kircher that included a three-dimensional model of his volcanic planet.

CHAPTER 2, CLIMATE AND ICE

On the climate system and current climate change: Richard B. Alley, *The Two-Mile Time Machine: Ice Cores, Abrupt Climate Change, and Our Future* (Princeton, NJ: Princeton University Press, 2000); David Archer, *The Long Thaw: How Humans are Changing the Next 100,000 Years of Earth's Climate* (Princeton, NJ: Princeton University Press, 2009); Tim Flannery, *The Weather Makers: How Man Is Changing the Climate and What It Means for Life on Earth* (New York: Grove Press, 2005); Robert Henson, *The Rough Guide to Climate Change*, 2nd ed. (London: Rough Guides, 2008); Elizabeth Kolbert, *Field Notes from a Catastrophe: Man, Nature, and Climate Change* (New York: Bloomsbury Press, 2006); Mark Lynas, *High Tide* (New York: Picador, 2004) and *Six Degrees: Our Future on a Hotter Planet* (Washingon, DC: National Geographic, 2008); Kerry Emanuel, *What We Know about Climate Change* (Cambridge, MA: MIT Press, 2007); and William F. Ruddiman, *Earth's Climate: Past and Future* (New York: W.H. Freeman, 2001).

On human evolution and dispersal, see Richard G. Klein, *The Human Career: Human Biological and Cultural Origins*, 2nd ed. (Chicago: University of Chicago Press, 1999); and David J. Meltzer, "Peopling of North America," *Developments in Quaternary Science 1* (2003): 539–563.

"Big-picture" books exploring possible connections between climate and human history: William J. Burroughs, *Climate Change in Prehistory: The End of the Reign of Chaos* (Cambridge: Cambridge University Press, 2005); Jared Diamond, *Collapse: How Societies Choose to Fail or Succeed* (New York: Viking, 2004); Brian Fagan, *The Great*

Warming: Climate Change and the Rise and Fall of Civilizations (New York: Bloomsbury Press, 2009), *The Long Summer: How Climate Changed Civilization* (New York: Basic Books, 2004), and *The Little Ice Age: How Climate Made History 1300–1850* (New York: Basic Books, 2000); Jean Grove, *Little Ice Ages: Ancient and Modern,* 2nd ed. (London: Routledge, 2004); Hubert Lamb, *Climate, History, and the Modern World,* 2nd ed. (London: Routledge, 1995); Emmanuel Le Roy Ladurie, *Times of Feast, Times of Famine: A History of Climate since the Year 1000,* trans. Barbara Bray (Garden City, NY: Doubleday, 1971); and William R. Ruddiman, *Plows, Plagues, and Petroleum: How Humans Took Control of Climate* (Princeton, NJ: Princeton University Press, 2005).

A classic account of the development of the sublime as an aesthetic for mountainous landscapes is Marjorie Hope Nicolson's *Mountain Gloom and Mountain Glory: The Development of the Aesthetics of the Infinite* (Ithaca, NY: Cornell University Press, 1959). Hans Neuberger's essay is "Climate in Art," *Weather* 25 (1970): 46–56.

On Alaskan ice: Julie Cruikshank, *Do Glaciers Listen? Local Knowledge, Colonial Encounters, and Social Imagination* (Vancouver: University of British Columbia Press, 2005); John Muir (and his story about the dog Stickeen), *Travels in Alaska* (orig. 1915; repr. San Francisco: Sierra Club Books, 1988); and Charles Wohlforth, *The Whale and the Supercomputer: On the Northern Front of Climate Change* (New York: North Point Press, 2004).

The National Park Service geologist with whom Ana Maria Spagna went hiking is Jon Riedel. His glacier monitoring project can be followed at www.nps.gov/noca/naturescience/glacial-mass-balance1.htm.

On climate-change literature and art: Marybeth Holleman, "What Happens when Polar Bears Leave," *ISLE: Interdisciplinary Studies in Literature and Environment* 14, no. 2 (Summer 2007): 183–194; the Cape Farewell Web site (www.capefarewell.com); James Balog's Extreme Ice Survey Web site (www.extremeicesurvey.org) and book, *Extreme Ice Now* (Washington, DC: National Geographic, 2009); and the exhibition catalog *Weather Report: Art and Climate Change,* curated by Lucy Lippard (Boulder, CO: Boulder Museum of Contemporary Art, 2007).

For lucid, up-to-date videos and annotations for other excellent sources on climate change as it is seen by scientists, social scientists, humanists, artists, and others, see the Web site 100 Views of Climate Change at http://changingclimates.colostate.edu.

CHAPTER 3, WET AND FLUID

On rivers, see Ellen Wohl's *A World of Rivers: Environmental Change on Ten of the World's Great Rivers* (Chicago: University of Chicago Press, 2010), *Disconnected Rivers: Connecting Rivers to Landscapes* (New Haven, CT: Yale University Press, 2004), and *Virtual Rivers* (New Haven, CT: Yale University Press, 2001). Also see Luna B. Leopold, *A View of the River* (Cambridge, MA: Harvard University Press, 1994); Robert Naiman, Henri Décamps, and Michael McClain, *Riparia: Ecology, Conservation, and Management of Streamside Communities* (Amsterdam: Elsevier, 2005); J. David Allan and Maria Castillo, *Stream Ecology: Structure and Function of Running Rivers* (Dordrecht: Springer, 2007); and Stanley Schumm, *The Fluvial System* (New York: Wiley, 1977). A lovely book written for nonspecialist readers (which we found too late to make full use of) is Jerry Dennis's *The Bird in the Waterfall: A Natural History of Oceans, Rivers, and Lakes* (New York: HarperCollins, 1996); the quotation from Leonardo da Vinci may be found there.

On desert waters, see the notes for the next chapter.

On wetlands: Ann Vileisis, *Discovering the Unknown Landscape: A History of America's Wetlands* (Washington, DC: Island Press, 1997); William J. Mitsch and James G. Gosselink, *Wetlands* (Hoboken, NJ: Wiley, 2007); David Carroll, *Swampwalker's Journal: A Wetlands Year* (Boston: Houghton Mifflin, 1999); S.M. Haslam, *Understanding Wetlands* (London: Taylor & Francis, 2003); and Michael Williams, ed., *Wetlands: A Threatened Landscape* (Oxford: Basil Blackwell, 1990). These Web sites are also useful: http://lakeaccess.org/ecology/lakeecologyprim7.html; www.waterencyclopedia.com/Hy-La/Lake -Formation.html; and www.biology.qmul.ac.uk/research/staff/s-araya/currents.htm#Surface%20waves.

An accessible edition of William Bartram's book is the one edited by Mark Van Doren and published by Dover in 1955 as *Travels of William Bartram;* the descriptions mentioned here are from chaps. 6 and 7. An interesting article about Coleridge and Bartram is by Ernest Hartley Coleridge (the poet's grandson), "Coleridge, Wordsworth, and the American Botanist William Bartram," *Transactions of the Royal Society of Literature of the United Kingdom* (1906), available online through Google Books.

Henri Bergson's words are from *Creative Evolution,* trans. Arthur Mitchell (New York: Henry Holt, 1911), p. 128. Aldo Leopold's words are from "The Green Lagoons," published in *A Sand County Alma-*

nac, with Essays on Conservation from Round River (orig. 1949; repr. Oxford and New York: Oxford University Press, 1966). Loren Eiseley's are from the essay "The Flow of the River," in *The Immense Journey* (New York: Vintage, 1957).

Philip Larkin's line is from "Water" (1954), published in *The Whitsun Weddings* (New York: Random House, 1964). The phrases quoted by Gerald Delahunty from Seamus Heaney's poems about bog bodies are from "Bogland" ("melting and opening underfoot"), "The Tollund Man" ("trove of the turf-cutters"), "The Grauballe Man" ("beauty and atrocity"), "Bog Oak" ("cobwebbed, black"), and "Punishment" ("tribal, intimate revenge"). Stevie Smith's poem "The Ass" is in *The Collected Poems of Stevie Smith* (London: Allen Lane, 1975), pp. 522–524.

Tolkien's passage is from the chapter called "The Passage of the Marshes," *The Lord of the Rings,* book 2: *The Two Towers,* 2nd ed. (Boston: Houghton Mifflin, 1982), p. 235. The comment that this passage owes "something to Northern France after the Battle of the Somme" is in Wayne G. Hammond and Christina Scull, *The Lord of the Rings: A Reader's Companion* (Boston: Houghton Mifflin, 2005), p. 453.

The David Blackbourn quotation is from *The Conquest of Nature: Water, Landscape, and the Making of Modern Germany* (New York: W. W. Norton, 2006), p. 49. The quotation from Rod Giblett about the horrifically and fascinatingly uncanny is from *Postmodern Wetlands: Culture, History, Ecology* (Edinburgh: Edinburgh University Press, 1996), p. 13. The quotation from Graham Swift is from *Waterland* (New York: Washington Square Press/Pocket Books, 1985), pp. 7–8.

Deborah Bird Rose's essay "At the Billabong," which was written for this book, was first published in *Philosophy Activism Nature* 5 (2008).

CHAPTER 4, DESERT PLACES, DESERT LIVES

For general overviews of desert landforms and ecology: Nathaniel Harris, *Atlas of the World's Deserts* (New York: Fitzroy Dearborn, 2003); Andrew S. Goudie, *Great Warm Deserts of the World: Landscapes and Evolution* (Oxford and New York: Oxford University Press, 2002); Tom L. McKnight and Darrel Hess, "The Topography of Arid Lands," in *Physical Geography: A Landscape Appreciation* (Upper Saddle River, NJ: Pearson, 2004); Sara Oldfield, *Deserts: The Living Drylands* (Cambridge, MA: MIT Press, 2004); "Global Desert Outlook," United Nations Environment Program (www.unep.org/geo/gdoutlook/index.asp); *Ecosystems of the World,* vols. 12A and 12B, *Hot Deserts* and *Arid Shrublands,* edited by Michael Evenari, Imanuel Noy-Meir, and David W. Goodall (Amsterdam: Elsevier, 1985).

For information on various plants and animals in the desert (as well as elsewhere on Earth), two Web sites are invaluable: Tree of Life Web Project (http://tolweb.org/tree/phylogeny.html) and E.O. Wilson's comprehensive Encyclopedia of Life (www.eol.org/).

On the Sahara, see "Sahara Desert" on the Web site The Encyclopedia of Earth, edited by Mark McGinley (www.eoearth.org/article/Sahara_desert).

On Australian deserts: C. Creagh, "Understanding Arid Australia," *Ecos* 73 (Spring 1992): 15–20; S.R. Morton, "Land of Uncertainty: The Australian Arid Zone," in H.F. Recher, D. Lunney, and I. Dunn, eds., *A Natural Legacy: Ecology in Australia* (Rushcutters Bay: Pergamon Press, 1986); and D.M. Stafford Smith and S.R. Morton, "A Framework for the Ecology of Arid Australia," in *Journal of Arid Environments* 18 (1990): 225–278. Pat Lowe and Jimmy Pike's *Jilji: Life in the Great Sandy Desert* (Broome, Western Australia: Magabala Books, 1990) provides wonderful insight into the desert world of the Walmajarri Aboriginal community. Sid Cowling's little field guide, *The Living Desert: The Nature of the Barrier Ranges* (Broken Hill, NSW: Broken Hill City Council, 1995), is a simple but excellent source for information on the ecology of western New South Wales. Jay Arthur's book *The Default Country: A Lexical Cartography of Twentieth-Century Australia* (Sydney: University of New South Wales Press, 2003) is specific to Australia, but its argument about the role of language in biasing our perception of deserts is applicable to all English-speaking societies.

On the deserts of the American Southwest, see Susan J. Tweit's three books, *Barren, Wild, and Worthless: Living in the Chihuahuan Desert* (Albuquerque: University of New Mexico Press, 1995), *The Great Southwest Nature Factbook* (Seattle: Alaska Northwest Books, 1992), and *Seasons in the Desert: A Naturalist's Notebook* (San Francisco: Chronicle Books, 1998). Gary Paul Nabhan's *Gathering the Desert* (Tucson: University of Arizona Press, 1985) is a treasure trove of information on a dozen desert plant species. Many of his numerous other books are also valuable. Ann Zwinger's *The Mysterious Lands* (New York: Dutton, 1989) is another and more wide-ranging guide to the plants of North America's deserts.

A good overview of global desert literature is Gregory McNamee's *The Desert Reader: A Literary Companion* (Albuquerque: University of New Mexico Press, 1995). Edward Abbey's *Desert Solitaire* (New York: Ballantine, 1968) remains the most influential introduction to the canyon-country deserts of North America. Tom Lynch's *Xerophilia: Ecocritical Explorations in Southwestern Literature* (Lubbock: Texas Tech University Press, 2008) is a readable literary study of desert writing from the American Southwest.

On desert spirituality: Douglas Burton-Christie, *The Word in the Desert: Scripture and the Quest for Holiness in Early Christian Monasticism* (Oxford and New York: Oxford University Press, 1993); and Lane Belden, *The Solace of Fierce Landscapes: Exploring Desert and Mountain Spirituality* (Oxford and New York: Oxford University Press, 1998).

The quotation from Georgia O'Keeffe is in her essay "About Myself," published in an exhibition catalog, *Georgia O'Keeffe: Exhibition of Oils and Pastels* (New York: An American Place, 1939); n.p. The observations of Navajo poet Luci Tapahonso are from her essay "Just Past Shiprock," published in *Sáanii Dahataal: The Women are Singing* (Tucson: University of Arizona Press, 1993), pp. 5–6. The comments of Aboriginal poet Jack Davis may be found in "From the Plane Window," *Black Life: Poems* (St. Lucia: University of Queensland Press, 1992), p. 73. The quotation from Rina Swentzell is from "An Understated Sacredness," *Connotations: The Island Institute Journal* (Spring 2007): 9–12 at p. 12, reprinted there from *Mass: Journal of the [University of New Mexico] School of Architecture and Planning* 3 (Fall 1985): 24–25. Edward Abbey's words are from these chapters of *Desert Solitaire:* "Down the River" (whirlwind and thorntree), "Cliffrose and Bayonets" ("shock," "child," "much to see"), "The First Morning" ("bare bones"), and "Bedrock and Paradox" ("at the center").

CHAPTER 5, THE COMPLEXITIES OF THE REAL

On dirt: William Bryant Logan, *Dirt: The Ecstatic Skin of the Earth* (New York: Riverhead Books, 1995); Yvonne Baskin, *Under Ground: How Creatures of Mud and Dirt Shape Our World* (Washington, DC: Island Press, 2005); and David R. Montgomery, *Dirt: The Erosion of Civilizations* (Berkeley: University of California Press, 2007).

On grasslands: Susan L. Woodward, *Grassland Biomes* (Westport, CT: Greenwood, 2008); Richard Manning, *Grassland: The History, Biology, Politics, and Promise of the American Prairie* (New York: Viking, 1995); Candace Savage, *Prairie: A Natural History* (Vancouver: Greystone Books, 2004); David J. Gibson, *Grasses and Grassland Ecology* (Oxford and New York: Oxford University Press, 2009); and Graham Harvey, *The Forgiveness of Nature: The Story of Grass* (London: Vintage, 2002). Aldo Leopold's *Sand County Almanac* also contains some valuable writing on the American prairie, as does Jeffrey Lockwood's *Grasshopper Dreaming: Reflections on Killing and Loving* (Boston: Skinner House Books, 2002).

On forests: For John Muir's comments, see *The Mountains of California* (orig. 1894; repr. Berkeley: Ten Speed Press, 1977), pp. 139–140 (overview), 167 (ponderosa pines), and 248ff. (windstorm). Nalini Nadkarni's brush-wielding tree is described in *Between Earth and Sky: Our Intimate Connections to Trees* (Berkeley: University of California Press, 2008), p. 172. For the quotations from Donald Culross Peattie's *A Natural History of Western Trees* (1950; repr. Boston: Houghton Mifflin, 1991), see pp. 17 (redwoods), 37 (limber pines), 319 (aspens), and 560 (mesquites).

Contributor Biographies

COAUTHORS

SueEllen Campbell teaches environmental literature and writing at Colorado State University. Her books include *Bringing the Mountain Home* and *Even Mountains Vanish: Searching for Solace in an Age of Extinction*. She is a past president of the Association for the Study of Literature and Environment (ASLE); has published numerous articles in literary ecocriticism and ecocritical theory and teaching; and is cofounder and codirector of Changing Climates @ Colorado State University (http://changingclimates.colostate.edu), an active multidisciplinary teaching and outreach initiative about global climate change (sponsored in part by an NSF-funded Science and Technology Center, CMMAP, whose work includes climate change research).

Alex Hunt teaches environmental and American literature at West Texas A&M University. He has written many articles about American environmental literature, edited *The Geographical Imagination of Annie Proulx: Rethinking Regionalism,* and co-edited *Postcolonial Green: Ecocritical Politics and World Narratives.*

Richard Kerridge teaches environmental literature and creative writing at Bath Spa University in England. He is a past member of the ASLE Executive Council and president of ASLE-UK; received the BBC Wildlife Award for Nature Writing in 1990 and 1991; co-edited *Writing the Environment,* the first collection of ecocritical essays published in Britain; has written *Beginning Ecocriticism* for the Manchester University Press Beginnings series edited by Peter Barry (forthcoming); and coauthored *Nearly Too Much: The Poetry of J. H. Prynne.*

Tom Lynch teaches environmental and American literature at the University of Nebraska–Lincoln. He is author of *Xerophilia: Ecocritical Explorations in Southwestern Literature* and articles on Australian desert literature and

American environmental justice, and he is currently working on an ecocritical and postcolonial reading of literature of the American West and the Australian Outback.

Ellen Wohl teaches geomorphology and geology at Colorado State University. She is author of many scientific papers and books, including *Of Rock and Rivers: Seeking a Sense of Place in the American West; Island of Grass; A World of Rivers: Environmental Change on Ten of the World's Great Rivers; Disconnected Rivers: Connecting Rivers to Landscapes;* and *Virtual Rivers.*

CONTRIBUTORS

Aishah Abdallah grew up in Saudi Arabia, where she is a wilderness leader and environmental educator. Working with the Girl Scouts, schools, and government agencies, she introduces women and girls to the natural world and shows them how to visit wildlands safely and responsibly. She is a member of the IUCN Commission on Education and Communication and is studying wilderness leadership and environmental education and interpretation at Prescott College.

John Calderazzo teaches creative nonfiction writing at Colorado State University. He is author of *Rising Fire: Volcanoes and Our Inner Lives; 101 Questions about Volcanoes; Writing from Scratch: Freelancing;* and many essays and articles about science and the environment. He is cofounder and codirector of Changing Climates @ Colorado State University.

Bruce Campbell is a mechanical engineer, physicist, and ergonomic tool specialist in Chicago and a frequent visitor to nearby prairie restorations and preserves.

Gerald Delahunty grew up in Ireland and teaches linguistics and Irish literature at Colorado State University. In addition to having written many articles, he is coauthor of *Communication, Language, and Grammar: A Course for Teachers* and of *The English Language: From Sound to Sense.*

Scott Denning is a climate scientist and teacher in Colorado State University's Department of Atmospheric Science. He is education and diversity director of the NSF-funded Science and Technology Center, CMMAP (Center for Multiscale Modeling of Atmospheric Processes) and author of many scientific papers about the atmosphere and earth-atmosphere carbon processes.

Mark Fiege teaches environmental and American history at Colorado State University, where he also directs the Public Lands History Center. He is author of *Irrigated Eden: The Making of an Agricultural Landscape in the American West* and a forthcoming environmental history of the United States.

Charles Goodrich is a poet, essayist, and the program director for the Spring Creek Project for Ideas, Nature, and the Written Word at Oregon State University. He is author of two books of poetry, *Going to Seed: Dispatches from the Garden* and *Insects of South Corvallis,* and a collection of essays, *The Practice of Home,* and was chief editor of *In the Blast Zone: Catastrophe and Renewal on Mount St. Helens.*

Julia Klein teaches grassland ecology in the Department of Forest, Rangeland, and Watershed Stewardship at Colorado State University and has done fieldwork for many years on the Tibetan Plateau involving land use, grassland, climate, and ecology.

Othman Llewellyn grew up in Colorado and has lived for thirty years in Arabia, where he works as an environmental planner for the Saudi Wildlife Commission. He is a member of the IUCN World Commission on Protected Areas and the IUCN Commission on Environmental Law, and he has written on the environmental ethics of Islam. He has prepared the forthcoming protected area system plan for Saudi Arabia and writes more lyrical work when he is able.

Kathleen Dean Moore teaches environmental humanities at Oregon State University, where she is a distinguished professor of philosophy. She is director of the Spring Creek Project for Ideas, Nature, and the Written Word. Her books include *Wild Comfort: The Solace of Nature; Riverwalking; Holdfast;* and *The Pine Island Paradox*. Her newest book, *Moral Ground: Ethical Action for a Planet in Peril,* is a call to honor obligations to the future.

Kate Rigby teaches at the Centre for Comparative Literature and Cultural Studies at Monash University in Australia. She is author of *Topographies of the Sacred: The Poetics of Place in European Romanticism,* founding president of ASLE-Australia/New Zealand, co-editor of the forthcoming *Ecocritical Theory: New European Approaches,* and co-editor of the magazine *PAN (Philosophy Activism Nature)*.

Deborah Bird Rose teaches at the Centre for Research on Social Inclusion, Macquarie University, Sydney. She is the author of several prize-winning books, including *Dingo Makes Us Human* and *Reports from a Wild Country*. Her work engages with Indigenous Australian and Western philosophies and focuses on entwined social and ecological justice. Her new book on extinctions and the moral imagination is *Wild Dog Dreaming: Love and Extinction,* published by the University of Virginia Press.

Ana Maria Spagna is a freelance essayist and author of *Now Go Home: Wilderness, Belonging, and the Crosscut Saw; Test Ride on the Sunnyland Bus: A Daughter's Civil Rights Journal;* and *Potluck: Community on the Edge of Wilderness*.

Fred Swanson is a geomorphologist with the United States Forest Service and long-time researcher at Oregon's H. J. Andrews Experimental Forest, part of the Long-Term Ecological Research network. He has written many papers about the effects of such biophysical processes as landslides, volcanic eruptions, wildfire, and windthrow on forested landscapes and streams. He was co-editor of *In the Blast Zone: Catastrophe and Renewal on Mount St. Helens*.

Diana Wall teaches biology at Colorado State University, where she is a university distinguished professor and directs the School of Global Environmental Sustainability. She was one of the lead authors of the Millennium Ecosystem Assessment and is a past president of the Ecological Society of America. She has written many papers about soil invertebrates, especially in grassland, desert, and Antarctic ecosystems.

Acknowledgments

I owe heartfelt thanks to the many, many people who helped with this book—who told me useful and wondrous scientific and cultural facts, answered questions, checked my explanations for accuracy and clarity, recommended sources, or put me up and gave me directions as I traveled to explore different kinds of landscapes.

Chief among these people—as the title page suggests—are Alex Hunt, Richard Kerridge, Tom Lynch, and Ellen Wohl, without whom this book would have been neither started nor finished. Thank you all for everything you gave, through conversations and on paper. Thanks to Richard for coming up with the idea of the on-the-spot pieces—and to the friends and colleagues who contributed their time and experiences of the world in these essays. Obviously, this book would be much poorer without your words. I'm also particularly grateful to the people who read the manuscript and made useful suggestions about it: John Calderazzo, Scott Denning, Mark Easter, Mark Fiege, Richard Kerridge, and Ellen Wohl, who read all or most of it; and Dan Binkley, Gene Kelly, and Fred Swanson, who read smaller parts. Richard and John helped edit the final version.

Kate Rigby gave me especially useful information about Australia, Douglas Burton-Christie about desert spirituality, Mica Glanz about human evolution, Larry Todd about human dispersal, Deb Anthony about the history of geological thinking, Charles Dawson about New Zealand, Jim Davidson and Rodney Ley about the effects of altitude on human bodies. Many others offered information, hospitality, or often both: Jan Alberghene and David Watters, Parker Huber, Cindy and Mitch Thomashow, and William Whalen in New England; Valerie and Michael Cohen, about the vocabulary of mountain climbers; Nina Bjornsson, Elisabet Dolinda Olafsdottir, and Anna Olafsdottir Bjornsson in Iceland; Laurence Wiland, Lisa Williamson, and Peter David in Wisconsin; Ben Brown, Christine Gardinier, Henry Williamson, Susan Laidlaw, and Doug Graham in the Southeast; Kathy and Frank Moore and Carolyn Servid in Alaska. Richard Kerridge and Tracy Brain were most hospitable during my visit to England; Scott Thomas put me up in London. Thank you all!

I am most grateful to the English Department and the College of Liberal Arts at Colorado State University for their material support (for travel funds and time released from teaching), and to the English Department chairman, Bruce Ronda, for his support and patience. And thanks to Charles Goodrich, Kathleen Dean Moore, and the Spring Creek Project for Ideas, Nature, and the Written Word for a very productive week of writing at the cabin at Shotpouch Creek.

For support for his travels to and researches in Australia, Tom Lynch thanks the Centre for Cross-Cultural Research in the Research School for the Humanities, Australian National University; the Fenner School of Environment and Society, Australian National University; and the University of Nebraska's Office of Research for a Layman award for a project titled "Outback/Out West."

Richard Kerridge thanks Dorothy Kerridge for contributing expert advice about geology and weather, and Chris Nicholson for giving ideas and support.

Finally, my gratitude to my family (especially my parents, Nancy and Laird Campbell) and my friends (especially Mary Lea Dodd, Stephanie G'Schwind, Louann Reid, and Walter Isle; and also my friends in ASLE, the Rocky Mountain Land Library, Changing Climates, and the Crestone and Blue River writing communities). And to my husband, John Calderazzo, for the many ways you've helped.

I must add this important caveat: although we have worked hard to be accurate, it is inevitable in a book of this nature and length that some errors remain. We regret them. It is certain that specialists will find statements with which they will wish to argue, as it is in the nature of specialists to do, and equally certain that new discoveries and thinking will displace some of what we have said. We hope that such matters spur readers on to further thinking—and to continuing the project of bringing together knowledge from many different fields. *SEC*

Index

TEXT
10/13 Sabon Open Type

DISPLAY
Din

COMPOSITOR
BookMatters, Berkeley

INDEXER
Marilyn Anderson

PRINTER AND BINDER
Maple-Vail Book Manufacturing Group

DATE DUE
